Radiocarbon and Climate Change

Edward A.G. Schuur · Ellen R.M. Druffel
Susan E. Trumbore
Editors

Radiocarbon and Climate Change

Mechanisms, Applications and Laboratory Techniques

Springer

Editors
Edward A.G. Schuur
Center for Ecosystem Science and Society, and Department of Biological Sciences
Northern Arizona University
Flagstaff, AZ
USA

Susan E. Trumbore
Max Planck Institute for Biogeochemistry
Jena
Germany

Ellen R.M. Druffel
Department of Earth System Science
University of California, Irvine
Irvine, CA
USA

ISBN 978-3-319-25641-2   ISBN 978-3-319-25643-6 (eBook)
DOI 10.1007/978-3-319-25643-6

Library of Congress Control Number: 2016930260

Printed on acid-free paper

This Springer imprint is published by SpringerNature
The registered company is Springer International Publishing AG Switzerland

Foreword

The content of this book originated as lectures and presentations for the short course "Radiocarbon in Ecology and Earth System Science" taught by many of the authors of this volume. This course was taught for the first time at the University of California, Irvine, in 2004, and was modeled after a stable isotope short course developed by Dr. Jim Ehleringer and his many colleagues. It was our intention to promote radiocarbon as a standard scientific tool available for ecology and Earth system science research, much as has happened with stable isotopes over the previous several decades. The development of this course was made possible by a generous grant from the W.M. Keck Foundation that built the W.M. Keck Carbon Cycle Accelerator Mass Spectrometer Facility in the Department of Earth System Science at UC Irvine. The establishment of this facility was led by Susan Trumbore, John Southon, and Ellen Druffel, with many of the authors of this volume also contributing much time and effort to its creation and to producing the science advances that have been enabled.

Radiocarbon has grown far beyond Willard Libby's initial conceptualization as a tool for age determination. At the same time, current uses of radiocarbon for understanding the Earth system build upon the solid foundation laid down by several generations of scientists who worked tirelessly using this unique isotope. This volume was intended to highlight accumulated knowledge from the short course in order to provide insight to the next generation of researchers in ecology and Earth system science. Increased numbers of radiocarbon measurements combined with more researchers skilled in the interpretation of radiocarbon data offer great promise toward addressing the human impact on global carbon cycling and its feedbacks to climate change.

While it is not possible to present the entire scope of radiocarbon research, presenting radiocarbon theory, history, applications, and analytical techniques in one volume builds a broad outline of the field of radiocarbon and its emergent role in defining changes in the global carbon cycle. The writing of this book was a collective effort by all of the authors, and also drew upon inspiration of many others working in this field. We wish to thank all of our mentors and colleagues who taught us, allowing the wide range of information in this book to be brought together in one volume. We especially thank Annett Börner for her excellent work on the illustrations as well as her tireless help in coordinating the many logistics of book publication. Finally, we also want to thank the many attendees of the short course who gave us helpful feedback for refining the material that is presented here. As authors, we hope that this book will engage curious minds to look further into the wealth of past radiocarbon research in the primary literature and also to uncover new uses for this 'mother of all isotopes.' By providing the basics on radiocarbon and the motivation to learn more, it is our hope that this book acts as a stepping stone toward greater understanding of the Earth system and of the way to live responsibly within it.

Edward A.G. Schuur
Ellen R.M. Druffel
Susan E. Trumbore

Contents

Chapter 1
Radiocarbon and the Global Carbon Cycle

E.A.G. Schuur, S.E. Trumbore, E.R.M. Druffel, J.R. Southon, A. Steinhof, R.E. Taylor and J.C. Turnbull

1.1 Introduction

There is an inextricable link between carbon (C) gas (CO_2, CH_4) concentrations and heat content in the atmosphere (IPCC 2013). This link arises from the ability of these gases to absorb long-wave radiation emitted from the Earth surface. Absorbed radiation is then re-emitted within the atmosphere rather than being lost to space. This so-called *greenhouse effect* of C and other atmospheric gases has the benefit of keeping Earth more than 30 °C warmer than it would otherwise be if the atmosphere had no heat-trapping capacity at all (World Meteorological Organization

E.A.G. Schuur
Center for Ecosystem Science and Society, and Department of Biological Sciences, Northern Arizona University, Flagstaff, AZ, USA

E.A.G. Schuur (✉)
Department of Biology, University of Florida, Gainesville, FL, USA

S.E. Trumbore · A. Steinhof
Max Planck Institute for Biogeochemistry, Jena, Germany

S.E. Trumbore · E.R.M. Druffel · J.R. Southon
Department of Earth System Science, University of California, Irvine, Irvine, CA, USA

R.E. Taylor
Department of Anthropology, University of California, Riverside, Riverside, CA, USA

R.E. Taylor
Cotsen Institute of Archaeology, University of California, Los Angeles, CA, USA

J.C. Turnbull
National Isotope Centre, GNS Science, Lower Hutt, New Zealand

J.C. Turnbull
Cooperative Institute for Research in Environmental Sciences, University of Colorado, Boulder, CO, USA

E.A.G. Schuur et al. (eds.), *Radiocarbon and Climate Change*,
DOI 10.1007/978-3-319-25643-6_1

1982). At the same time, changes in atmospheric gas concentrations observed today as a result of human activity, and as had happened in the past due to other mechanisms, can alter the radiation balance of the Earth surface with large consequences. Greenhouse gas concentrations have fluctuated during the Ice Ages of the recent past 1.5 million years (Myr), accompanied by ice sheets that persisted for thousands of years, covering vast areas and lowering sea levels worldwide (Augustin et al. 2004; Petit et al. 1999). Much further back in time, C gas concentrations were much higher than at present and forests grew in places like Antarctica and Greenland, which today are covered by ice (Rothman 2002; Dahl-Jensen et al. 2013).

Changes in atmospheric C gases arose in part by processes that control the distribution and movement of C within the Earth system. This global C cycle is controlled by biologic and physical factors that move C among the major reservoirs of the Earth: atmosphere, hydrosphere, lithosphere, and biosphere. Because C is one of the essential building blocks of all life on Earth, enzymatic processes are continuously acting upon this element in most of the reservoirs where it exists. Plants convert inorganic CO_2 from the atmosphere into organic C using sunlight and then rely on the chemical energy stored in C bonds for metabolism and growth. The vast majority of other organisms in turn depend directly on plant C for energy, either consuming plants directly or eating other animals that are eating plants. This cycle returns organic C back to CO_2, or to CH_4 in anoxic environments, or converts it to other non-living organic C forms. Microbial decomposers complete this cycle by obtaining C bond energy from organisms after death. Ecological processes drive the activity of organisms that are obtaining energy from C bonds, and C transformations are changing the form and location of C (Fig. 1.1). As a result, life on Earth leaves a strong imprint on the global C cycle, which would appear remarkably different if subject to physical processes alone.

The rates and processes that control C within the Earth system reservoirs are the subject of chapters contained within this book. This overview chapter begins by summarizing recent changes in the global C cycle, contrasting patterns that exist today with that of the past several hundred years. This period of time roughly defines the Anthropocene, the Earth era dominated by humans (Crutzen 2002; Lewis and Maslin 2015). The global C cycle of the future is likely to continue to shift dramatically along with projected changes in human population and development (Fig. 1.2). Clues to this future can be, in part, gained from observations of

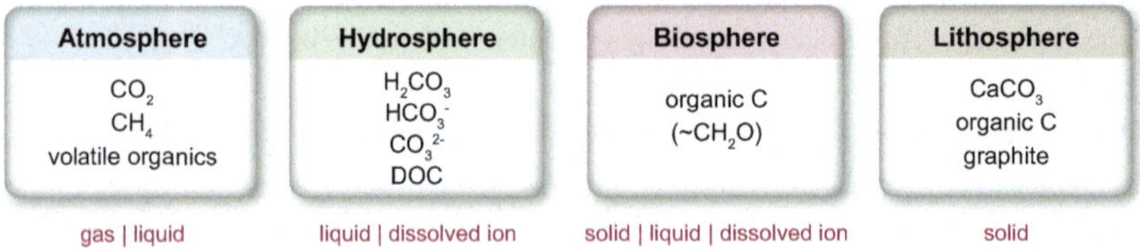

Fig. 1.1 Forms and locations of C in the Earth system

Fig. 1.2 Aspects of the global C cycle. Image sources from *top left* to *bottom right*: Unnamed photograph by Xenja Santarelli, retrieved from https://www.flickr.com/photos/78752351@N03/8464430910 (*); cirrus clouds by Alana Sise, retrieved from https://www.flickr.com/photos/alanasise/7889379304 (*); Douglas Complex fires 2013: Rabbit Mountain by Marvin Vetter, Oregon Department of Forestry, retrieved from https://www.flickr.com/photos/oregondepartmentofforestry/10479092053 (*); surface coal mine detail, Gillette, Wyoming/2008 by Greg Goebel, retrieved from https://www.flickr.com/photos/37467370@N08/7639182256 (**); waves in the sunlight by Susanne Nilsson, retrieved from https://www.flickr.com/photos/infomastern/16312966961 (**); forest light by Scott Wylie, retrieved from https://www.flickr.com/photos/scotbot/14122145262 (*); Moody Wheat Fields—Vale of Pewsey by Ben Cremin, retrieved from https://www.flickr.com/photos/bencremin/14309123919 (*); aerial view of the Amazon rainforest by Neil Palmer (CIAT), retrieved from https://www.flickr.com/photos/ciat/5641016391 (**); slash and burn agriculture in the Amazon by Matt Zimmermann, retrieved from https://www.flickr.com/photos/16725630@N00/1524189000 (*). *Note* (*) CC BY 2.0 (https://creativecommons.org/licenses/by/2.0/legalcode). (**) CC BY-SA 2.0 (https://creativecommons.org/licenses/by-sa/2.0/legalcode)

the past, and a summary of the global C cycle of the last Ice Age provides an important frame for understanding the redistribution of C among reservoirs of the Earth system. Understanding past C cycle changes in context with scenarios of future change helps to build scenarios of anticipated changes to both climate and the world in which we live.

Within the broader context of climate change, this book is focused on the use of radiocarbon (^{14}C) to study the global C cycle (Libby 1952). Radiocarbon is an isotope of C. It has six protons and electrons and therefore overall the same chemistry as ^{12}C, the most abundant form of C. The two additional neutrons in ^{14}C and one in ^{13}C give each isotope different mass. As a result, molecules that contain these isotopes react at slightly differing rates and form bonds with slightly differing strength. Variation in reaction rates and bond strengths leads to differences in the distribution of isotopes as C moves among the Earth reservoirs. This chapter examines the overall distribution of C isotopes as a framework for understanding the global C cycle and the changes that are happening to it.

The distribution of ^{14}C is also affected by radioactive decay, a process whereby the instability of the ^{14}C nucleus eventually destroys this isotope. Radioactive decay and loss do not affect ^{12}C and ^{13}C. They are stable isotopes that are neither created nor destroyed as they circulate among the Earth reservoirs. The loss of ^{14}C atoms by radioactive decay is balanced by the production of new ^{14}C atoms in the upper atmosphere, and this important feature gives rise to the unique attribute of ^{14}C. Not only is it redistributed among reservoirs like a stable isotope, but it also acts as a clock due to the regularity of radioactive decay. This seminal discovery by Willard Libby and colleagues led to a Nobel Prize in 1960. Perhaps no other single discovery has so profoundly affected so many different disciplines of science (Taylor 2014). The history of the development of ^{14}C dating described in detail in Chap. 2 provides an important backdrop to all of the science that forms the content of this book.

Just as human activities have changed the global C cycle, ^{14}C has been altered even more dramatically. Development and testing of thermonuclear weapons in the early 1960s almost doubled the atmospheric content of ^{14}C atoms over the level sustained by natural background production rates (Levin and Kromer 2004; Hua et al. 2013; Manning et al. 1990). Changing production rates created problems for ^{14}C dating, but also presented new opportunities for studying C cycling on shorter timescales as compared to the time it takes ^{14}C atoms to radioactively decay. Atmospheric weapons testing ceased, and the resulting decline in atmospheric ^{14}C atoms has been used as a global-scale pulse-chase labeling experiment for Earth's C cycle (Levin and Hesshaimer 2000). This, in combination with the continued burning of ^{14}C-free fossil fuel, continues to shift the distribution of ^{14}C atoms in the Earth system into patterns that have never been observed over the entire 50-kyr time record of ^{14}C (Suess 1955). As a result, new applications of ^{14}C that have not been invented as of the writing of this book are likely to be discovered by energetic new scientists educated in the science of ^{14}C. These important features are the themes introduced in this chapter and followed in more detail throughout other chapters of this book, giving insight into C cycling from small to large scales and through all the important Earth reservoirs.

1.2 Carbon Reservoirs in the Earth System

1.2.1 Pre-1850 Pools and Fluxes

The preindustrial period, defined conventionally as pre-1850, represents an extended period (~10 kyr) where atmospheric CO_2 concentration was relatively stable compared to today. Variations in CO_2 were less than 20 ppm once the Earth had finally emerged from the Last Glacial Maximum (LGM) around ~20 kya. In the preindustrial period, the atmosphere contained approximately 590 Petagrams ($Pg = 1 \times 10^{15}$ g) C, mainly as CO_2, while terrestrial vegetation and soils to 3 m depth contained 450–650 Pg C and 3500 Pg C, respectively, in a diverse array of organic forms (Fig. 1.3) (Ciais et al. 2013; IPCC 2013; Jobbágy and Jackson 2000; Hugelius et al. 2014; Schuur et al. 2015). The ocean contained a much larger pool of C, about 38,000 Pg as dissolved inorganic carbon (DIC), but only a small fraction of this (900 Pg C) was contained in the surface ocean (~100 m depth) that

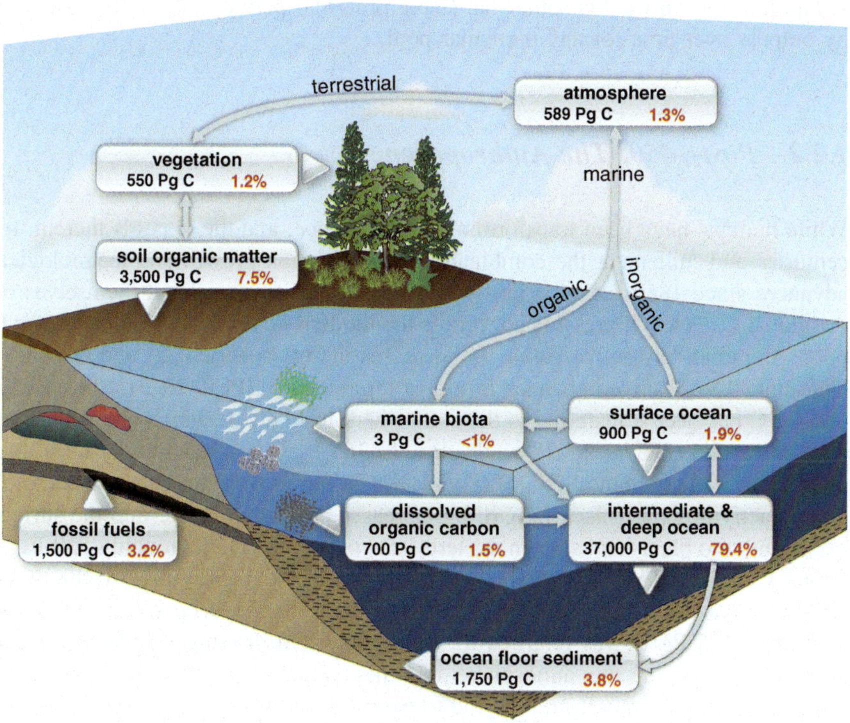

Fig. 1.3 Major C reservoirs in the Earth system shown in Pg C (*black numbers*) and as fraction of the total C represented here (*red numbers*). Number shown here represents the midpoints of ranges published in Fig. 6.1 of IPCC (2013) with the exception of the soil organic matter pool that harmonizes new numbers for permafrost with global soil organic C inventories (Schuur et al. 2015)

comes into rapid (~annual) contact with the atmosphere. Another 700 Pg C was contained as dissolved organic carbon (DOC) in the ocean with another 150 Pg C in organic forms in the top 1 m of ocean sediments (Emerson and Hedges 1988). Ocean vegetation and detritus (particulate organic carbon (POC)) contained a relatively small 3 Pg C. Rock and soil carbonates contain millions of Pg C, but inputs and outputs are small, only a fraction of 1 Pg C $year^{-1}$ (Schlesinger 1982). In contrast, recoverable fossil organic C (coal, oil, gas) comprised 1500 Pg (IPCC 2013), which until the modern exploitation of fossil fuel was largely isolated from the atmosphere but is now being mined out of this pool in significant quantities. Some of the large fluxes of C from one reservoir to another in the preindustrial period included the movement of atmospheric C into the land and ocean reservoirs via photosynthesis (159 Pg C $year^{-1}$) (Beer et al. 2010), much of which was balanced by a return flux from respiration and fires. Less than half that amount (60 Pg C $year^{-1}$) entered the surface ocean as direct dissolution, and this too was balanced by CO_2 release to the atmosphere of a similar magnitude to uptake. Despite large quantities of C moving among land, ocean, and atmosphere driven largely by biologic and physical processes, these pools were thought to be roughly in equilibrium during this somewhat stable pre-1850 period, with inputs balanced by outputs over time for any particular pool.

1.2.2 Post-1850: The Anthropocene

While humans have been transforming the landscape, and the C pools therein, for centuries and millennia, the combination of population growth and technological advances since 1850 fueled explosive changes that have imprinted themselves on the global C cycle. Of pressing concern is the modern increase in the atmospheric C pool. As radiatively active gases, the increases in atmospheric CO_2 and CH_4 have been linked to observed changes in global temperature (IPCC 2013). Since 1850, 156 Pg C has been transferred to the atmosphere from deforestation and soil tilling (Houghton and Hackler 2002), while close to 365 Pg C was released by fossil fuel combustion (Marland et al. 2006) (Fig. 1.4). Carbon emitted to the atmosphere by human activities was offset, in part, by increases in natural uptake into the surface ocean and into plants and soils in the terrestrial biosphere. Of the roughly 520 Pg of excess C delivered into the atmosphere, about 155 Pg C rapidly dissolved into the surface ocean and then slowly started to move into the deep ocean, while an additional 125 Pg C was returned to the land system, leaving behind about 240 additional Pg C in the atmosphere that has raised CO_2 concentrations from ~280 ppm in the preindustrial period to over 400 ppm as of 2013 (Fig. 1.5). The current annual imbalance in the global C cycle driven by emissions from human

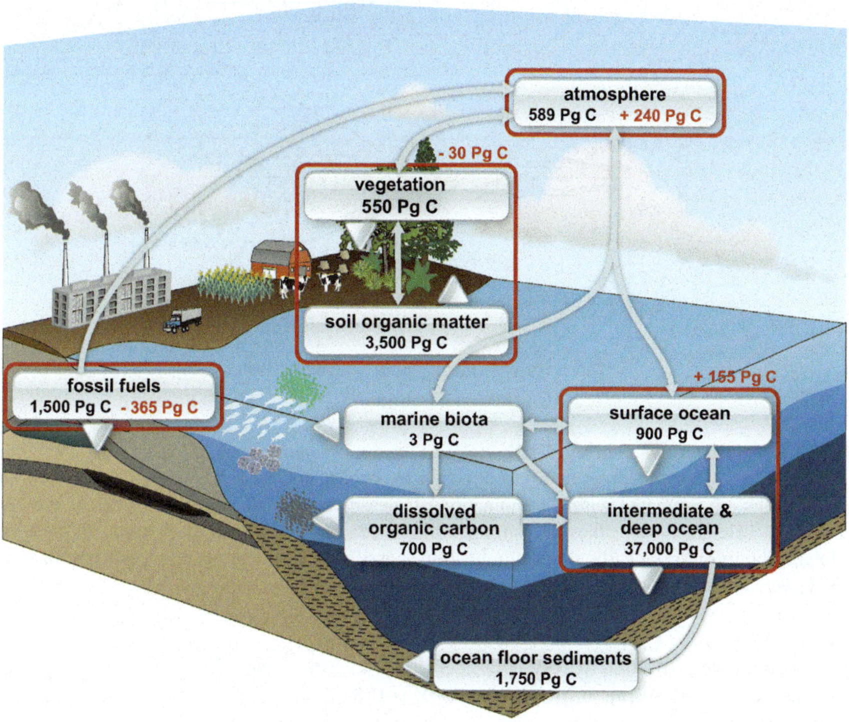

Fig. 1.4 Major additions and subtractions to the C reservoirs over the past 150 years as a direct or indirect result of human activity (*red numbers*). Units are Pg C

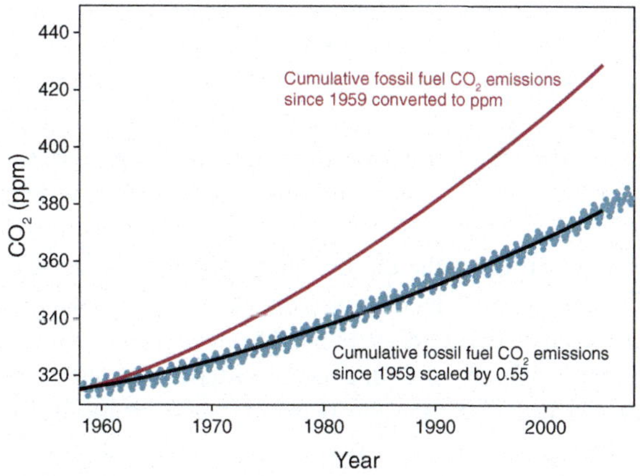

Fig. 1.5 Atmospheric observations of CO_2 shown in *blue* for the past 50 years. The *red line* depicts the concentration of CO_2 that would result if all of known additions from fossil fuel burning remained in the atmosphere. The *black line* represents fossil fuel emissions scaled by 0.55, suggesting that about 45 % of these emissions have moved from the atmosphere into other Earth system reservoirs. Data adapted from Keeling et al. (2014)

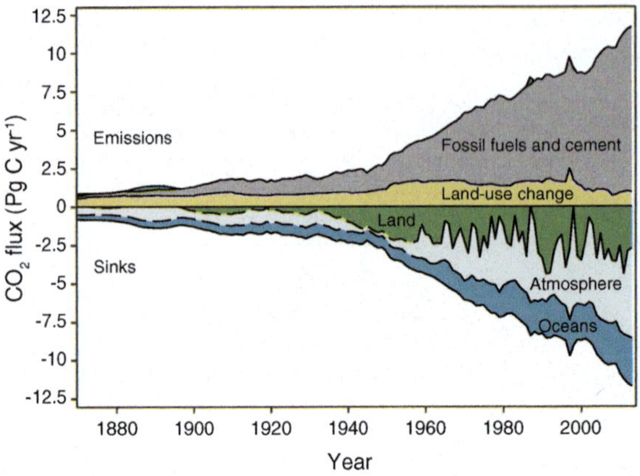

Fig. 1.6 Emissions of C from fossil fuel and land-use change over the past century. These emissions have been incorporated into atmosphere, ocean, and land sinks. The atmosphere and ocean sinks are estimated with a combination of measurements and modeling, whereas the land sink is calculated as the remaining difference. Figure modified from http://www.globalcarbonproject.org/carbonbudget/index.htm, based on Le Quéré et al. (2014) [CC BY 3.0 (https://creativecommons.org/licenses/by/3.0/legalcode)]

activities was estimated to be 10.1 ± 0.5 Pg C year^{-1} (in 2014) from fossil fuels and cement production, with another 0.9 ± 0.5 Pg C year^{-1} (2003–2012 average) from land-use change (Le Quéré et al. 2014) (Fig. 1.6). Total C emissions were 42 % higher in 2013 compared to 1990, largely driven by increases in fossil fuel use. Emissions from land-use change have declined only slightly in absolute terms, but proportionally went from 36 % of the total in 1960 to 19 % in 1990 all the way to only 8 % in 2013 as fossil fuel emissions increased. Averaged over the decade from 2004 to 2013, 8.9 Pg C year^{-1} and 0.9 Pg C year^{-1} were released from fossil fuel combustion and land-use change. Of this amount released into the atmosphere, 2.6 Pg C year^{-1} (26 % of inputs) dissolved into the ocean, 2.9 Pg C year^{-1} (30 % of inputs) was absorbed by terrestrial ecosystems via plant photosynthesis, while the remaining 4.3 Pg C year^{-1} (44 % of inputs) accumulated in the atmosphere (Le Quéré 2014).

While a large proportion of the modern atmospheric C increase was due to direct human activities, the future trajectory of atmospheric C also depends in part on the response of terrestrial and ocean systems (Canadell et al. 2007; Le Quere et al. 2007). Because natural fluxes of C in and out of terrestrial and ocean reservoirs are an order of magnitude larger than the perturbation from fossil fuels and land-use change, relatively small changes in ocean dissolution or biologic C cycling processes such as photosynthesis and decomposition can have a large impact on the size of the atmospheric pool. Indeed, there is some evidence that the capacity for both the oceans and the land to absorb excess atmospheric CO_2 is decreasing (Le Quere et al. 2007) and that the airborne fraction (proportion of C inputs that remain in the atmosphere) is increasing, although this result is under debate (Ballantyne et al. 2012; Knorr 2009; Raupach et al. 2014). Because of the sensitivity to global changes, the responses of land and ocean uptake processes to changes in temperature, precipitation, and CO_2 levels have received significant attention in the scientific literature. Attention has also focused on processes and pools in the global C cycle that are perhaps less understood, but could be of equal importance in the potential feedback to atmospheric CO_2 and the climate system within this century (Gruber et al. 2004). Vulnerable C pools and processes include thawing permafrost C, wetland drying, fires/land-use change, CH_4 hydrates, and the ocean biologic pump (Field and Raupach 2004; Davidson and Janssens 2006; Schuur et al. 2008, 2015; Ciais et al. 2012). These pools and processes have been identified to be potentially susceptible to change from the direct and indirect effects of climate change, but that the level of risk and timescale of change is highly uncertain. Incomplete knowledge of all of these processes contributes to large model uncertainty in estimating the future sink strength of ocean and land sinks (Fig. 1.7) (Friedlingstein et al. 2014).

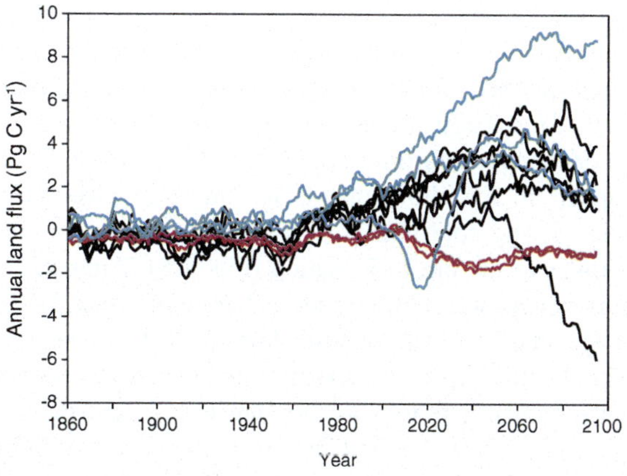

Fig. 1.7 Earth system models (ESM) are used to forecast changes in global C pools under scenarios of climate change. Each line represents a different ESM projection of past and future C uptake by the terrestrial biosphere under business-as-usual warming scenarios. Differences in model formulations and responses lead to projections of the terrestrial biosphere acting either as a source or as a sink of atmospheric C by 2100. Figure modified from Friedlingstein et al. (2014)

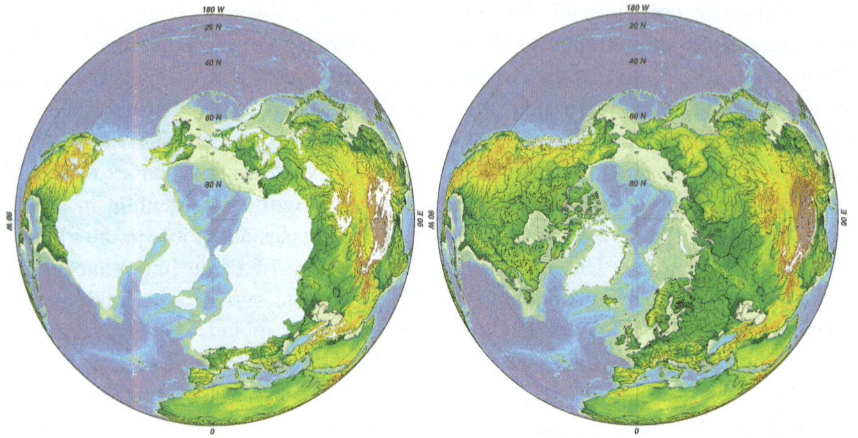

Fig. 1.8 Glacial periods during the Pleistocene featured a much colder climate with widespread ice sheets in the Northern Hemisphere and lower sea levels (*left panel*). This climate corresponded with shifts in the global C cycle including lower atmospheric C and lower C in terrestrial vegetation and soils. Excess dissolved C was thought to be stored in the deep ocean. Figure modified from Hannes Grobe/AWI via Wikimedia Commons [CC BY 3.0 (https://creativecommons.org/licenses/by/3.0/legalcode)]

1.2.3 Carbon Pools During Glacial Times

It is useful to compare pre- and post-1850 C pools to those that have been estimated for recent glacial periods when the climate was vastly different than today. Atmospheric CO_2 for a large fraction of the Pleistocene era (1.5 Myr) has been recorded in ice cores collected from Greenland and Antarctica. Gases analyzed from air bubbles trapped in ice show that CO_2 concentrations cycled between two quasi-stable states during cold glacial ($\sim$180 ppm) and warm interglacial periods ($\sim$280 ppm) similar to the pre-1850 period (Fig. 1.8) (Augustin et al. 2004; Petit et al. 1999). This represents an atmospheric change of about 210 Pg C, which is a similar net change than has been recorded from 1850 to the present. While the magnitude is similar, the processes driving this change in atmospheric CO_2 were not dominated by human activities as they are today. Reorganization of C pools during glacial periods is discussed in Chap. 7, but is generally linked to changes in both land and ocean C. Based on pollen reconstruction of vegetation in combination with estimates of ice sheet distribution, land vegetation and soils to 1 m depth are thought to have contained 700 Pg C, which is less than the estimates of the terrestrial pool pre-1850. Lower C storage in both the atmosphere and the terrestrial biosphere means that additional C was likely stored in the ocean during cold glacial periods. Stable ^{13}C isotope evidence from the shells of organisms (foraminifera) in ocean sediment cores points to an additional quantity of C dissolved into the ocean DIC pool (Bird et al. 1996; Kaplan et al. 2002). The movement of C among the atmosphere, land, and oceans between glacial and interglacial periods is thought to have been first triggered by changes in solar radiation inputs to earth that then caused feedbacks within the global C cycle and amplified this redistribution (Shakun et al. 2012). Less atmospheric CO_2 contributed, in part, to the colder climate of the Ice Ages. While these glacial and interglacial periods extend back more than 1 Myr, Chap. 7 focuses on changes during a single glacial–interglacial period contained within the last 50 kyr. This timescale is defined by the detection limit of ^{14}C using accelerator mass spectrometry (AMS) technology.

1.3 Radiocarbon Distribution Across Reservoirs

We now move from the discussion of C pools and fluxes to focus on the distribution of ^{14}C atoms within these pools. Because ^{14}C is produced in the atmosphere and also undergoes radioactive decay, the rate of C mixing from the atmosphere into the land biosphere and into the ocean can be estimated by measuring ^{14}C within the Earth system reservoirs. The overview of the distribution of ^{14}C is presented here for pre- and post-1850, with some aspects covered in greater detail in Chap. 3, while a discussion of ^{14}C distribution over the glacial–interglacial periods is left to Chap. 7. Radiocarbon is continually produced in the upper atmosphere as cosmic rays interact with nitrogen gas (N_2), primarily consisting of ^{14}N atoms (Lal and Suess 1968). Collisions produce ^{14}C with six protons and eight neutrons, which are then ultimately destroyed by radioactive β^- decay with a measured half-life of 5730 years (Karlen et al. 1968). This time frame is long enough for ^{14}C atoms to move through many reservoirs of the Earth system as part of the global C cycle.

Radiocarbon atoms produced in the upper atmosphere are quickly oxidized to carbon monoxide and CO_2 in a matter of a few weeks and distributed throughout the atmosphere via mixing of the upper and lower atmosphere. As is discussed in Chap. 4, mixing of the upper and lower atmosphere, as well as the mixing of the atmosphere both latitudinally and longitudinally, are not instantaneous processes. But, mixing within the atmosphere occurs much faster than the mixing of the ocean and land C reservoirs (Broecker and Peng 1982). Despite the continual production of ^{14}C in the stratosphere, the atmosphere pre-1850 contained only about 2 % of all ^{14}C atoms on Earth (Fig. 1.9). More than two-thirds of the total ^{14}C atoms were contained in the deep ocean, with an additional 8–10 % in the surface ocean and 2–18 % in marine sediments (to 1 m depth) (Bolin et al. 1979; Levin and Hesshaimer 2000; Damon and Sternberg 1989). The distribution of ^{14}C atoms reflected both the rates of C transfer between Earth system C reservoirs and the radioactive decay rate. Reservoirs such as the deep ocean that contain a large fraction of the ^{14}C atoms do so in part because this pool mixes slowly with other reservoirs. Slow mixing is reflected in the ratio of ^{14}C atoms in the deep ocean as compared to the modern prebomb atmosphere, which has been defined as equal to 1.00 fraction modern (F) (the specific standard is wood from the year 1890) (Fig. 1.9). Ratios less than 1.00, such as 0.84 for the deep-ocean DIC and 0.60 for the deep ocean DOC, show that significant time has elapsed such that radioactive decay has depleted ^{14}C atoms in these

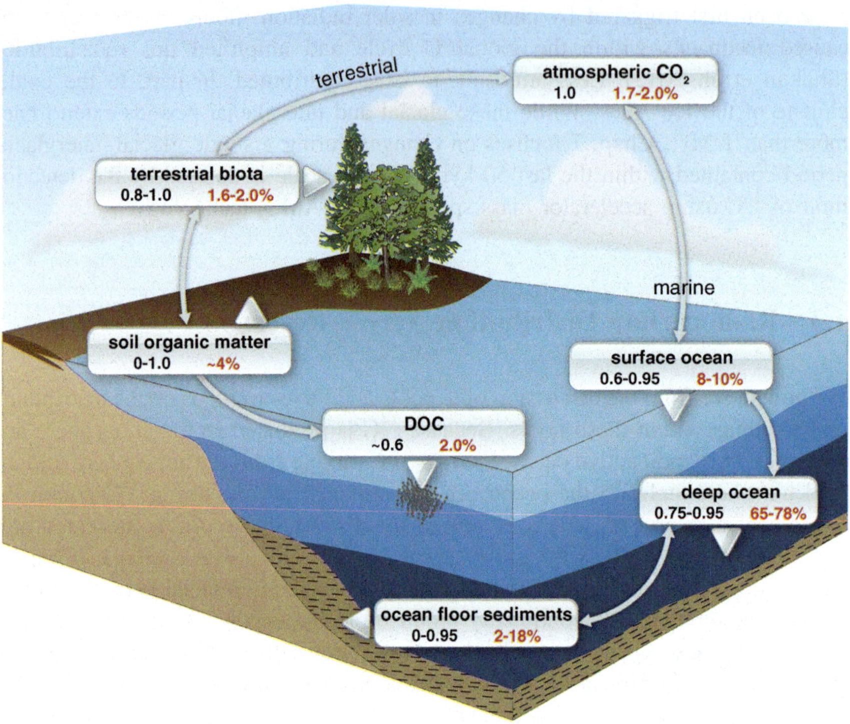

Fig. 1.9 The range of ^{14}C isotope ratios (*black numbers*) among different reservoirs of the Earth system, expressed relative to the atmospheric ^{14}C content defined in the preindustrial as 1.0, specifically based on wood from 1890. The percent of total ^{14}C atoms in the Earth system (*red numbers*) is a combination of the total C pool size and the ^{14}C age of the pool

reservoirs. Surface ocean DIC and the land biosphere have ratios somewhat closer to those of the atmosphere. While there is some depletion of ^{14}C atoms by radioactive decay, this is balanced by faster rates of replenishment of new ^{14}C atoms from the atmosphere as compared to rates of replenishment for the deep ocean. Processes that move ^{14}C atoms through the ocean are covered in more detail in Chap. 5, while Chap. 6 examines C cycling in terrestrial systems.

The distribution of ^{14}C atoms in the Earth system before 1850 was initially assumed to have been relatively constant through past time (Arnold and Libby 1949). Constant production, however, turned out not to be the case as changes in solar radiation had influenced the production of ^{14}C atoms in the stratosphere that then influenced the abundance of ^{14}C atoms in other Earth system reservoirs (Stuiver and Quay 1980; Stuiver and Braziunas 1989; Laj et al. 2002). The assumption of constant ^{14}C production is introduced in the historical overview presented in Chap. 2 and examined in much greater detail in Chap. 7. After 1850, changes in the distribution of ^{14}C atoms are linked to human activities and parallel the changes that occurred to the global C cycle described previously. Between the

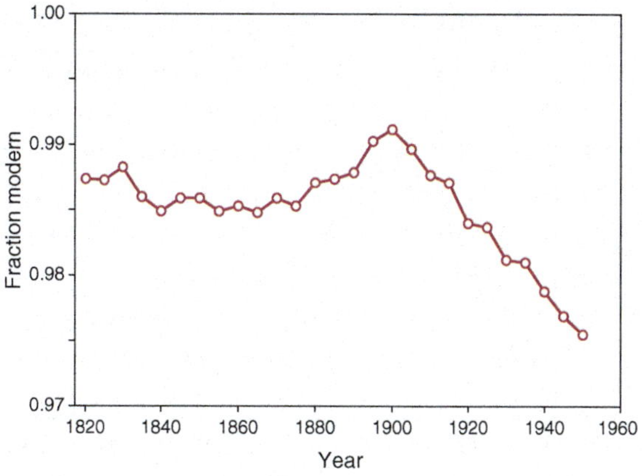

Fig. 1.10 Fraction modern (F) from tree rings in the industrial period pre-1950. The decline in ^{14}C reflects the dilution of atmospheric ^{14}C from fossil fuel burning that increases total atmospheric C with ^{14}C-free CO_2. Fraction modern values were defined as 1.00 for the preindustrial atmosphere, specifically based on wood from 1890. Data from Reimer et al. (2013)

early 1800s and 1945, the atmospheric ratio of ^{14}C compared to the modern atmosphere decreased below 1.00 F (Suess 1958). In this case, it was not radioactive decay in the atmosphere that depleted ^{14}C atoms. Instead, the input of C from land-use change and fossil fuel combustion, in particular, moved old C into the atmosphere that was already depleted in ^{14}C atoms (Fig. 1.10). While the pool of C was increasing in the atmosphere due to human activities, the pool of ^{14}C atoms was not increasing at the same rate, causing the ratio of ^{14}C to ^{12}C to decline.

Fig. 1.11 Hydrogen bomb testing in the Pacific Ocean in the early 1960s was the source of the majority of ^{14}C atoms in the atmosphere that were derived from atmospheric weapons testing

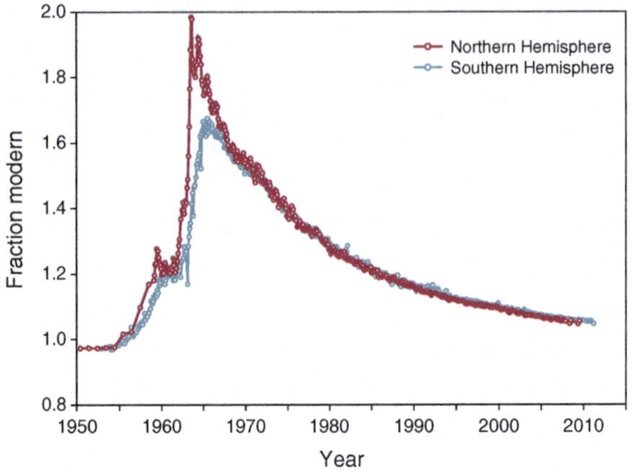

Fig. 1.12 Atmospheric ^{14}C doubled in the Northern Hemisphere, where most of the nuclear weapons testing took place. Peak ^{14}C values were lower in the Southern Hemisphere and occurred slightly later due to the time of atmospheric mixing (data from Hua et al. 2013)

The decline in the ratio of ^{14}C atoms between 1850 and 1950 was only a modest change compared to what happened next. Development and aboveground testing of thermonuclear weapons created a new source of ^{14}C atoms in the stratosphere through similar processes as described for solar radiation. The period of aboveground weapons testing was short and most intense during hydrogen bomb testing in the early 1960s (Fig. 1.11). It had the effect of almost doubling the number of ^{14}C atoms in the Northern Hemisphere atmosphere where most of the tests were conducted (Fig. 1.12). Peak ^{14}C values appeared roughly one year later in the Southern Hemisphere and were lower, because it takes about one year for air to mix across the equator and ^{14}C was diluted by that process (Levin and Hesshaimer 1996; Hua et al. 2013; Manning et al. 1990). Testing was subsequently confined belowground after 1963 with the exception of a few limited aboveground tests that occurred thereafter. These bomb ^{14}C atoms provided a 50-year record of global C cycling, as they mixed through the atmosphere, dissolved in the oceans, and cycled through the terrestrial biosphere. This timescale of years to decades that is different than the timescale of radioactive decay provides valuable information on the exchange of C between these reservoirs. Processes defined by timescales of years to decades, reflected by both the pre-1950 atmospheric decline and the subsequent bomb period, are addressed in more detail for the atmosphere in Chap. 4, oceans in Chap. 5, and the terrestrial biosphere in Chap. 6.

1.4 Scope of This Book

Cosmic radiation, the dilution of the pre-1950 atmosphere, and the production of bomb ^{14}C atoms are all processes that affected the entire Earth system. On much more limited spatial scales, humans deliberately or inadvertently have introduced

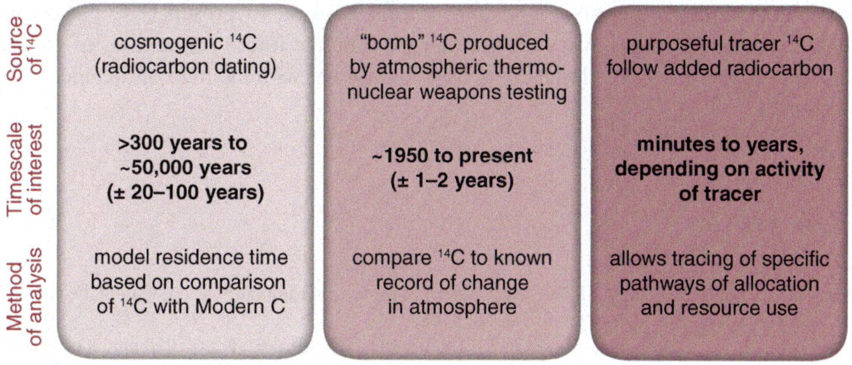

Fig. 1.13 Timescales of interest and science approaches used in ^{14}C studies

^{14}C atoms to trace the allocation of C within organisms or ecosystems on timescales of years or less. Together, tracking ^{14}C in various forms is useful on timescales ranging from minutes to about 50 kyr, which approaches the detection limit of AMS (Fig. 1.13). As will be illustrated throughout the chapters in this book, ^{14}C can be used in four general ways to understand the cycling of C atoms through ecosystems and the Earth system. (1) For *closed systems* that take up C from the atmosphere or ocean and then do not exchange C further, the amount of radioactive decay determines the ^{14}C age of an object. A calibrated age measured in calendar years can then be estimated from the ^{14}C age. This nomenclature and methods for ^{14}C dating are covered in Chaps. 3 and 7. (2) For closed systems containing C fixed during the post-1950 bomb period, a comparison of ^{14}C values with the changing atmospheric ^{14}C record (Fig. 1.12) produces a calendar year for the fixation of C. Ages for closed systems are used to provide timescales for climate proxies and ages for artifacts and to determine the ages and growth rates of trees or plants that have no annual markers such as tree rings. (3) However, most global or ecosystem C reservoirs are not isolated from continued exchange with the atmosphere. In these cases, C is constantly added and removed, and the ^{14}C value of a reservoir can provide an estimate of the rate of exchange with the atmosphere. Models to estimate this rate of exchange are described in Chap. 3 and applications to ocean and terrestrial C reservoirs in Chaps. 5 and 6. (4) Lastly, ^{14}C can be used analogous to stable isotopes as source pool tracers using mixing models described in Chap. 3. As source tracers, they do not rely so much on the radioactive decay of ^{14}C or the rate of decline of bomb ^{14}C in the atmosphere, but instead use the fact that different C pools have different characteristic ages and therefore different ^{14}C values. These differences can be used to estimate the relative contributions of individual sources with different ^{14}C values to a mixed C pool or flux. These four general ways of using ^{14}C to understand the global C cycle extend well beyond the method of ^{14}C dating invented by Libby. Chapters in this book detail how these ways of using ^{14}C have been applied to answer specific questions about ecosystem and global-scale C cycling in different Earth system reservoirs, providing insight into these important processes that would have remained hidden otherwise. The aim of this book is to present ^{14}C basics and applications in a single overview in order to stimulate future generations of curious students to see what novel questions they can answer with the help of this incredible isotope.

References

Arnold, J., and W. Libby. 1949. Age determinations by radiocarbon content—Checks with samples of known age. *Science* 110: 678–680.

Augustin, L., C. Barbante, P. Barnes, J. Barnola, M. Bigler, E. Castellano, O. Cattani, J. Chappellaz, D. DahlJensen, B. Delmonte, G. Dreyfus, G. Durand, S. Falourd, H. Fischer, J. Fluckiger, M. Hansson, P. Huybrechts, R. Jugie, S. Johnsen, J. Jouzel, P. Kaufmann, J. Kipfstuhl, F. Lambert, V. Lipenkov, G. Littot, A. Longinelli, R. Lorrain, V. Maggi, V. Masson-Delmotte, H. Miller, R. Mulvaney, J. Oerlemans, H. Oerter, G. Orombelli, F. Parrenin, D. Peel, J. Petit, D. Raynaud, C. Ritz, U. Ruth, J. Schwander, U. Siegenthaler, R. Souchez, B. Stauffer, J. Steffensen, B. Stenni, T. Stocker, I. Tabacco, R. Udisti, R. van de Wal, M. van den Broeke, J. Weiss, F. Wilhelms, J. Winther, E. Wolff, M. Zucchelli, E.C. Members, and E.C. Members. 2004. Eight glacial cycles from an Antarctic ice core. *Nature* 429(6992): 623–628.

Ballantyne, A., C. Alden, J. Miller, P. Tans, and J. White. 2012. Increase in observed net carbon dioxide uptake by land and oceans during the past 50 years. *Nature* 488: 70–72.

Beer, C., M. Reichstein, E. Tomelleri, P. Ciais, M. Jung, N. Carvalhais, C. Rodenbeck, M. Arain, D. Baldocchi, G. Bonan, A. Bondeau, A. Cescatti, G. Lasslop, A. Lindroth, M. Lomas, S. Luyssaert, H. Margolis, K. Oleson, O. Roupsard, E. Veenendaal, N. Viovy, C. Williams, F. Woodward, and D. Papale. 2010. Terrestrial gross carbon dioxide uptake: Global distribution and covariation with climate. *Science* 329: 834–838.

Bird, M.I., J. Lloyd, and G.D. Farquhar. 1996. Terrestrial carbon storage from the last glacial maximum to the present. *Chemosphere* 33: 1675–1685.

Bolin, B., E.T. Degens, S. Kempe, and P. Ketner. 1979. *The global carbon cycle*. Chichester: Wiley.

Broecker, W.S., and T.-H. Peng. 1982. *Tracers in the Sea*. Lamont-Doherty Geol Observatory, Palisades: Eldigio Press.

Canadell, J., C. Le Quere, M. Raupach, C. Field, E. Buitenhuis, P. Ciais, T. Conway, N. Gillett, R. Houghton, and G. Marland. 2007. Contributions to accelerating atmospheric CO_2 growth from economic activity, carbon intensity, and efficiency of natural sinks. *Proceedings of the National Academy of Sciences of the United States of America* 104: 18866–18870.

Ciais, P., A. Tagliabue, M. Cuntz, L. Bopp, M. Scholze, G. Hoffmann, A. Lourantou, S. P. Harrison, I.C. Prentice, D. Kelley, and C. Koven. 2012. Large inert carbon pool in the terrestrial biosphere during the Last Glacial Maximum. *Nature Geoscience* 5(1): 74–79. doi:10.1038/ngeo1324.

Ciais, P., T. Gasser, J.D. Paris, K. Caldeira, M.R. Raupach, J.G. Canadell, A. Patwardhan, P. Friedlingstein, S.L. Piao, and V. Gitz. 2013. Nature climatic change. Published online: 14 July 2013; doi:10.1038/nclimate1942.

Crutzen, P. 2002. Geology of mankind. *Nature* 415: 23.

Dahl-Jensen, D., M. Albert, A. Aldahan, N. Azuma, D. Balslev-Clausen, M. Baumgartner, A. Berggren, M. Bigler, T. Binder, T. Blunier, J. Bourgeois, E. Brook, S. Buchardt, C. Buizert, E. Capron, J. Chappellaz, J. Chung, H. Clausen, I. Cvijanovic, S. Davies, P. Ditlevsen, O. Eicher, H. Fischer, D. Fisher, L. Fleet, G. Gfeller, V. Gkinis, S. Gogineni, K. Goto-Azuma, A. Grinsted, H. Gudlaugsdottir, M. Guillevic, S. Hansen, M. Hansson, M. Hirabayashi, S. Hong, S. Hur, P. Huybrechts, C. Hvidberg, Y. Iizuka, T. Jenk, S. Johnsen, T. Jones, J. Jouzel, N. Karlsson, K. Kawamura, K. Keegan, E. Kettner, S. Kipfstuhl, H. Kjaer, M. Koutnik, T.

Kuramoto, P. Kohler, T. Laepple, A. Landais, P. Langen, L. Larsen, D. Leuenberger, M. Leuenberger, C. Leuschen, J. Li, V. Lipenkov, P. Martinerie, O. Maselli, V. Masson-Delmotte, J. McConnell, H. Miller, O. Mini, A. Miyamoto, M. Montagnat-Rentier, R. Mulvaney, R. Muscheler, A. Orsi, J. Paden, C. Panton, F. Pattyn, J. Petit, K. Pol, T. Popp, G. Possnert, F. Prie, M. Prokopiou, A. Quiquet, S. Rasmussen, D. Raynaud, J. Ren, C. Reutenauer, C. Ritz, T. Rockmann, J. Rosen, M. Rubino, O. Rybak, D. Samyn, C. Sapart, A. Schilt, A. Schmidt, J. Schwander, S. Schupbach, I. Seierstad, J. Severinghaus, S. Sheldon, S. Simonsen, J. Sjolte, A. Solgaard, T. Sowers, P. Sperlich, H. Steen-Larsen, K. Steffen, J. Steffensen, D. Steinhage, T. Stocker, C. Stowasser, A. Sturevik, W. Sturges, A. Sveinbjornsdottir, A. Svensson, J. Tison, J. Uetake, P. Vallelonga, R. van de Wal, G. van der Wel, B. Vaughn, B. Vinther, E. Waddington, A. Wegner, I. Weikusat, J. White, F. Wilhelms, M. Winstrup, E. Witrant, E. Wolff, C. Xiao, J. Zheng, N. Community, and N. Community. 2013. Eemian interglacial reconstructed from a Greenland folded ice core. *Nature* 493: 489–494.

Damon, P., and R. Sternberg. 1989. Global production and decay of radiocarbon. *Radiocarbon* 31: 697–703.

Davidson, E.A., and I.A. Janssens. 2006. Temperature sensitivity of soil carbon decomposition and feedbacks to climate change. *Nature* 440: 165–173.

Emerson, S., and J.I. Hedges. 1988. Processes controlling the organic carbon content of open ocean sediments. Paleoceanography 3.

Field, C.B., and M.R. Raupach. 2004. *The global carbon cycle: Integrating humans, climate, and the natural world.* Washington D.C.:Island Press.

Friedlingstein, P., M. Meinshausen, V. Arora, C. Jones, A. Anav, S. Liddicoat, and R. Knutti. 2014. Uncertainties in CMIP5 climate projections due to carbon cycle feedbacks. *Journal of Climate* 27: 511–526.

Gruber, N., P. Friedlingstein, C.B. Field, R. Valentini, M. Heimann, J.E. Richey, P. Romero-Lankao, D. Schulze, and C.-T.A. Chen. 2004. The vulnerability of the carbon cycle in the 21st century: An assessment of carbon-climate-human interactions. In *The global carbon cycle: Integrating humans, climate, and the natural world*, ed. C.B. Field, and M.R. Raupach, 45–76. Washington, D.C: Island Press.

Houghton, R.A., and J.L. Hackler. 2002. Carbon flux to the atmosphere from land-use changes. In *Trends: A compendium of data on global change*. Carbon Dioxide Information Analysis Center, Oak Ridge National Laboratory, U.S. Department of Energy, Oak Ridge, TN.

Hua, Q., M. Barbetti, and A. Rakowski. 2013. Atmospheric radiocarbon for the period 1950–2010. *Radiocarbon* 55: 2059–2072.

Hugelius, G., J. Strauss, S. Zubrzycki, J.W. Harden, E.A.G. Schuur, C.L. Ping, L. Schirrmeister, G. Grosse, G.J. Michaelson, C. Koven, J. O'Donnell, B. Elberling, U. Mishra, P. Camill, Z. Yu, J. Palmtag, and P. Kuhry. 2014. Improved estimates show large circumpolar stocks of permafrost carbon while quantifying substantial uncertainty ranges and identifying remaining data gaps. 11:4771–4822.

IPCC. 2013. Climate change 2013: The physical science basis. In: Stocker, T.F., D. Qin, G.-K. Plattner, M. Tignor, S.K. Allen, J. Boschung, A. Nauels, Y. Xia, V. Bex and P.M. Midgley (eds.) *Contribution of working group I to the fifth assessment report of the intergovernmental panel on climate change*. Cambridge: Cambridge University Press, 1535 pp, doi:10.1017/CBO9781107415324.

Jobbágy, E.G., and R.B. Jackson. 2000. The vertical distribution of soil organic carbon and its relation to climate and vegetation. *Ecological Applications* 10: 423–436.

Kaplan, J.O., I.C. Prentice, W. Knorr, and P.J. Valdes. 2002. Modeling the dynamics of terrestrial carbon storage since the Last Glacial Maximum. *Geophysical Reseach Letters* 29(22): 2074. doi:10.1029/2002GL015230.

Karlen, I., I.U. Olsson, P. Kllburg, and S. Kilici. 1968. Absolute determination of the activity of two 14C dating standards. *Arkiv Geofysik* 4: 465–471.

Keeling, R.F., S.C. Piper, A.F. Bollenbacher, and S.J. Walker. 2014. *Scripps CO_2 program.* Scripps Institution of Oceanography, University of California, La Jolla, CA.

Knorr, W. 2009. Is the airborne fraction of anthropogenic CO_2 emissions increasing? Geophysical Research Letters 36.

Laj, C., C. Kissel, A. Mazaud, E. Michel, R. Muscheler, and J. Beer. 2002. Geomagnetic field intensity, North Atlantic deep water circulation and atmospheric delta C-14 during the last 50 kyr. *Earth and Planetary Science Letters* 200: 177–190.

Lal, D., and H.E. Suess. 1968. The radioactivity of the atmosphere and hydrosphere. *Annual Review of Nuclear Science* 18.1: 407–434.

Le Quere, C., C. Rodenbeck, E. Buitenhuis, T. Conway, R. Langenfelds, A. Gomez, C. Labuschagne, M. Ramonet, T. Nakazawa, N. Metzl, N. Gillett, and M. Heimann. 2007. Saturation of the Southern Ocean CO_2 sink due to recent climate change. *Science* 316: 1735–1738.

Le Quéré, C., G.P. Peters, R.J. Andres, R.M. Andrew, T.A. Boden, P. Ciais, P. Friedlingstein, R.A. Houghton, G. Marland, R. Moriarty, S. Sitch, P. Tans, A. Arneth, A. Arvanitis, D.C.E. Bakker, L. Bopp, J.G. Canadell, L.P. Chini, S.C. Doney, A. Harper, I. Harris, J.I. House, A.K. Jain, S. D. Jones, E. Kato, R.F. Keeling, K. Klein Goldewijk, A. Körtzinger, C. Koven, N. Lefèvre, F. Maignan, A. Omar, T. Ono, G. Park, B. Pfeil, B. Poulter, M.R. Raupach, P. Regnier, C. Rödenbeck, S. Saito, J. Schwinger, J. Segschneider, B.D. Stocker, T. Takahashi, B. Tilbrook, S. Van Heuven, N. Viovy, R. Wanninkhof, A. Wiltshire, and S. Zaehle. 2014. Global carbon budget 2013. *Earth System Science Data* 6: 235–263.

Levin, I., and V. Hesshaimer. 1996. Refining of atmospheric transport model entries by the globally observed passive tracer distributions of (85)krypton and sulfur hexafluoride (SF6). *Journal of Geophysical Research-Atmospheres* 101: 16745–16755.

Levin, I., and V. Hesshaimer. 2000. Radiocarbon—A unique tracer of global carbon cycle dynamics. *Radiocarbon* 42: 69–80.

Levin, I., and B. Kromer. 2004. The tropospheric (CO_2)-C-14 level in mid-latitudes of the Northern Hemisphere (1959–2003). *Radiocarbon* 46: 1261–1272.

Lewis, S., and M. Maslin. 2015. Defining the anthropocene. *Nature* 519: 171–180.

Libby, W.F. 1952. *Radiocarbon dating*, 2nd ed. The Chicago: University of Chicago Press.

Manning, M., D. Lowe, W. Melhuish, R. Sparks, G. Wallace, C. Brenninkmeijer, and R. Mcgill. 1990. The use of radiocarbon measurements in atmospheric studies. *Radiocarbon* 32: 37–58.

Marland, G., T.A. Boden, and R.J. Andres. 2006. Global, regional and national CO_2 emissions. In: *Trends: A compedium of data on global change*. Carbon Dioxide Information Analysis Center, Oak Ridge National Laboratory, US Department of Energy, Oak Ridge, TN.

Petit, J., J. Jouzel, D. Raynaud, N. Barkov, J. Barnola, I. Basile, M. Bender, J. Chappellaz, M. Davis, G. Delaygue, M. Delmotte, V. Kotlyakov, M. Legrand, V. Lipenkov, C. Lorius, L. Pepin, C. Ritz, E. Saltzman, and M. Stievenard. 1999. Climate and atmospheric history of the past 420,000 years from the Vostok ice core, Antarctica. *Nature* 399: 429–436.

Quéré, C.L. 2014. Global carbon budget 2014. *Earth System Science Data.*

Raupach, M.R., M. Gloor, J.L. Sarmiento, J.G. Canadell, T.L. Frölicher, T. Gasser, R.A. Houghton, C. Le Quéré, and C.M. Trudinger. 2014. The declining uptake rate of atmospheric CO_2 by land and ocean sinks. *Biogeosciences* 11: 3453–3475. doi:10.5194/bg-11-3453-2014.

Reimer, P., E. Bard, A. Bayliss, J. Beck, P. Blackwell, C. Ramsey, C. Buck, H. Cheng, R. Edwards, M. Friedrich, P. Grootes, T. Guilderson, H. Haflidason, I. Hajdas, C. Hatte, T. Heaton, D. Hoffmann, A. Hogg, K. Hughen, K. Kaiser, B. Kromer, S. Manning, M. Niu, R. Reimer, D. Richards, E. Scott, J. Southon, R. Staff, C. Turney, and J. van der Plicht. 2013. IntCal13 and Marine13 radiocarbon age calibration curves 0–50,000 Years cal BP. *Radiocarbon* 55: 1869–1887.

Rothman, D. 2002. Atmospheric carbon dioxide levels for the last 500 million years. *Proceedings of the National Academy of Sciences of the United States of America* 99: 4167–4171.

Schlesinger, W. 1982. Carbon storage in the caliche of arid soils—A case-study from ARIZONA. *Soil Science* 133: 247–255.

Schuur, E.A.G., J. Bockheim, J.G. Canadell, E. Euskirchen, C.B. Field, S.V. Goryachkin, S. Hagemann, P. Kuhry, P.M. Lafleur, H. Lee, G. Mazhitova, F.E. Nelson, A. Rinke, V.E. Romanovsky, N. Shiklomanov, C. Tarnocai, S. Venevsky, J.G. Vogel, and S.A. Zimov. 2008.

Vulnerability of permafrost carbon to climate change: Implications for the global carbon cycle. *BioScience* 58: 701–714.

Schuur E.A.G., A.D. McGuire, G. Grosse, J.W. Harden, D.J. Hayes, G. Hugelius, C.D, Koven, P. Kuhry, D.M. Lawrence, S.M. Natali, D. Olefeldt, V.E. Romanovsky, C. Schädel, K. Schaefer, M. Turetsky, C. Treat, and J.E. Vonk. 2015. Climate change and the permafrost carbon feedback. *Nature* 520:171–179.

Shakun, J., P. Clark, F. He, S. Marcott, A. Mix, Z. Liu, B. Otto-Bliesner, A. Schmittner, and E. Bard. 2012. Global warming preceded by increasing carbon dioxide concentrations during the last deglaciation. *Nature* 484: 49–54.

Stuiver, M., and T. Braziunas. 1989. Atmospheric C-14 and century-scale solar oscillations. *Nature* 338: 405–408.

Stuiver, M., and P. Quay. 1980. Changes in atmospheric C-14 attributed to a variable sun. *Science* 207: 11–19.

Suess, H. 1955. Radiocarbon concentration in modern wood. *Science* 122: 415–417.

Suess, H. 1958. Radioactivity of the atmosphere and hydrosphere. *Annual Review of Nuclear Science* 8: 243–256.

Taylor, R.E. 2014. *Radiocarbon dating*, 2nd ed. Walnut Creek: Left Coast Press.

World Meteorological Organization 1982. *WMO Global Ozone Research and Monitoring Project, Rep. No. 14 of the Meeting of Experts on Potential Climatic Effects of Ozone and Other Minor Trace Gases*. Geneva: World Meteorological Organization.

Chapter 2
Radiocarbon Dating: Development of a Nobel Method

R.E. Taylor

2.1 Discovery of Radiocarbon

In later retrospectives, Willard Frank Libby [1909–1980], the American nuclear chemist responsible for the development of the ^{14}C dating method (Fig. 2.1), noted his indebtedness to his collaborators, James Richard Arnold [1923–2012] and Ernest Carl Anderson [1920–2014]. He expressed the view that "Certainly, nothing would have been done without them" (Libby 1952).

Libby's later recall of how the ^{14}C dating discovery process unfolded appears not to have been entirely accurate in each and every detail. This is understandable since, as both a review of his publications for this period and more recent discussions with contemporary colleagues and students reveal, during the time that he and the group was engaged in the research that established ^{14}C as a dating isotope, Libby was supervising a large cohort of graduate students and postdoctoral fellows who were pursuing a wide range of research topics, as was his custom throughout most of his scientific career. For a complete and accurate reconstruction of the history of the early development of ^{14}C dating, the only dating method whose development was recognized with a Nobel Prize, the information provided by James Arnold and Ernest Anderson has been absolutely essential.

R.E. Taylor (✉)
Department of Anthropology, University of California, Riverside, Riverside, CA, USA

R.E. Taylor
Cotsen Institute of Archaeology, University of California, Los Angeles, Los Angeles, CA, USA

R.E. Taylor
Keck Carbon Cycle Accelerator Mass Spectrometry Laboratory, University of California, Irvine, Irvine, CA, USA

E.A.G. Schuur et al. (eds.), *Radiocarbon and Climate Change*,
DOI 10.1007/978-3-319-25643-6_2

Fig. 2.1 Willard Frank Libby (1909–1980). 1960 Nobel Laureate in chemistry for the development of ^{14}C dating. *Source* US Atomic Energy Commission photograph taken in 1952

As was the case with a number of isotopes, radiocarbon (^{14}C) was produced in a particle accelerator prior to being identified as a naturally occurring radioisotope. Both Libby (1967) and Martin Kamen [1913–2002], a physical chemist who had a research position at the University of California, Berkeley (UCB), Radiation Laboratory in the mid-1930s (Kamen 1963), have noted that the first published suggestion of the possibility of the existence of ^{14}C was reported by UCB physicist Franz Newell Devereux Kurie [1907–1972] in a one-page letter in the June 15, 1934, issue of *Physical Review* (Fig. 2.2). The experiment Kurie carried out photographed particle tracks created when neutrons interacted with supercooled nitrogen and oxygen gases contained in a Wilson cloud chamber. These neutrons—uncharged particles found in atomic nuclii—had been produced when beryllium was bombarded by deuterons in the Berkeley 27-cm (11-in.) cyclotron.

In reporting on this experiment, Kurie listed six possible reactions to account for the observed tracks—three for nitrogen and three for oxygen. For those involving nitrogen, he listed ^{14}C, ^{13}C, and ^{12}C. He noted that ^{14}C was improbable since this isotope does not correspond "to a known isotope of these elements … Whether or not N … is made radioactive by neutrons has not been conclusively established." He argued that either ^{12}C or ^{13}C as stable isotopes would be favored (Kurie 1934).

Fig. 2.2 Franz N.D. Kurie (1907–1972). Research Fellow, University of California, Berkeley, Radiation Laboratory, 1932–1939. While at Berkeley, he conducted experiments which resulted in the first published suggestion of the possible existence of ^{14}C. *Source* University of California, Lawrence Berkeley National Laboratory

2.2 "Wild Bill" Libby at Berkeley: 1927–1941

Kurie's paper appeared the year following Libby's receipt of his doctoral degree in chemistry at UCB. Libby's 26-page dissertation was entitled "Radioactivity of Ordinary Elements, especially Samarium and Neodymium: Method of Detection" filed in the Spring of 1933 (Libby 1933). Libby had also been an undergraduate student at Berkeley, completing his baccalaureate degree in chemistry in 1931. He had acquired the sobriquet and perhaps epithet of "Wild Bill Libby" based, at least originally, on his high school football reputation which had followed him to Berkeley. However, his very forceful personality, the manner in which he approached his laboratory experiments, and certain other aspects of both his later personal life and professional career reinforced his lifelong reputation as "Wild Bill," at least in the eyes of some of his contemporaries and later in the view of both some of his supporters and antagonists. James Arnold would much later comment that those "meeting him for the first time tended to be either attracted or repelled. He made both friends and enemies easily" (Arnold 1992).

While still a UCB undergraduate, Libby had built for his Senior Honors Thesis what almost certainly was the first Geiger–Müller (GM) ionizing radiation detector assembled in the USA (Libby 1964b, 1981, Libby nd, de Messières 2001). The basic features of the GM tube design had been first published in 1928 in a German

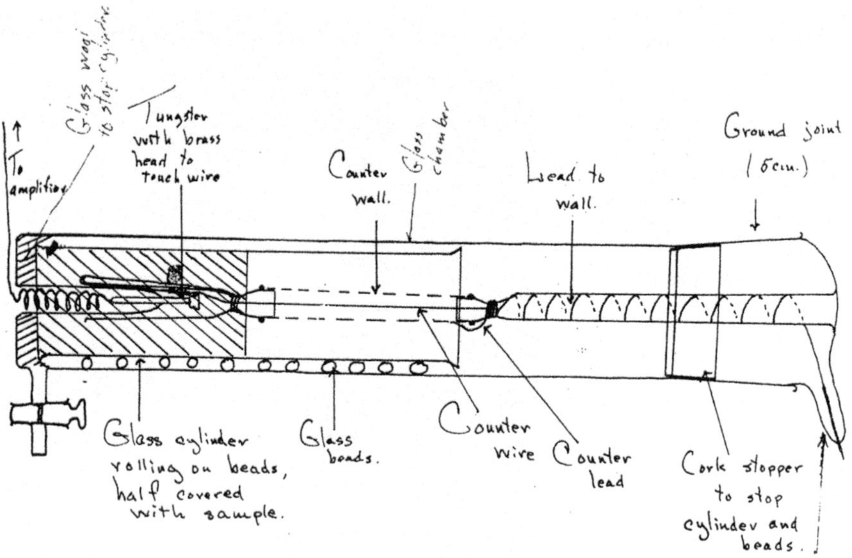

Fig. 2.3 Drawing by Libby of the design of the screen wall counter used in his dissertation research at the University of California, Berkeley, in 1932–1933. *Source* Libby (1933)

scientific periodical by Hans Geiger [1882–1945] and his student, Walther Müller [1905–1979] (Geiger and Müller 1928). The GM detector that Libby constructed consisted of a brass tube down the length of which was stretched an iron wire. It used air at a reduced pressure as the counting gas. A meter that could measure milliampere-level pulses in the detector current was connected to the output of the grid element of a two-stage amplifier circuit. These pulses recorded the effects of a cascade of electrons produced when ionizing radiation was present in the detector. A description of the circuit of this amplifier constituted Libby's first scientific publication (Libby 1932).

In Libby's dissertation research, his GM counter envelope was fabricated from glass rather than metal (Fig. 2.3). A wire grid or "screen wall" was substituted as the cathode of the counter in place of the more typical solid metal wall. His samples of rare earth elements were mounted on the inside surface of a glass sleeve rolling on spherical glass beads acting as ball bearings which allowed the sleeve to be moved in and out of the active region of the counter where the effects of ionizing radiation could be detected. This permitted measurements to be alternated between the sample and the blank or background count levels in the detector without changing the counter gas (Libby 1933; Libby and Lee 1939). The reason for this arrangement was that the count rate of his samples resulting from beta decay was very close to the counter's background rate (i.e., the activity exhibited in the counter with no sample present). With some relatively minor modifications, the general design of the counter used in Libby's dissertation research served as the model of the type of screen wall counters that he would employ more than a decade later in all of his ^{14}C experiments using solid carbon (C) counting at the University of Chicago.

Fig. 2.4 Ernest Orlando Lawrence (1901–1958). 1939 Nobel Laureate in physics for the invention of the cyclotron. Organized Radiation Laboratory (now Lawrence Berkeley Laboratory), University of California, Berkeley. *Source* University of California, Lawrence Berkeley National Laboratory

Ernest Orlando Lawrence [1901–1958], who would receive the physics Nobel prize in 1939 for the invention and development of the cyclotron, had arrived at Berkeley from Yale University in 1928 (Fig. 2.4). He began his research with a 27-cm (11-in.) cyclotron instrument that he constructed in 1931. This was the instrument that had been employed to produce neutrons for Kurie's experiment. Throughout the decade of the 1930s, Lawrence built larger and larger diameter cyclotrons that permitted the production of particles of higher and higher energies.

In an attempt to determine whether ^{14}C was the isotope being produced from nitrogen interaction with neutrons in a cloud chamber, Samuel Ruben [1913–1943], then a Berkeley chemistry graduate student, was set to work by Lawrence to produce ^{14}C by irradiating ammonium nitrate in Lawrence's 68-cm (27-in.) cyclotron. At that time, ^{14}C was assumed to have a half-life of not more than a few months based on the energy of its decay products which were similar to a radioactive isotope of sulfur, ^{35}S. The experiment, undertaken on the assumption of a short half-life, failed to produce any measurable amounts of ^{14}C. No studies focused on ^{14}C were immediately undertaken in view of its presumed short half-life. This was because the work of Lawrence's laboratory between 1937 and 1939 was largely focused on the production of isotopes for biomedical applications. In this pre-US National Science Foundation or National Institutes of Health era,

Lawrence believed that this promised the best chance of more stable, long-term funding to support the operation and expansion of cyclotron research at Berkeley (Kamen 1963; Libby 1967, 1979).

In late 1939, a question was raised concerning the practical worth of radioisotopes in biochemical or biomedical research. For example, ^{11}C, which was used as a tracer in the study of photosynthesis, had a half-life of only 20 min. As a result, Lawrence initiated a maximum effort to determine definitively whether any longer-lived radioactive isotopes existed for any of the biologically important elements, which included hydrogen, nitrogen, oxygen, and C (Kamen 1963). The work on C was undertaken by Ruben and Kamen who began by exposing graphite to deuteron bombardment in a 94-cm (37-in.) cyclotron. Following the bombardment, the graphite was burned to CO_2, precipitated out as $CaCO_3$, and used to coat the inside of a screen wall counter designed by Libby. The counter registered the presence of artificial ^{14}C, and after several months, when there was no decrease in the activity of ^{14}C, the researchers came to realize that the half-life had to be orders of magnitude in excess of that previously assumed.

This initial experiment had been based on the assumption that the reaction of deuterons with ^{13}C rather than neutrons with ^{14}N would be the *most likely* reaction in the formation of ^{14}C. To exclude the possibility of a favored reaction of neutrons with nitrogen, an experiment was devised in which a solution of saturated ammonium nitrate was irradiated with neutrons. The expectation was that no detectable amount of ^{14}C would be produced. Instead, a relatively small amount of precipitate paralyzed the screen wall counter because of the high ^{14}C count rate. Within a few months in early February 1940, despite strong theoretical arguments to the contrary, Kamen and Ruben had experimentally demonstrated that the ("slow") neutron mode of ^{14}C production was heavily favored and that the half-life had to be somewhere in the range of 10^3–10^5 (1000–100,000) years (Ruben and Kamen 1941; Kamen 1985).

Libby later reported that sometime in the mid-1930s, he had become aware of the work of Serge A. Korff [1906–1983] who was then a cosmic ray physicist at the Bartol Research Foundation. Korff had been measuring the increase of cosmic rays with increasing altitude in the atmosphere by sending GM counters aloft in balloons (Heusser 1990). As we have noted, neutrons had been produced in the laboratory, but it was not known whether they existed in natural radiation received by the earth. Korff set to work to build a counter that would be sensitive only to neutrons.

Sending these counters up in balloons, Korff found an increase of neutrons with altitude up to about 16 km (10 miles) after which there was a rapid decrease (Fig. 2.5). These data were interpreted to indicate that neutrons existed naturally but were secondary radiation, a product of the collision of high-energy cosmic rays with nuclei of the gaseous components of the earth's atmosphere. The decrease at the top of the atmosphere was attributed to the escape into space of some of the neutrons so formed (Libby 1952). Korff noted that the neutrons that remained would be slowed down by collisions with atmospheric gas nuclei and that these "slow" or thermal neutrons would disappear by being captured by nitrogen to form ^{14}C. This was apparently the first published prediction that ^{14}C should exist *in nature* (Korff and

Fig. 2.5 Serge Alexander Korff (1906–1983) New York University [on *right* in photograph], preparing a balloon to lift counter to measure neutrons at various altitudes. *Source* Bartol Research Foundation, Franklin Institute

Danforth 1939; Montgomery and Montgomery 1939; Korff 1940; Libby n.d.). Four decades later, Libby stated that the origin of his conceptualization of ^{14}C dating involved his reading of the Korff and Danforth (1939) publication which reported finding neutrons in the atmosphere: "As soon as I read Korff's paper … that's carbon dating" (Libby 1979).

2.3 Libby at the Manhattan Project: 1941–1946

The entrance of the USA into World War II quickly redirected the activities of a large percentage of the US scientific community into a wide variety of applied science and engineering projects directed toward enhancing the pursuit of the war effort. US scientists involved in nuclear physics or nuclear chemistry research were quickly recruited to pursue the development of a nuclear weapon based on fears, which turned out to be correct, that Nazi-ruled Germany had a similar project underway.

In 1941, Libby was at Princeton as a Guggenheim Fellow on his first sabbatical leave from Berkeley. He was one of the first scientists recruited by Nobel chemistry laureate Harold C. Urey [1893–1981] to join the Manhattan Project (officially, the Manhattan Engineer District) at Columbia University in New York, one of more than thirty sites in the USA, Canada, and England which were established to work on the World War II development of a nuclear weapon (Fig. 2.6). Urey's group had the task of developing a means of undertaking uranium isotope separations on a large scale by gaseous diffusion (Arnold 1981).

Libby's most important Manhattan Project scientific research involved the development of a material that could withstand the extremely corrosive effects of uranium hexafluoride (UF_6) as part of a technique to accomplish the gaseous

Fig. 2.6 Libby and Harold C. Urey (1893–1981) at an award ceremony for Libby. *Source* University of Chicago Regenstein Library, Special Collections, Photographic archive

thermal diffusion enrichment of ^{235}U in its separation from ^{238}U. Interestingly, he later expressed the view that his research to develop a thermal diffusion barrier was "better than my carbon 14 dating" (Libby 1978).

2.4 Libby at Chicago: 1946–1954

In 1945, at the conclusion of World War II, looking forward to continuing his relationship with Urey, Libby accepted an appointment as a professor in the Department of Chemistry and Institute for Nuclear Studies (later renamed the Enrico Fermi Institute for Nuclear Studies) at the University of Chicago rather than returning to Berkeley. He arrived to take up his position in October 1945 with a number of research projects in mind. Libby later commented that he initially determined to pursue his ^{14}C project "in secret." In a 1979 interview, when later asked to identify the most difficult and critical part of the development of the technique, Libby responded: "Being smart enough to keep it secret until it was in hand … I don't care who you are. You couldn't get anybody to support it. It's obviously too crazy." Libby stated that only Urey initially knew of the goal of the research (Libby 1979). Neither of his collaborators at the time, James Arnold nor Ernest Anderson, recalled that they were ever informed by Libby, or were cognizant on their own, that the research they were undertaking was being conducted in a secretive manner (James Arnold and Ernest Anderson, personal communications 2001).

The first published hint of the direction in this thinking appeared in a short paper that appeared in *Physical Review* in June 1946. In this paper, he noted that if the half-life of ^{14}C was much greater than 1000 years, then a balance between ^{14}C production and decay would exist in living organics and predicted what the specific ^{14}C activity of modern C would be. Most importantly, however, he predicted that there would be a significant difference in the ^{14}C activity between living biological and fossil C (Libby 1946).

In his autobiography, Emilio Segrè, an Italian physicist resident at the University of Chicago at this time, notes parenthetically that he had "been told that the suggestion [for ^{14}C dating] came from [Enrico] Fermi in a Chicago seminar" (Segrè 1991). Segrè's (1970) earlier biography of Fermi made no mention of this rumor. Contemporaries report that the informal exchange of ideas was a constant feature of academic life at both Berkeley and Chicago. If Libby's later statement that the basic idea was formed in his mind in 1939 is not considered credible, this rumor might reflect the views of some that Libby, like some of his colleagues, was not adverse to taking and rushing quickly to do "quick and dirty" experiments to determine the feasibility of some offhanded comment expressed by a colleague.

What ever was its ultimate source, the kernel of the conception of using ^{14}C as a dating isotope is without question contained in Libby's 1946 publication. Interestingly, two reviewers of the paper recommended rejection on the grounds of "insufficient general interest" (Arnold 1981). As far as can be determined, this paper contained the earliest use of the term "radiocarbon," and although no contemporary statement has been found that confirms this, the later oral statements of his contemporary collaborators lead to the conclusion that Libby himself was the first to employ this term for ^{14}C.

At a party (he later thought perhaps a Christmas party) late in 1946, Arnold, now a postdoctoral fellow who, since February 1946, had worked with Libby and two other professors at Chicago and reports that he became cognizant of Libby's serious intention to use ^{14}C as a means of dating. Arnold had been able to undertake his dissertation research as part of the Manhattan Project efforts, while he had been a Princeton graduate student during the years 1943–1945. For a short period in the Spring of 1946, he was associated with Libby at Chicago working with reactor-produced ^{14}C. Employing his wartime connections, Libby had had a sample of barium carbonate inserted into a reactor at the Oak Ridge National Laboratory in Tennessee that was producing a high flux of neutrons. This bombardment yielded about a millicurie of ^{14}C. This amount of ^{14}C was many orders of magnitude in excess of what had ever been, to that date, produced and was sufficient to undertake the experiment that would produce a more accurate and precise ^{14}C half-life measurement (Marlowe 1980, 1999; Arnold, personal communication 2001).

During the Christmas holidays of that year, Arnold mentioned Libby's work to his father, who although a corporate attorney by profession, was very knowledgeable in Egyptian archaeology, having served for many years as the American Secretary for the British-based Egypt Exploration Society with responsibility for fund raising (Arnold 1992). The elder Arnold communicated the news of this potential new dating technique to Ambrose Lansing [1891–1959], then a curator in the Egyptology Department at the Metropolitan Museum in New York. As a result of this conversation, Lansing sent a set of ten samples from various Egyptian sites to James Arnold at Chicago (Marlowe 1980, 1999). That package contained what was to become the first sample to be dated by ^{14}C (C-l), cypress wood from the tomb of the Third Dynasty Egyptian king Zoser or Djoser (Netjerikhet). Lansing did not communicate the age of any of the samples, but did indicate the provenience of each. Since James Arnold had grown up in a household where Egyptian archaeology was talked about often and at length, he was knowledgeable enough to be able to deduce the age on his own (Arnold, personal communication 1986). The arrival of these samples was certainly highly premature for as yet there was no experimental confirmation of any of Libby's assumptions concerning the distribution of natural ^{14}C in nature. However, Arnold later reported that Libby took the package containing the samples and placed it on a shelf behind his desk. Arnold recalls thinking at that time that Libby was really serious about his ^{14}C dating idea (Arnold, personal communication 1986; Arnold 1992).

2.5 The Critical Experiments: 1946–1948

In May 1946, Ernest Anderson joined Libby's group, originally as a laboratory assistant and then as his first graduate student at the University of Chicago (Anderson, personal communication 1986; Marlowe 1999). During the war years, Anderson had worked on the Manhattan Project at the Los Alamos Scientific Laboratory. However, unlike Arnold, he had not been able to utilize these years to obtain his Ph.D. In Libby's group, he was first given the responsibility by Libby to fabricate the counting apparatus (Fig. 2.7). This began a three-year period [1946–1948] during which three critical experiments were undertaken which empirically examined the implications of the predictions Libby had published in his short 1946 *Physical Review* paper.

Fig. 2.7 Libby and Ernest Carl Anderson (1920–2013) taken in 1948 in Libby's laboratory in Jones Hall, University of Chicago. In the background are parts of the instrumentation used in making ^{14}C measurements. *Source* University of Chicago Regenstein Library, Special Collections, Photographic archive

The first experiment was designed to determine whether there was a measurable difference in ^{14}C content between a contemporary or modern C sample and a sample of fossil C of great geological age. A modern sample would establish the equivalent of a "zero ^{14}C age" or 0 BP value on the ^{14}C timescale. A sample of fossil C, in which it would be reasonably expected that its ^{14}C concentration would have decayed below detectable levels, would establish an infinite or "greater than" ^{14}C age. If no difference between the ^{14}C activity in modern and fossil C could be measured, this clearly would constitute a catastrophic flaw in Libby's original conceptualization of using ^{14}C as a dating isotope.

For the modern sample, biological CH_4 gas ("biomethane") from a sewage disposal plant in Baltimore, Maryland, was employed. The sample of great geological age was petroleum CH_4 ("petromethane") obtained from a Sun Oil Company refinery. In the absence at that time of any reliable technology that could directly measure the differences in natural ^{14}C content of these samples to a reasonable level of precision, the two CH_4 gas samples were enriched to the same degree in thermal diffusion columns used to make ^{13}C for tracer biomedical applications. The ^{14}C activity in both of these enriched CH_4 samples was then counted in a conventional GM counter (Libby 1970).

It was determined that whereas the ^{14}C activity of the modern biomethane increased in direct proportion to the measured amount of ^{13}C enrichment, there was no significant increase in the ^{14}C activity of the petromethane. Clearly, the ^{14}C in the CH_4 derived from the fossil source had decayed to a level below detection limits. There was thus no measurable ^{14}C to enrich in the petromethane (Anderson et al. 1947a, b).

The focus of much of the subsequent research on the ^{14}C dating project was to develop a relatively routine method to directly measure natural ^{14}C levels without resorting to time-consuming and energy-intensive and thus costly gaseous enrichment procedures using a thermal diffusion column (Libby 1970). The use of proportional gas counters using CH_4 or CO_2 was briefly considered (Libby 1947; Anderson, personal communication 1987). However, gas counting for ^{14}C was finally rejected by Libby in favor of a modification of his dissertation screen wall counter design using solid or elemental C.

A screen wall counter using solid C rather than a gas system was chosen by Libby on the basis of several considerations. The principal problem was to fabricate a detector that could obtain the maximum and most stable count rates with the minimum physical size (Anderson 1949, 1953). The physical size of the counters was crucial since, if all other factors remain constant, the background count rate—in this case, the counting rate of a detector absent a ^{14}C-containing carbonaceous sample—is proportional to its size. It was also extremely important to employ a detector that yielded stable count rates over relatively long periods since attempts would be made to measure ^{14}C at natural concentrations close to the background levels of the counters.

A detailed comparison of gas and screen wall detector technologies available at that time determined that the counting sensitivities of the two alternative approaches were not significantly different for counter volumes and pressures that were then

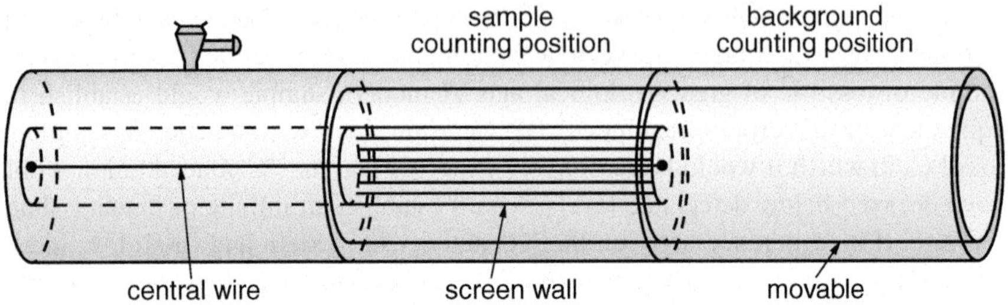

Fig. 2.8 Chicago solid C counting instrumentation. Simplified diagram of major features of the screen wall counter showing the sample counting position of sleeve containing solid sample and the background counting position of sleeve when the screen wall would be moved to the right side of the counter. Figure modified from Taylor (1987: Fig. 6.2)

considered practical. The use of larger counters and higher pressures together with more reliable electronics and stable power supplies would later shift the balance in favor of gas counters. The relatively low efficiency of a screen wall counter, e.g., the percentage of the decay events that occurred within the counter that was actually detected, which in this case was about 5 %, was offset by the fact that it could contain about 20 times more C than a gas counter of similar size (Anderson 1953; Anderson, personal communication 1987, 2002).

The major design features of the counter assembly used at the Chicago laboratory are shown in Fig. 2.8. The effective or active volume of the detector represented about one-third of the total length of the cylinder as defined by the wire grid (Libby 1952, 1955). A highly significant outcome of this arrangement, as with the original Berkeley design, was to permit the background of the detector and radioactivity emanating from the sample to be measured repeatedly without changing the counting gas filling (Libby 1967). This characteristic was a critically important consideration in the decision to use the screen wall counter design since it increased the chances of obtaining reproducible counting rates (Anderson, personal communication 1986). Counting background and ^{14}C activity in the sample were accomplished in the same manner as in Libby's pre-World War II instrument by having a sleeve that could be moved in and out of the sensitive volume of the counter.

Samples intended for solid C counting after chemical pretreatment were converted first to CO_2 by combustion or acidification. Following chemical steps to eliminate trace impurities in the sample gas, the CO_2 was reduced by reaction with magnesium at high temperature. One of the reaction products was elemental C or "lamp black." The magnesium oxide was separated from the C by treating the reaction product with concentrated hydrochloric acid. Unfortunately, not all of the magnesium oxide—termed "ash" in Libby's publications—could be removed even with repeated acid treatments. The amount of ash remaining in the sample had to be determined very carefully, since it would reduce the measured ^{14}C specific activity (Libby 1955).

The inside surface of the sample sleeve, which was constructed of metal rather than glass as was the case in Libby's Berkeley instrument, was coated with the C powder to which had been added, at various times, water, ethyl alcohol, and/or agar to form a paste. Spreading the paste containing the C powder on the inside surface of the metal sleeves was reported by those who were involved in solid C counting to be a very challenging task which required a great detail of dexterity and perseverance. Once the liquid was evaporated and the coating thoroughly dried, considerable care had to be exercised to avoid dislodging the coating from the walls of the cylinder. Arnold later reported that this feature of the solid C counting instrument caused him to regard the technology as being "developed in Hell" due to the frequency with which the coating became dislodged. Interestingly, Anderson was one of the few individuals involved in early ^{14}C studies who did not have highly negative memories of having to work with solid C counting technology.

The air used for evaporating off the water or alcohol contained traces of radon, an isotope in the uranium decay chain, which resulted in unwanted radioactivity being picked up in the C sample. Fortunately, radon has a half-life of a few days. By waiting a few weeks to pass allowed the radon to decay below the level where the beta daughters of the alpha radiation from the radon could be detected. This step became a major problem as fallout from atmospheric nuclear testing rapidly increased in the early 1950s, causing solid C counting to become an obsolete technology. For counting, the C-coated sleeve was fitted into the screen wall counter assembly. The cylinder was sealed, originally using a specially prepared black wax compound and later with o-rings, evacuated, and a counting gas was then introduced. Alternate sample and background counts were obtained by sliding the sleeve containing the sample coating in and out of the sensitive volume of the counter as defined by the wire grid.

The most crucial technical element in the development of routine low-level ^{14}C work in decay counting was the application of the anticoincidence principle to reduce background count rates. This involved surrounding the central detector containing the sample with a ring of Geiger counters and connecting their electronic output circuits together. With such an arrangement, pulses coming from the sample counter could be electronically compared as to coincidence or non-coincidence (anticoincidence) with pulses coming from one or more of the "guard" counters. The circuitry in the instrument was arranged so as to count pulses from the central counter *only* if these pulses were *not* accompanied by an essentially simultaneous pulse from the surrounding guard counters (Fig. 2.9). If a pulse from the sample and guard counters arrived within a very small time window (several milliseconds), it was assumed that thc ionization event occurring *inside* the sample counter had been caused by the same event that triggered the guard counter, was thus *not* the result of a ^{14}C-decay event inside the sample counter, and thus would not be counted as a ^{14}C-decay-caused event. Figure 2.10 is the sample counter surrounded by the ring of GM counters used in the Chicago laboratory.

Although Libby was not sure of the source of this idea to reducing background counts (cf. Libby 1970, 1979), the principle of *coincidence* counting had been a standard method employed in cosmic ray studies since the 1930s (Anderson 1953). Arnold expressed his view that the concept was "borrowed by Libby from the cosmic ray physicists and put together in the first crude form by Anderson" (Arnold

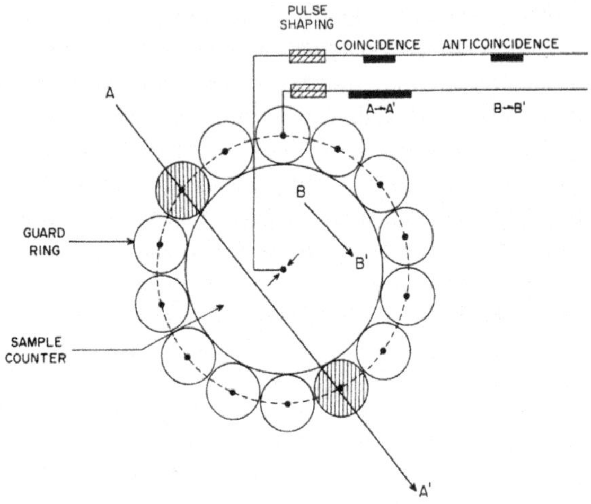

Fig. 2.9 Principle of anticoincidence counting used to reduce background counts in Libby's solid C counters. *Source* Taylor (1987: Fig. 4.7)

Fig. 2.10 Chicago solid C counting instrumentation. Screen wall counter used for ^{14}C measurements (center counter) surrounded by a ring of GM tubes. *Source* R. Berger and L.M. Libby, University of California, Los Angeles (USA), and Isotope Foundation, Santa Monica, California

1992). Several others questioned by the author had the impression that Anderson was the source of the idea. When queried about this much later, Anderson did not take credit for it, stating that the idea was "kind of in the air." Anderson later discovered that Danish scientists in the 1930s had employed the anticoincidence principle for low-level counting applications. This technology was not known to Libby or his coresearchers when ^{14}C dating was being developed at Chicago because it had been published in a Dutch language journal (Libby nd, Anderson, personal communication 1987, 2001).

Throughout the first part of 1948, problems with the counting instrumentation were slowly and painstakingly solved by Anderson and Arnold to the point that reasonably reliable counting data could be obtained (Anderson 1949). At first, only two or three days of acceptable counting runs in an entire month could be obtained (Marlowe 1999, information confirmed by Arnold 2006). By the early Spring of 1948, Anderson's and Arnold's painstaking work permitted estimates of the specific ^{14}C activity in a sample to a precision of about ± 10 %. This was sufficient to undertake the second and third critical experiments resulting in the establishment of the use of ^{14}C as a dating isotope.

The second critical experiment constituted the dissertation research of Anderson. His task was to determine whether the natural ^{14}C activity in *living* organics varied as a function of the geographic location on Earth. Since ^{14}C production in the atmosphere at the geomagnetic poles was about five times that at the equator, the worldwide comparability of ^{14}C-inferred age values would require the relatively rapid and complete mixing of cosmogenic ^{14}C into air masses near the Earth's surface on a worldwide basis. This would result in ^{14}C concentrations in living organics to be approximately equal no matter their geographical locality. If this turned out not to be the case, it would seriously violate one of the prime assumptions on which the ^{14}C dating model rested.

To test this assumption, Anderson measured the ^{14}C specific activity expressed in terms of counts (or disintegrations) per minute per gram of C (cpm g^{-1} or dpm g^{-1}) in a series of mostly wood samples from locations situated from 65° north (Sweden) to 45° south (Australia and Tierra del Fuego). In several cases, the specific ^{14}C activity of samples from high altitudes (2750–3000 m) was also obtained. The result of these experiments revealed that the values of widely geographically separated modern organics ranged from 11.81 ± 0.60 to 13.33 ± 0.43 cpm g^{-1} of C. Anderson summarized his results by stating that the "average of three equatorial samples is 12.32 ± 0.35 [cpm g^{-1}] and that of the six samples from magnetic latitudes greater than 40° [North or South] is 12.54 ± 0.20 [cpm g^{-1}]. The difference is 0.22 ± 0.40, which is not statistically significant" (Anderson 1949).

The average terrestrial biospheric specific activity of ^{14}C reported in Anderson's dissertation was 12.50 ± 0.2 cpm g^{-1} of C. This same average value was also reported in Libby et al. (1949) although there is some variation in the values cited for individual samples. A higher value of 15.3 ± 0.1 cpm g^{-1} of C reported in Anderson and Libby (1951) and incorporated in Libby's *Radiocarbon Dating* volume (Libby 1952, 1955) was a result of the later recalibration of the counter data of Anderson to account for the effect of differences in ash content in samples and counter efficiency (Libby 1952; confirmed by Anderson, personal communication 1987). Interestingly enough, however, even this first set of results revealed variations in ^{14}C activity of marine shell and in samples from the higher latitude regions. Both of these problems have continued to be studied down to the present day.

The third critical set of ^{14}C measurements was obtained on a suite of presumed known-age samples. The first reported ^{14}C age determination (C-l) was on the piece of cypress wood sent by Ambrose Lansing to Libby in January 1947 from the tomb of Djoser (Zoser), whose reign is thought to have occurred from about 2665 to 2650 BC (Shaw 2000). The age on this sample was calculated in Libby's laboratory by Arnold on July 12, 1948 (Arnold and Anderson, personal communication), and published in March 1949 (Libby et al. 1949). If one wanted to identify a specific day as the single "birthday" of radiocarbon dating, then this date might be used. In publishing this result, the age of a wood of assumed similar age from the tomb of Snofru (Sneferu), the first king of the Egyptian Fourth Dynasty who was considered the father of Khufu (Cheops to the Greeks), the builder of the first or great Giza pyramid, was also measured. In combining the ages obtained by a weighted average expressed as specific activity in counts per second per gram of C from multiple measurements on both samples, the final result was reported as agreeing "with the expected value within 1 standard deviation unit" (Libby et al. 1949).

The second measured sample was expected to have a known age about one-half that of the Zoser–Sneferu samples. However, the result of this measurement was never reported in print until 1967 (Libby 1967). This is because its count rate was essentially the same as that of the modern or contemporary samples measured by Anderson. What had transpired was that John Wilson, then the director of the University of Chicago Oriental Institute and a highly respected Egyptologist, had been asked to supply a piece of wood dating from the Hellenistic period (323 BC–14 AD). Unfortunately, the sample he selected had been obtained from a "reputable Cairo antiquities dealer" a phrase well known to be oxymoron among Egyptologists. When measured, its ^{14}C activity was statistically indistinguishable from that obtained on biomethane. It was a modern piece of wood and therefore obviously a fake (James Arnold in Marlowe 1980).

Since Arnold reports that this affair ruined his Christmas (December 1948), one wonders what would have happened if the wood supplied by John Wilson would have been the first "known-age" sample to be measured or if the sample supplied by Lansing would have been a modern fake. Libby later commented that fortunately their early studies encountered few fakes "as otherwise faith in radiocarbon dating would have been rapidly shaken and the research abandoned …" He later publicly mused that much of his ^{14}C research benefited from what he called "good fortune," which he suggested bordered on the "miraculous" (Libby 1970; Deevey 1984).

The capstone to the more than fifteen months of intensive work was "Age Determinations by Radiocarbon Content: Checks with Samples of Known Age," published in the December 23, 1949, issue of *Science* (Arnold and Libby 1949). In this paper, the first "Curve of Knowns" was presented with six data points spanning a period from about AD 600 to 2700 BC (Fig. 2.11). Earlier in that year, Libby had organized a measurement of the ^{14}C half-life by a group at the Argonne National Laboratory located near Chicago (Engelkemeir et al. 1949). Their value of

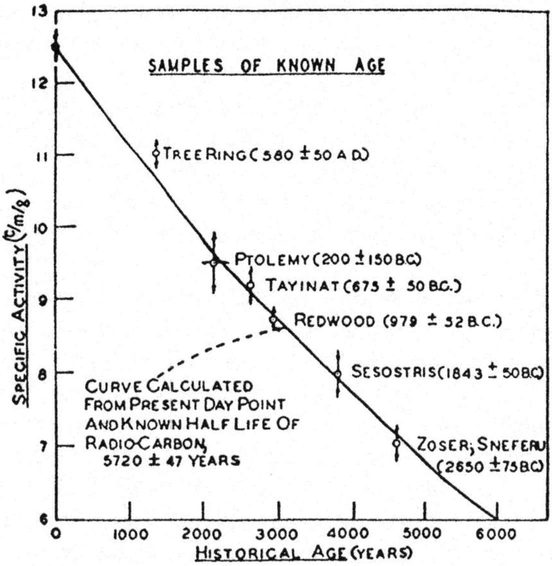

Fig. 2.11 The first of Libby's "Curve of Knowns" (Arnold and Libby 1949: Fig. 1; Libby 1952: Fig. 1). This plot employed the first ^{14}C half-life used by Libby, 5730 ± 40 years, and expressed ^{14}C activity in samples in terms of counts of ^{14}C per minute per gram of C (c/m/g). Libby's second "Curve of Knowns" substituted 5568 ± 30 years, the so-called Libby half-life and expressed ^{14}C activity in samples in terms of percentage of modern ^{14}C activity. *Source* Rainer Berger, University of California, Los Angeles (USA)

5720 ± 47 year was used to convert the sample count rate of each sample to its ^{14}C age. For Libby, the reasonably close (±10 %) correspondence between the "known-age" and "^{14}C age" of these seven samples indicated that the major assumptions on which the dating model rested were valid for at least the last 5000 years to a first-order approximation based on the then typical counting precision of ±10 %.

2.6 Dating Unknown-Age Samples: 1949–1954

By early 1949, Libby had been confident enough to move into what he called the "great unknown periods of prehistory." Since there were no fixed chronometric points with which the ^{14}C values could be directly compared, it was anticipated that validation would be accomplished by the degree of "internal consistency from a wide variety of samples and in a wide variety of problems" (Arnold and Libby 1949).

During the period of the Chicago laboratory's routine dating operation from about March 1949 until early 1954, ^{14}C values on more than 500 samples were obtained. About two-thirds of these determinations were on samples of archaeological significance, most of which had been submitted by or through the collaborators. The first listing of "provisional" ^{14}C dates was prepared in mimeograph form on January 1, 1950, with a supplement dated in April. Both lists were sent principally to the collaborators. A number of these values were subsequently modified, and these revised values appeared in a booklet published by the Institute of Nuclear Studies and dated September 1, 1950. This booklet contained 148 ^{14}C values and carried a notation in the preface that the "list itself is not for publication in its present form, though the dates themselves may be quoted freely" (Arnold and Libby 1950; Johnson 1951).

The text of the booklet with some modifications constituted the first Chicago list of dates, which appeared in the journal *Science* in February 1951 (Arnold and Libby 1951). This inaugurated the custom of laboratories publishing their ^{14}C values in the form of "date lists." For US laboratories, the earliest date lists continued to appear in the journal *Science*. The half-life chosen by Libby to calculate the ^{14}C age determinations beginning with the Chicago date list booklet in 1950 was 5568 ± 30 year. This value replaced the earlier 5720 ± 47 year that had been used to calculate the first "Curve of Knowns" (Arnold and Libby 1949). The 5568 ± 30 year value was a weighted average of three published measurements, 5580 ± 45, 5589 ± 75, and 5513 ± 165 year that had been obtained by Libby and his collaborators and two other groups (Engelkemeir et al. 1949). The 5568 years value became known as the "Libby half-life" and has continued to be used to calculate "conventional" ^{14}C age values, as later explicitly defined by Stuiver and Polach (1977), even though it now is clear that the half-life Libby originally used (5720 ± 47 year) was closer to the correct value. With only a few exceptions, e.g., some French laboratories, the "Libby half-life" has been used to calculate all conventional ^{14}C values published in the journal *Radiocarbon.*

The Chicago laboratory ceased operation with the departure of Libby in October 1954 to accept an appointment by President Eisenhower to become a commissioner of the US Atomic Energy Commission (AEC) [now the US Department of Energy]. He held this position until June 1959 when he joined the faculty of the University of California, Los Angeles. At the AEC, Libby was a major supporter of the growth of the electric nuclear power industry as well as the development of nuclear weapons. He also directed "Project Sunshine," a study of worldwide stratospheric mixing using radioactive fallout from the testing of nuclear weapons (Buck 1983; de Messières 2001). This project continued the work he had begun during the Manhattan Project and he had maintained direct involvement in US atomic energy development at the AEC while at the University of Chicago from the time the AEC had been established, serving as a member of its General Advisory Committee. It was his access to classified information concerning the magnitude of the fallout from "bomb" ^{14}C that alerted him to the problem that this fallout would create for solid C counting. Together with his colleague at Chicago, physicist Edward Teller, Libby was an outspoken advocate of the development of nuclear weapons. As the result, both were considered controversial figures in the public discourse concerning the uses of atomic energy.

2.7 Radiocarbon Dating Comes of Age: 1954–1960

The acceptance by the vast majority of scientists from a number of disciplines of the overall, general validity of the ^{14}C method is reflected in the rapid establishment of laboratories to perform ^{14}C analyses. Solid C counting instrumentation was installed at eight institutions in the early 1950s. These pioneering laboratories included the University of Arizona (Wise and Shutler 1958), University of Copenhagen, Denmark (Anderson et al. 1953), Lamont Geological Observatory, Columbia University (Kulp et al. 1951), University of Michigan (Crane 1956), New Zealand Institute of Nuclear Studies (Fergusson and Rafter 1953), University of Pennsylvania (Ralph 1955), University of Saskatchewan (McCallum 1955), and Yale University (Blau et al. 1953). In at least three instances at US universities (Arizona, Michigan, and Pennsylvania), archaeologists were instrumental in establishing the laboratories. Some thought at that moment that they stood "before the threat of the atom in the form of [^{14}C] dating. This may be our last chance for old-fashioned, uncontrolled guessing" (Phillips et al. 1951).

Low-level ^{14}C counting using Libby's solid C approach required an extraordinary attention to detail in laboratory procedure. Many who attempted to duplicate the technique experienced moderate to severe difficulties in obtaining reproducible values. Most frustrating was the problem of avoiding contamination of the solid C samples with nuclear fission fallout products resulting from the testing of thermonuclear devices in the atmosphere beginning in the middle of the 1950s. As a result of these and other problems, the laboratories previously noted, which had begun operations with solid C counting, converted over to some type of gas counting by the end of the decade. Meanwhile, other early laboratories had commenced operations with gas counting using CO_2 or acetylene or, in one case, with the use of liquid scintillation technology. These included the facilities at the University of Groningen, the Netherlands (de Vries and Barendsen 1954), the University of Heidelberg, Germany (Munnich 1957), and the US Geological Survey in Washington, D.C. (Suess 1954).

By the inauguration of the journal *Radiocarbon* in 1959, more than twenty ^{14}C laboratories were actively pursuing a wide range of research topics utilizing ^{14}C. *Radiocarbon* (initially the *Radiocarbon Supplement* to the *American Journal of Science*) was inaugurated to solve the problem of the increasing number of ^{14}C laboratory date lists which could no longer be accommodated in the journal *Science* as well as to provide an archive of primary data on individual ^{14}C dates (Deevey 1984). This later function was of particular concern at that time since, in some cases, ^{14}C values were being cited without laboratory number or without descriptions of any kind (Deevey and Flint 1959). Rapid development during this period also took place particularly in Europe (Waterbolk 1960) as well with laboratories being built in England (at the British Museum and Cambridge), Sweden (Stockholm and Uppsala), France (Gif and Saclay), Ireland (Dublin), Germany (Cologne and Hanover), Belgium (Louvain), Russia (Moscow), Italy (Pisa and Rome) as well as in Switzerland (Bern) and Norway (Trondheim). In the USA during this period, the first commercial ^{14}C laboratory was opened and several oil companies briefly supported ^{14}C facilities.

Initial ^{14}C data that yielded age estimates at dramatic variance with the views of individual archaeologists or geologists (e.g., Neustupný 1970; Antevs 1957; Lee 1981) generated discussions that questioned the validity of the ^{14}C method in general (Barker 1958; Johnson, personal communication 1986). However, with the rapidly mounting evidence of the general validity of the ^{14}C model in broad outline, discussions quickly turned to questions of accuracy of ^{14}C values from specific archaeological or geological contexts or geochemical environments (Broecker and Kulp 1956).

2.8 Nobel Prize: 1960

The capstone at the end of the first decade of ^{14}C dating was the award of the 1960 Nobel Prize in chemistry to Libby for the development of the method (Fig. 2.12). Translated from Swedish, the citation reads "At its meeting of November 3, 1960, the Royal Swedish Academy of Science has decided, in conformity with the terms of the November 27, 1895 will of Alfred Nobel, to award the prize to be given this

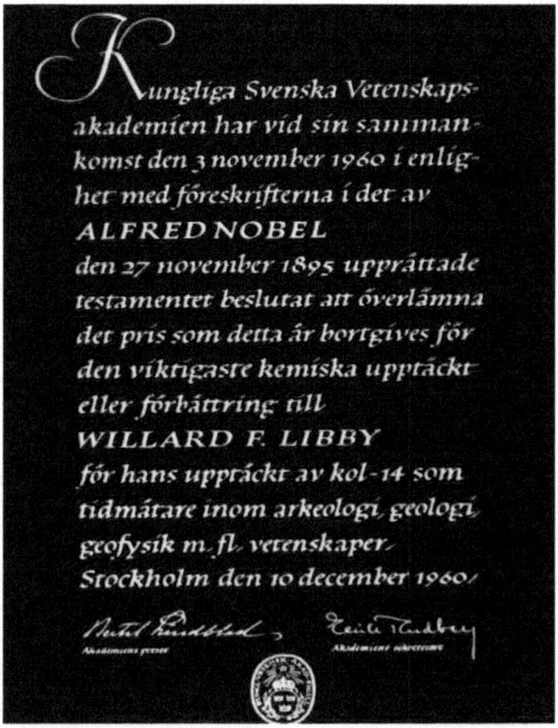

Kungliga Svenska Vetenskaps-
akademien har vid sin samman-
komst den 3 november 1960 i enlig-
het med föreskrifterna i det av
ALFRED NOBEL
den 27 november 1895 upprättade
testamentet beslutat att överlämna
det pris som detta år bortgives för
den viktigaste kemiska upptäckt
eller förbättring till
WILLARD F. LIBBY
för hans upptäckt av kol-14 som
tidmätare inom arkeologi, geologi,
geofysik m. fl. vetenskaper.
Stockholm den 10 december 1960.

Fig. 2.12 Text of the 1960 Nobel Prize in chemistry awarded to Willard F. Libby for the development of the ^{14}C dating method. *Source* Image made available for reproduction by L.M. Libby, University of California, Los Angeles (USA)

year for the most important chemical discovery or improvement to Willard F. Libby for his method to use Carbon-14 for age determinations in archeology, geology, geophysics, and other sciences." Interestingly, when President Eisenhower was informed that Libby had been awarded the Nobel Prize in chemistry, he expressed surprise since he had not been aware that Libby was a scientist. Eisenhower had known him only in his role as a member of the Atomic Energy Commission (Libby 1981).

At the time the Nobel Laureate was awarded, Libby had the year previously arrived at the University of California, Los Angeles (UCLA), to take up a faculty appointment as professor of chemistry and, two years later, as the director of the Institute of Geophysics and Planetary Physics (IGPP). In 1961, Libby brought to UCLA Gordon J. Fergusson [1922–2009], formerly of the New Zealand Institute of Nuclear Studies ^{14}C laboratory, to build an Isotope Laboratory for the IGPP and collaborate in geophysical research. In the 1960s, the author of this chapter served as a graduate research assistant in the UCLA laboratory while a graduate student in the UCLA Department of Anthropology specializing in archaeology and archaeometry/archaeological science (Taylor 1970). In 1968, Rainer Berger [1930–2003] succeeded Fergusson as the director of the UCLA laboratory having faculty appointments in the UCLA departments of anthropology, geography, as well as the IGPP. Libby's UCLA laboratory, having operated for 34 years, was closed when Rainer Berger retired from UCLA in 1995.

2.9 Afterward

Willard Libby retired from UCLA in 1976 and died in Los Angeles four years later at the age of 71 (Arnold 1981). His wife, Leona Marshall Libby, also a noted chemist, noted that "Until the day of his death, he was carrying on his research at top speed as he always did. One week before [his death], he had completed a seminal paper on the geometrical theory of catalysis of light hydrocarbon reactions, important in oil refineries and on the role of electron tunneling therein" (Libby 1981).

The Analy High School in Sebastopol, California, from where Libby graduated in 1926, named a Willard F. Libby Science Classroom Wing in his honor and a large mural with his likeness, has been painted on an exterior wall of the school. A Sebastopol park was named the Willard Libby Memorial Park, and an 8 km (5 mile) segment of a California State highway between Sebastopol and an adjacent town became the Willard F. Libby Memorial Highway.

During the period of his collaboration with Libby, James Arnold pioneered in the development of liquid scintillation technology for low-level counting (Arnold 1954). After leaving the ^{14}C dating project at Chicago in 1955, he served as the faculty of Princeton University. While at Princeton, he discovered the presence of ^{10}Be in nature (Arnold 1956). In 1958, he was asked to be the founding chairman of the University of California, San Diego (UCSD), Department of Chemistry. He spent the remainder of his illustrious scientific career at UCSD, a campus located at La Jolla, California, overlooking the Pacific Ocean. Arnold was elected to the US National Academy of Science in 1964 and was appointed as the first holder of UCSD Harold C. Urey Professorship in Chemistry. Arnold retired as professor emeritus from UCSD in 1993 and died at the age of 88 on January 6, 2012.

Upon completion of his Ph.D. at Chicago, Ernest Anderson returned to the University of California, Los Alamos Scientific Laboratory in Los Alamos, New Mexico, where he spent the remainder of his scientifically productive career. He continued to be involved in the development of low-level counting (Anderson and Haynes 1956) and assisted in the development of the ^{14}C laboratory at the University of Copenhagen in the early 1950s (Anderson et al. 1953). While in Denmark, he was surprised to learn that the principle of anticoincidence counting for low-level radioactivity measurement had been developed in the 1930s by Dutch scientists. In 1966, he received the Ernest O. Lawrence Award from the US Department of Energy for his contributions to nuclear technology. He retired from Los Alamos in 1995 and died at the age of 92 on May 20, 2013.

For a more detailed discussion of the scientific background to and initial development of ^{14}C dating with a particular focus on its application in archaeology, interested readers are directed to Chap. 8 in the 2nd edition of *Radiocarbon Dating: An Archaeological Perspective* (Taylor and Bar-Josef 2014).

Acknowledgements Portions of this chapter were written during a sabbatical year provided by the University of California, Riverside with supplemental support from the Gabriel R. Vierra Memorial Fund. The support of Ellen Druffel, Sue Trumbore, and John Southon of the University of California, Irvine W. M. Keck Carbon Cycle Accelerator Mass Spectrometry Laboratory is very much appreciated as well as the comments of Ellen Druffel on a draft of this chapter.

References

Anderson, E.C. 1949. *Natural radiocarbon*. Ph.D. dissertation, University of Chicago.

Anderson, E.C. 1953. The production and distribution of natural radiocarbon. *Annual Review of Nuclear Science* 2: 63–89.

Anderson, E.C., and R.N. Hayes. 1956. Recent advances in low level counting techniques. *Annual Review of Nuclear Science* 6: 303–323.

Anderson, E.C., and W.F. Libby. 1951. World-wide distribution of natural radiocarbon. *Physical Review* 81: 64–69.

Anderson, E.C., W.F. Libby, S. Weinhouse, A.F. Reid, A.D. Kirshenbaum, and A.V. Grosse. 1947a. Radiocarbon from cosmic radiation. *Science* 105: 576.

Anderson, E.C., W.F. Libby, S. Weinhouse, A.F. Reid, A.D. Kirshenbaum, and A.V. Grosse. 1947b. Natural radiocarbon from cosmic radiation. *Physical Review* 72: 931–936.

Anderson, E.C., H.H. Levi, and H. Tauber. 1953. Copenhagen natural radiocarbon measurements, I. *Science* 118: 6–9.

Antevs, E. 1957. Geological tests on the varve and radiocarbon chronologies. *The Journal of Geology* 65: 129–148.

Arnold, J.R. 1954. Scintillation counting of natural radiocarbon I: the counting method. *Science* 119: 155–157.

Arnold, J.R. 1956. Beryllium-10 produced by cosmic rays. *Science* 124: 584–585.
Arnold, J.R. 1981. Willard F. Libby (1908–1980). *The American philosophical society yearbook 1980*. Philadelphia: The American Philosophical Society, pp 608–612.
Arnold, J.R. 1992. (with Schuch RL) The early years with Libby at Chicago: a retrospective. In *Radiocarbon after four decades an interdisciplinary perspective*, ed. R.E. Taylor, A. Long, and R.S. Kra, 3–10. New York: Springer-Verlag.
Arnold, J.R., and W.F. Libby. 1949. Age determinations by radiocarbon content: checks with samples of known age. *Science* 10: 678–680.
Arnold, J.R., and W.F. Libby. 1950. *Radiocarbon dates (September 1, 1950)*. Institute for Nuclear Studies: The University of Chicago.
Arnold, J.R., and W.F. Libby. 1951. Radiocarbon dates. *Science* 113: 111–120.
Barker, H. 1958. Radio carbon dating: its scope and limitations. *Antiquity* 32: 253–263.
Blau, M., E.S. Deevey Jr, and M.S. Gross. 1953. Yale natural radiocarbon measurements, I. pyramid valley, New Zealand and its problems. *Science* 118: 1–6.
Broecker, W.S., and J.L. Kulp. 1956. The radiocarbon method of age determination. *American Antiquity* 22: 1–11.
Buck, A.L. 1983. *A history of the atomic energy commission*. Washington DC: U.S. Department of Energy.
Crane, H.R. 1956. University of Michigan radiocarbon dates I. *Science* 124: 664–672.
de Messières, N. 2001. Libby and the interdisciplinary aspect of radiocarbon dating. *Radiocarbon* 43: 1–5.
de Vries, H., and G.W. Barendsen. 1954. Measurements of age by the carbon-14 technique [Groningen I]. *Nature* 174: 1138–1141.
Deevey Jr, E.S. 1984. Zero BP plus 34: 34 years of radiocarbon. *Radiocarbon* 26: 1–6.
Deevey Jr, E.S., and R.F. Flint. 1959. Preface. *Radiocarbon* 1: 1.
Engelkemeir, A.G., W.H. Hamill, M.G. Inghram, and W.F. Libby. 1949. The half-life of radiocarbon (C^{14}). *Physical Review* 75: 1825–1833.
Fergusson, G.J., and T.A. Rafter. 1953. New Zealand ^{14}C age measurements I. *New Zealand Journal of Science and Technology, Section B*. 35: 127–128.
Geiger, H., and W. Müller. 1928. Elektronenzählrohr zur Messung schwächster Aktivitäten [Electron Tube to Measure Weak Radiation]. *Naturwissenschaften* 16(31): 617–618.
Heusser, C.J. 1990. Serge Alexander Korff (1906–1989). Radiocarbon 32: iv.
Johnson, F. 1951. Introduction. In *Radiocarbon dating*, vol 8, 1–3. Memoirs of the Society for American Archaeology. (*American Antiquity* 17(1)[part 2]: 1–3).
Kamen, M.D. 1963. Early history of carbon-14. *Science* 140: 584–590.
Kamen, M.D. 1985. *Radiant science, dark politics, a memoir of the nuclear age*. Berkeley: University of California Press.
Korff, S.A. 1940. On the contribution to the ionization at sea-level produced by the neutrons in the cosmic radiation. *Terrestrial Magnetism and Atmospheric Electricity* 45: 133–134.
Korff, S.A., and W.E. Danforth. 1939. Neutron measurements with boron-trifluoride counters. *Physical Review* 55: 980.
Kulp, J.L., H.W. Feely, and L.E. Tryon. 1951. Lamont natural radiocarbon measurements, I. *Science* 114: 565–568.
Kurie, F.N.D. 1934. A new mode of disintegration induced by neutrons. *Physical Review* 45: 904–905.
Libby, L.M. nd. *The isotope people*. Unpublished book manuscript.
Lee, R.E. 1981. Radiocarbon: ages in error. *Anthropological Journal of Canada* 19: 9–29.
Libby, W.F. 1932. Simple amplifier for Geiger-Muller counters. *Physical Review* 42: 440–441.
Libby, W.F. 1933. Radioactivity of ordinary elements, especially Samarium and Neodymium: method of detection. PhD dissertation, University of California, Berkeley.
Libby, W.F. 1946. Atmospheric Helium three and radiocarbon from cosmic radiation. *Physical Review* 69: 671–672.
Libby, W.F. 1947. Measurement of radioactive tracers particulary C^{14}, S^{35}, T and other longer-lived low-energy activities. *Analytical Chemistry* 19: 2–6.

Libby, W.F. 1952. *Radiocarbon dating*. Chicago: University of Chicago Press.
Libby, W.F. 1955. *Radiocarbon dating*, 2nd ed. Chicago: The University of Chicago Press.
Libby, W.F. 1964. Berkeley radiochemistry. *Annual Review of Physical Chemistry* 15: 7–12.
Libby, W.F. 1967. History of radiocarbon dating. *Radioactive dating and methods of low level counting*, 3–25. Vienna: International Atomic Energy Agency.
Libby, W.F. 1970. Radiocarbon dating. *Philosophical Transactions of the Royal Society of London* 269A: 1–10.
Libby, W.F. 1978. *Interview with Willard F. Libby by Mary Terrall*. Los Angeles: Oral History Program, University of California.
Libby, W.F. 1979. *Interview with Willard F. Libby by Greg Marlowe*. New York: American Institute of Physics.
Libby, L.M. 1981. Willard Frank Libby 1908–1980. In: *Willard F. Libby collected papers,* Berger, R. Libby, L.M. ed. vol. 1, pp 5–6. Tritium and radiocarbon. Santa Monica: Geo Science Analytical.
Libby, W.F., and D.D. Lee. 1939. Energies of the soft beta-radiations of Rubidium and other bodies. Method for their determination. *Physical Review* 55: 245.
Libby, W.F., E.C. Anderson, and Jr Arnold. 1949. Age determination by radiocarbon content: world wide assay of natural radiocarbon. *Science* 109: 227–228.
Marlowe, G. 1980. W.F. Libby and the archaeologists, 1946–1948. *Radiocarbon* 22: 1005–1014.
Marlowe, G. 1999. Year One: radiocarbon dating and American archaeology, 1947–1948. *American Antiquity* 64: 9–31.
McCallum, K.J. 1955. Carbon-14 age determinations at the University of Saskatchewan (I). *Transactions of the Royal Society of Canada* 49(4): 31–35.
Montgomery, C.G., and D.D. Montgomery. 1939. The intensity of neutrons of thermal energy in the atmosphere at sea level. *Physical Review* 56: 10–12.
Munnich, K.O. 1957. Heidelberg natural radiocarbon measurements I. *Science* 126: 194–199.
Neustupný, E. 1970. The accuracy of radiocarbon dating. In *Radiocarbon variations and absolute chronology*, ed. I.U. Olsson, 23–34. Almqvist and Wiksell: Stockholm.
Phillips, P., Ford, J., Griffin, J.B. 1951. Archaeological survey in the lower Mississippi alluvial valley 1940–1947. *Papers of the Peabody Museum of American Archaeology and Ethnology*, Vol. 25. Cambridge: Peabody Museum.
Ralph, E.K. 1955. University of Pennsylvania radiocarbon dates I. *Science* 121: 149–151.
Rubin, S., and M. Kamen. 1941. Long-lived radioactive carbon: C_{14}. *Physical Review* 59: 349–354.
Segrè, E. 1970. *Enrico fermi physicist*. Chicago: University of Chicago Press.
Segrè, E. 1991. *A mind always in motion the autobiography of Emilio Segrè*. Berkeley: University of California Press.
Shaw, I. 2000. *The Oxford history of ancient Egypt*. Oxford: Oxford University Press.
Stuiver, M., and H. Polach. 1977. Discussion: reporting of ^{14}C data. *Radiocarbon* 19: 355–363.
Suess, H.E. 1954. U. S. Geological survey radiocarbon dates I. *Science* 120:467–473.
Taylor, R.E. 1970. *Chronological problems in western Mexican archaeology: a dating systems approach to archaeological research.* PhD dissertation, University of California, Los Angeles.
Taylor, R.E., and O. Bar-Josef. 2014. *Radiocarbon dating: an archaeological perspective*, 2nd ed. Walnut Creek: Left Coast Press.
Waterbolk, H.T. 1960. The 1959 carbon-14 symposium at Groningen. *Antiquity* 34: 14–18.
Wise, E.N., and D. Shutler Jr. 1958. University of Arizona radiocarbon dates (I). *Science* 127: 72–73.

Chapter 3
Radiocarbon Nomenclature, Theory, Models, and Interpretation: Measuring Age, Determining Cycling Rates, and Tracing Source Pools

S.E. Trumbore, C.A. Sierra and C.E. Hicks Pries

3.1 Introduction

This chapter serves several purposes. It expands on the introduction to isotopes and the processes that distribute carbon (C) isotopes differently within the Earth system. It briefly discusses stable C isotopes and the processes responsible for mass-dependent fractionation of isotopes (Sect. 3.2). This mass-dependent fractionation affects radiocarbon (^{14}C) but is predictable from the distribution of the stable isotope ^{13}C and the fact that the mass difference between ^{14}C and ^{12}C is approximately twice than that between ^{13}C and ^{12}C. To isolate the effects that influence the ^{14}C isotope value alone, ^{14}C data are normally reported in a way that corrects for mass-dependent fractionation using the measured ^{13}C value.

The reporting of ^{14}C data, discussed in Sect. 3.3, differs according to application. We introduce all the regularly used nomenclatures, with a focus on the ones most commonly used in C cycle research, including Fraction Modern (*F*), ^{14}C age, and measures used to track bomb ^{14}C ($\Delta^{14}C$). Section 3.4 continues the link between nomenclature and data reporting by introducing the theory and models commonly used for interpreting ^{14}C data in terms of its three major uses: (1) closed systems, using the calibration curves to convert ^{14}C age into the time elapsed since C in the sample was fixed from the atmosphere; (2) open systems, estimating the rate of exchange of C between reservoirs;

S.E. Trumbore (✉) · C.A. Sierra
Max Planck Institute for Biogeochemistry, Jena, Germany

S.E. Trumbore
Department of Earth System Science, University of California Irvine, Irvine, CA, USA

C.E. Hicks Pries
Department of Biology, University of Florida, Gainesville, FL, USA

C.E. Hicks Pries
Climate and Ecological Sciences Division, Lawrence Berkeley National Laboratory, Berkeley, CA, USA

E.A.G. Schuur et al. (eds.), *Radiocarbon and Climate Change*,
DOI 10.1007/978-3-319-25643-6_3

and (3) mixing models for estimating the contributions of different C sources. Deciding which model applies to the system being studied is the most important step in making valid interpretations of ^{14}C measured in a given compound or reservoir of C. The remaining chapters of the book will build on these general approaches using examples specific to tracing C through the different 'spheres' of the Earth system.

3.2 Carbon Isotope Basics: General Nomenclature and Isotope Fractionation

Atomic nuclei with the same number of protons (atomic number) but different numbers of neutrons are called *isotopes*. Isotopes of a particular element behave the same chemically because they have the same number of protons (and therefore electrons). However, the different isotopes have different masses, and therefore, the bonds they form have slightly different strengths.

Carbon has three naturally occurring isotopes. The most abundant is ^{12}C with 6 neutrons and 6 protons in its nucleus, representing 98.89 % (Hoefs 2009) of the C atoms on Earth. The other 1.11 % of all C is ^{13}C with 6 protons and 7 neutrons. Both ^{12}C and ^{13}C are *stable* isotopes created originally by nucleosynthesis in stars. The abundance of these stable isotopes remains constant through time, unchanged since their synthesis. However, individual C atoms are constantly being transferred between different forms (organic and inorganic) of C and among the various C reservoirs (atmosphere, biosphere, hydrosphere, lithosphere) in the Earth system. This movement of atoms can change the relative abundance of ^{12}C and ^{13}C in any particular reservoir.

In this book, we will discuss stable C isotopes briefly because the processes that fractionate them also affect the distribution of ^{14}C. For a more thorough treatment of stable isotopes and their applications in ecology and the C cycle, the reader is referred to some of the excellent texts available [Hoefs (2009), Faure (1986), Mook (2000), Kendall and McDonnell (1998), and Rundel et al. (1989)].

Radiocarbon (^{14}C) is the radioactive isotope of C, with 6 protons and 8 neutrons. Radiocarbon is a *cosmogenic radionuclide*, which means it is constantly being created by the interaction of cosmic rays with the atmosphere and Earth's surface. The nucleus of ^{14}C is unstable and will spontaneously emit a β particle (electron):

$$^{14}_{6}C \rightarrow {}^{14}_{7}N + \beta^{-} + \gamma(\text{energy}) \tag{3.1}$$

In between its production and decay, the ^{14}C nucleus associates first with oxygen to form ^{14}CO, which is in turn oxidized to $^{14}CO_2$ within a matter of a few months. The $^{14}CO_2$ mixes into the troposphere and enters the global C cycle that continuously exchanges C between atmosphere, land, and ocean C reservoirs (Fig. 3.1). The balance of production and decay are such that the natural abundance of ^{14}C was

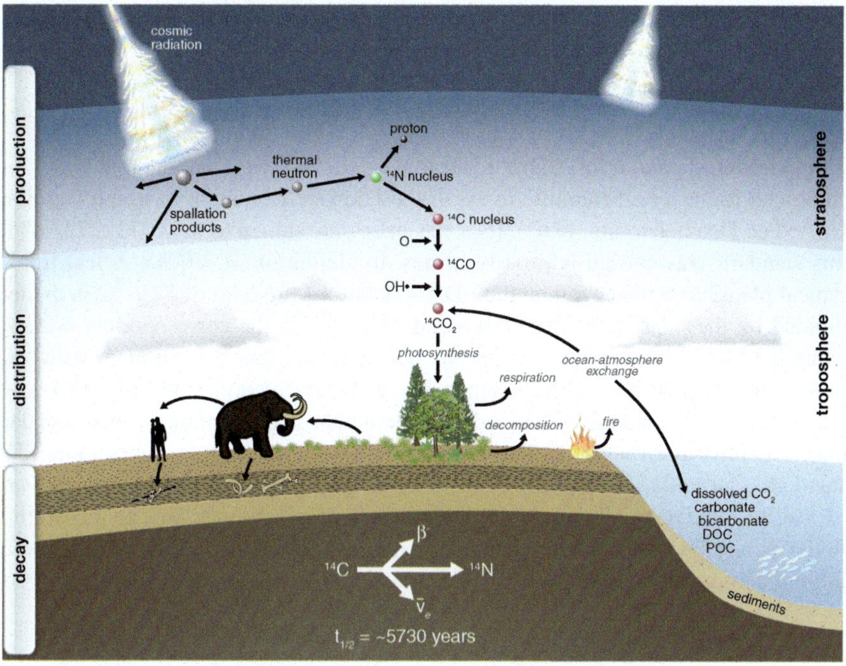

Fig. 3.1 Radiocarbon is produced in the stratosphere as a spallation product when cosmic rays interact with the Earth's atmosphere. During the time between production and decay, it is oxidized to $^{14}CO_2$, mixed through the atmosphere, and enters the Earth's C cycle

roughly one in every trillion (10^{12}) C atoms (an abundance of $\sim 10^{-10}\%$) in the preindustrial atmosphere. Radiocarbon is also produced in very small amounts by cosmic rays that interact with oxygen in rocks at the Earth's surface and in nuclear reactions mediated by humans. Another radioactive isotope, ^{11}C, is human-produced but very short-lived and is used for tracer studies. It will not be discussed in any more detail in this book.

It is difficult to measure the absolute abundances of isotopes, so mass spectrometric techniques instead measure relative isotope abundance. Isotopic composition is therefore expressed as follows:

$$R = \frac{\text{rare}}{\text{abundant}} = \frac{^{13}C}{^{12}C} \tag{3.2}$$

where R is the ratio of the rare isotope to the abundant isotope. The ratio R in a sample is compared to that of a known standard. Because the differences found in nature between sample and standard are relatively small, the isotopic value is expressed using the delta (δ) notation as the deviation in parts per thousand (per mil) of R for the sample compared to that of a universally agreed upon standard:

$$\delta^{13}\mathrm{C} = \left[\frac{\left(\frac{^{13}\mathrm{C}}{^{12}\mathrm{C}}\right)_{\mathrm{sample}}}{\left(\frac{^{13}\mathrm{C}}{^{12}\mathrm{C}}\right)_{\mathrm{standard}}} - 1 \right] \times 1000 \tag{3.3}$$

The original standard for stable C isotope reporting was based on the calcium carbonate found in a belemnite fossil in the Pee Dee formation in South Carolina, USA (Pee Dee belemnite, PDB). Its R has been determined to be 0.0112372. While this standard was chosen arbitrarily, it has, by definition, $\delta = 0$ ‰. A leaf from a typical plant has a lower R than the PDB standard; if the ratio of $^{13}C/^{12}C$ in the leaf divided by the $^{13}C/^{12}C$ of the PDB standard is 0.972, the corresponding $\delta^{13}C_{PDB}$ value is $(0.972 - 1) \times 1000 = -28$ ‰. This value is typical for a plant with a C_3 photosynthetic pathway with enzymes that favor the incorporation of $^{12}CO_2$ over $^{13}CO_2$. In fact, the actual PDB standard material has been used up, so new standard materials have been adopted that are themselves related back to the original PDB standard. Currently, $\delta^{13}C$ values of all C-bearing materials should be measured and expressed relative to Vienna PDB (VPDB, a replacement standard of the original PDB) on a scale normalized by assigning consensus values of −46.6 ‰ to L-SVEC lithium carbonate and +1.95 ‰ to NBS 19 calcium carbonate (Coplen et al. 2006).

3.2.1 Mass-Dependent Fractionation of Isotopes

Isotopes distribute themselves differently in the environment due to the two main processes—*equilibrium* and *kinetic fractionation*. Equilibrium fractionation processes reflect the fact that isotopes will partition differently among phases for the same molecule (e.g., for H_2O as vapor, liquid, or ice phases) or between different types of chemical bonds (in chemical reactions) for a system at chemical equilibrium. The underlying reason has to do with the energetic states of bonds in molecules and with molecules in their environment (degrees of freedom or entropy).

3.2.2 Equilibrium Exchange Reactions

At equilibrium when reactions are completely reversible, isotopes of an element will distribute themselves among the reactant and products according to the lowest energy state of the overall system. The energy states of molecular bonds are quantized, and the energy of the total molecule depends on the degree of activation of its rotational, translational, and vibrational modes. At room temperatures, translational and rotational modes are fully activated, and the most important factor that varies for different isotopes is the energy associated with vibration.

For C isotopes, the most common application of equilibrium fractionation is with the inorganic C cycle such as the precipitation of calcium carbonate from water (Mook et al. 1974). The heavier isotope will partition preferentially into the molecule in which it has a stronger bond, or into the phase with less entropy (e.g., a solid versus a liquid versus a gas). Thus, the δ^{13}C of C in calcium carbonate, which has fewer degrees of freedom because it is a solid, is more enriched (by 9–10 ‰) than that of atmospheric CO_2 in equilibrium with the water from which the $CaCO_3$ is precipitated (Mook 2000; Fig. 3.2). The degree of fractionation will also be

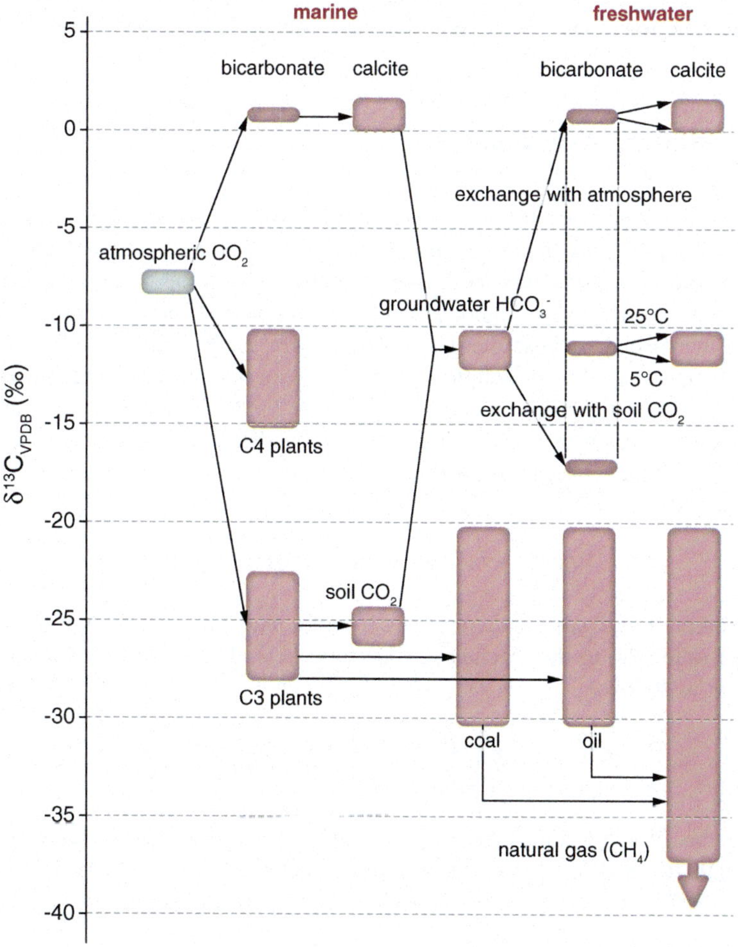

Fig. 3.2 Schematic survey of δ^{13}C variations found in the Earth's major C reservoirs. Figure modified from IAEA (2001)

temperature dependent; at higher temperatures, the differences in energy between reactants and products will be lower, and therefore, isotopic fractionation will be smaller.

It should be noted that the equilibrium condition rarely is used to interpret the distribution of stable isotopes of C. For example, the dissolution of CO_2 in seawater is controlled by the rate of diffusion of CO_2 through the air–water interface, rather than the equilibration between gas and hydrated (dissolved gas) phases (Wanninkhof 1985). In any case, the equilibrium constants for C isotopes in the dissolved inorganic C (DIC) system are difficult to determine experimentally because many phases coexist simultaneously [gaseous and dissolved CO_2, HCO_3^- (bicarbonate ion) CO_3^{2-} (carbonate ion) and calcium carbonate; (Mook 2000)].

3.2.3 Kinetic Fractionation

Kinetic fractionation occurs in non-equilibrium conditions such as when a reaction is not reversible. This fractionation reflects the fact that the lighter isotope ^{12}C will react and diffuse faster than the heavier isotope ^{13}C—in the first case because the bond strengths are lower (higher ground-state vibrational energy) for the ^{12}C bond and in the second because the gas molecule with smaller mass moves faster at a given temperature. For dissolution of CO_2 in seawater, the observed kinetic fractionation (determined experimentally) is −2.4 ± 2.0 ‰, because $^{13}CO_2$ is diffusing across the air–water interface slower than $^{12}CO_2$ (Wanninkhof 1985).

A prominent example of stable C isotope fractionation is the process of photosynthesis:

$$6CO_2 + 6H_2O + \text{light} \rightarrow C_6H_{12}O_6 + 6O_2$$

The isotopic *fractionation factor*, alpha (α), of a reaction is the ratio of the product isotope ratio (R_P) to the reactant isotope ratio (R_R). The value of α for photosynthesis is as follows:

$$\alpha = \frac{R_P}{R_R} = \frac{0.9724}{0.9926} = 0.9796 \quad (3.4)$$

where the $\delta^{13}C$ of atmospheric CO_2 (the reactant) is −7.4 ‰ (corresponding $R_R = 0.9926$) and the $\delta^{13}C$ of the plant material (the product) is −27.6 ‰ ($R_p = 0.9724$). This can also be expressed as an *enrichment factor*:

$$\varepsilon = (\alpha - 1) \times 1000 \quad (3.5)$$

The enrichment factor is used in geochemical and atmospheric chemistry applications. In the case of the plant material compared to the atmosphere, $\varepsilon = -20.35$ ‰. In other words, the isotopic value of the product plant material is ~20 ‰ lower or more depleted in ^{13}C than the reactant CO_2. Another way to express this is the isotope discrimination (Δ) defined as follows:

$$\Delta_{RP} \approx \delta_{reactant} - \delta_{product} \sim -7.4\ \text{‰} - (-27.6\ \text{‰}) = 20.2\ \text{‰} \tag{3.6}$$

Although generally small, differences in nomenclature used by biologists versus geochemists should be noted. The values of Δ and ε vary with the environmental factors (temperature) and biological factors that may reflect the balance of processes such as photorespiration and competition for Rubisco in photosynthesis.

Figure 3.2 provides a schematic with the approximate ranges of $\delta^{13}C$ values found in the Earth's major C reservoirs. More ^{13}C-depleted (more negative $\delta^{13}C$) values are associated with the organic matter produced by C_3 photosynthesis. Methane produced by fermenting such organic matter is even lower in $\delta^{13}C$, because the microbes that produce it discriminate against ^{13}C in the already $\delta^{13}C$-depleted substrate organic material. Dissolved bicarbonate ions and calcium carbonate are enriched compared to the source CO_2, which can vary in $\delta^{13}C$ depending on the degree to which atmospheric CO_2 is affected by CO_2 derived from the decomposition of plant materials in groundwater recharge zones (soils). Hence, soil carbonates can have lower $\delta^{13}C$ values compared to marine carbonates.

3.2.4 *Isotopes in the Global Carbon Cycle*

The three most abundant isotopes of C, ^{12}C, ^{13}C, and ^{14}C distribute themselves differently among the C reservoirs in the Earth system (Fig. 3.3). These distributions reflect the biogeochemistry of C (Fig. 3.3a), such that most of it is dissolved in the oceans. Mass-dependent fractionation of isotopes during the transfer of C from one reservoir to another means that the distribution of ^{13}C does not directly mirror that of ^{12}C—proportionally more ^{13}C ends up in the atmosphere and ocean and less in organic C reservoirs because of the discrimination against the heavy isotope in biological processes (Fig. 3.3b). The distribution of ^{14}C reveals, in addition, the time C resides in the different reservoirs. For example, reservoirs like the terrestrial biosphere have ^{14}C values close to that in the atmosphere. In contrast, the deep ocean has lower ^{14}C values that reflect its isolation from exchange with the atmosphere over many centuries (Fig. 3.3c). Changes in the distribution of C among the major inorganic and organic reservoirs in the past should thus be associated with changes in the distribution of isotopes, the subject of Chap. 7.

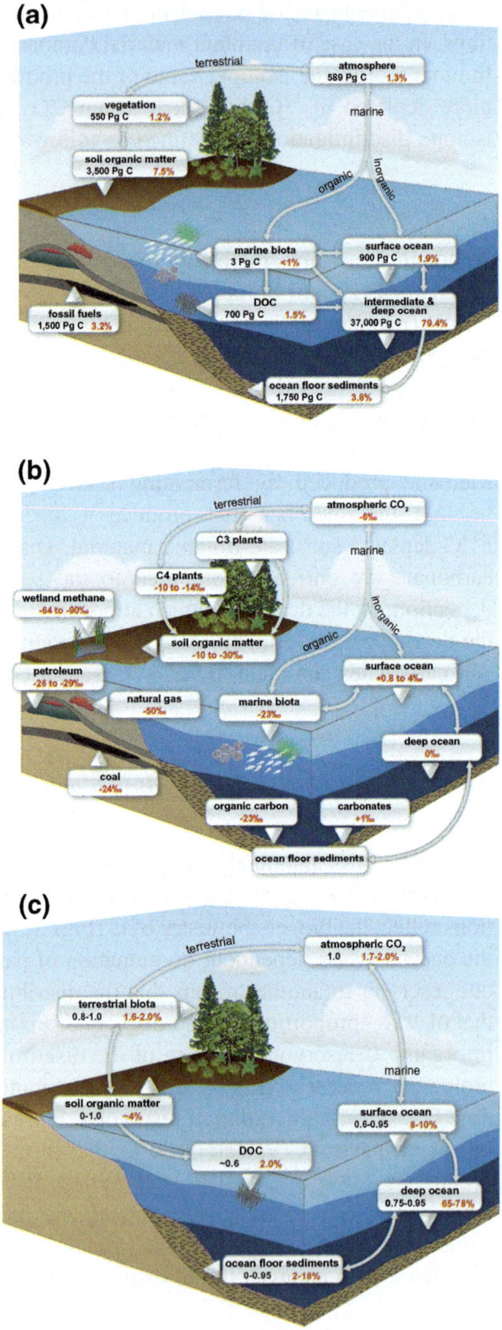

Fig. 3.3 The distribution of C in petagram C (PgC) and percent of Earth's total active C pool (**a**); the range of $\delta^{13}C$ values (**b**); and the range of ^{14}C values (expressed relative to the atmospheric ^{14}C content, where atmospheric ^{14}C in the preindustrial is defined as 1.0) among different reservoirs of the Earth system (**c**). For simplicity, the ^{14}C distributions shown here have been corrected for isotope discrimination and reflect just the effects of radioactive decay (see Sect. 3.3.2)

3.3 Reporting Radiocarbon Data: Radiocarbon Nomenclature and Calculations

There are several ways to report ^{14}C data, depending on the application. As with stable isotopes, the reporting is based on comparing the $^{14}C/^{12}C$ of the sample to that of a universal standard. However, for ^{14}C there are two extra factors. First, the $^{14}C/^{12}C$ ratio will reflect mass-dependent fractionation in addition to the signal (mixing, radioactive decay) that we want to know. Normally ^{14}C data are corrected for mass-dependent isotope fractionation using the sample's $\delta^{13}C$ value. Second, the standard against which all laboratories report their measurements will inevitably change over time as the ^{14}C it contains is subjected to radioactive decay—every year there is a little less ^{14}C in the standard than the year before. This is easily corrected for by selecting a specific time (in the case of ^{14}C it is 1950; Chap. 2) and decay correcting to 'put back' the ^{14}C lost between 1950 and the year the standard was measured. The different ways to report ^{14}C depend on whether or not these corrections for mass-dependent isotope fractionation (based on ^{13}C) and decay corrections have been made. For completeness, we will go over the nomenclature as originally defined by Stuiver and Polach (1977). While the nomenclature of reporting ^{14}C data gets confusing, the basic reason ^{14}C is expressed in various ways has to do with its very different uses as an age dating tool or as a tracer.

3.3.1 *Raw Data: Ratios of Unknowns to Standards*

Because measurement of ^{14}C has been accomplished by two methods, decay counting and accelerator mass spectrometry (AMS), there are two ways in which raw data are used to calculate the expressions by which ^{14}C data are reported. The first relies on the *activity* (radioactivity of a sample, A_S), measured as decays per unit time for a given number of atoms of C (Stuiver and Polach 1977). In decay counting laboratories, this was determined by counting decays for the prepared standard (OX-I, see below, measured either as a gas (CO_2, acetylene) or a liquid (benzene); Chap. 2). The second method uses data generated by AMS and relies on the measured ratio of ^{14}C atoms detected to the coulombs of integrated ^{12}C current at the high-energy end of the mass spectrometer (R_S, the ratio of ^{14}C atoms/^{12}C current; see also Chap. 8). In some AMS laboratories, the ratio of ^{14}C atoms to ^{13}C current is measured instead due to the very high currents generated by ^{12}C. The ^{12}C is calculated from the ^{13}C current, and the $^{13}C/^{12}C$ ratio is based on the measured or assumed $\delta^{13}C$ value.

The two methods of reporting are related by the radioactive decay equation that defines activity (decays per unit time) as the product of the number of atoms and the probability that each will undergo radioactive decay in a given time. This probability is the decay constant, λ. Both decay counting and AMS measurements have artifacts that may cause the measured activity or isotope ratio to differ from the actual value. Such systematic errors can include counter efficiency and quenching (for decay counting) and uncertainties in the calibration of the current measurement device (for AMS). These errors make it very difficult to accurately measure activity or isotope ratio, which is why ^{14}C measurements rely on the parallel measurement of an internationally accepted standard, just as is done for the stable isotope measurements. The assumption is that any systematic alteration of the isotopes by sample preparation or decay counting methods will be the same for the standard and the sample. In other words, the ratio of the sample to standard will be measured accurately, even if the absolute value of activity (or $^{14}C/^{12}C$) is not accurately determined. This assumption can be tested if there is a range of materials called secondary standards that have known activity or isotopic ratio compared to the primary standard.

The principal modern ^{14}C standard is NIST (National Institute of Standards and Technology) Oxalic Acid I or OX-I ($C_2H_2O_4$), made from a crop of 1955 sugar beets (Olsson 1970). The absolute ^{14}C standard is ninety-five percent of the activity of OX-I, measured in the year 1950. This activity (also referred to as 'Modern') is equivalent to a decay rate of 13.56 decays per minute per gram of C (decay counting) or 1.18×10^{-12} $^{14}C/^{12}C$ (AMS), measured in 1950. This is equal to the measured activity of 1890 wood corrected for radioactive decay to 1950. The wood standard was chosen to represent the pre-industrial atmospheric $^{14}CO_2$. The absolute activity or isotope ratio value is not as important as the fact that laboratories agree on the standard material and report their results relative to that standard. By convention, 1950 is year 0 BP (before present) in the ^{14}C dating literature. The year 1950 was chosen to honor the publication of the first ^{14}C dates calculated in December 1949 (Chap. 2). Raw data are thus reported as either the activity or ratio of the sample (A_S or R_S) divided by 0.95 times the activity or ratio of the OX-I standard (A_O or R_O).

3.3.2 Corrections for Mass-Dependent Isotope Fractionation

Just like the stable isotope ^{13}C, ^{14}C is affected by mass-dependent fractionation in the environment, and this effect needs to be removed for meaningful ^{14}C age calculations. For example, the $\delta^{13}C$ difference between atmospheric CO_2 and most tree cellulose (discrimination, defined above) is about 20 ‰. The fractionation of ^{14}C is roughly twice that of ^{13}C, since the mass difference between 12 and 14 is twice than that between 12 and 13, so the ^{14}C content of cellulose will be lower than that of the $^{14}CO_2$ from which it is fixed by $\sim$40 ‰. Even if the leaf C and the air from which that leaf was made have the same ^{14}C 'age,' the leaf will appear to be $\sim$330 years old due to the fractionation.

To remove this fractionation effect, the sample is corrected so that all data are expressed as if they had a constant $\delta^{13}C$ value of −25 ‰. In terms of ratios:

$$\left[\frac{^{14}C}{^{12}C}\right]_{sample[-25\,‰]} = \left[\frac{^{14}C}{^{12}C}\right]_{sample[\delta‰]} \left[\frac{1+\frac{-25}{1000}}{1+\frac{\delta}{1000}}\right]^2 \quad (3.7)$$

where δ is the ^{13}C value of the sample. Alternatively:

$$R_{SN} = R_S \left[\frac{0.975}{1+\frac{\delta}{1000}}\right]^2 \quad (3.8)$$

where R_{SN} is the sample ratio normalized to a ^{13}C of −25 ‰, and R_S is the measured isotope ratio of the sample. The terminology of ^{14}C was developed before the advent of AMS, and most of the earlier texts and papers discussing the nomenclature of ^{14}C are written in terms of radioactivity, or the number of decays per unit time, rather than the ratios of atoms. Activity (A) is just the number of ^{14}C atoms multiplied by the decay constant (e.g., $A_S = \lambda R_S$). One often sees this simplified approximation of Eq. (3.8) for the ^{13}C correction in the literature:

$$A_{SN} = A_S\left(1 - 2\left[\frac{25+\delta}{1000}\right]^2\right) \quad (3.9)$$

where A_{SN} is the normalized activity of the sample, A_S is the actual activity of the sample, and δ is the ^{13}C value of the sample in permil (‰). By convention, the OX-I standard is corrected to its actual measured $\delta^{13}C$ value of −19 ‰ not −25 ‰ (Stuiver and Polach 1977).

The basis for reporting ^{14}C is thus the ratio (or activity) of the measured sample, normalized to a $\delta^{13}C$ value of −25 ‰, divided by 0.95 times the measured ratio (or activity) of the OX-I standard measured at the same time (with $\delta^{13}C$ of −19 ‰). The resulting value is known as Fraction Modern (F):

$$F = \frac{R_{SN}}{R_{ON}} = \frac{R_S\left(\frac{0.975}{1+\delta/1000}\right)^2}{0.95R_{O,-19}} \quad (3.10)$$

In AMS, samples and standards can be measured many times, and corrections for mass-dependent fractionation may be done at different stages of the measuring and reporting process depending on the laboratory. Some AMS laboratories, for example, measure the $^{14}C/^{13}C$ ratio only and use separately the measured values of $^{13}C/^{12}C$ to correct for mass-dependent fractionation. Details of the calculation of ^{14}C data in such cases can be found in Donahue et al. (1990).

Most AMS laboratories will report data as F (Fraction Modern). Other ways to report ^{14}C data, which will vary depending on the application, can all be derived from F (Table 3.1). These other ways were defined originally by Stuiver and Polach (1977); however, they actually do not use the F notation and instead use percent

Table 3.1 The various ways to express ^{14}C information. Where given, *y* indicates the year of measurement

Notification	Name	Definition	^{13}C normalized	Decay-corrected	
				Standard since 1950	Known-age sample
F	Fraction Modern	(See text)	Yes	No	
pM	percent Modern	$100 \times F$	Yes	No	
^{14}C age	Conventional radiocarbon age	$-8033\ln(F)$	Yes	No	
D	Delta	$1000 \times (F - 1)$	Yes	No	
Absolute expressions					
pM_{abs}	Absolute percent Modern	$F \times \exp((1950 - y)/8267)$	Yes	Yes	
$\Delta^{14}C$	Delta ^{14}C	(See text)	Yes	Yes	
Δ	Delta	(See text)	Yes	Yes	Yes

Modern (pM), which is $F \times 100$. In the following sections, we will review the complete nomenclature defined by Stuiver and Polach (1977). We will emphasize a subset of the most commonly used terms in C cycle research.

3.3.3 Radiocarbon Age

Because of the relatively rapid cycling of C between the atmosphere and living biomass, most fast-growing tissues of plants growing in pre-industrial times maintained a $^{14}C/^{12}C$ value equal to that of atmospheric CO_2, once corrected for mass-dependent isotope fractionation effects. Similarly, animals reflect the $^{14}C/^{12}C$ value of the plants or animals they consume, with little time for radioactive decay to occur during the life span of the organism. Upon the death of an organism, the ^{14}C in its tissues is no longer replenished from its food, and the total number of ^{14}C atoms in its tissues decreases due to radioactive decay. If the C in tissues remains intact and isolated from exchange with the environment, the $^{14}C/^{12}C$ ratio measured at some later time, combined with knowledge of the radioactive decay rate of ^{14}C, can be used to indicate time since the death of the organism. This is the basis for ^{14}C dating, with the 'clock' provided by the predictable radioactive decay of ^{14}C. The number of atoms that disintegrate per unit of time (dN/dt) is proportional to the number of atoms N present at a time t. Mathematically, this equation is simply:

$$\frac{dN}{dt} = -\lambda N \tag{3.11}$$

where λ is the radioactive decay constant. The solution to this differential equation is given by the following:

$$N(t) = N_0 \exp(-\lambda t) \tag{3.12}$$

where N_0 represents the number of atoms present at an initial time $t = 0$ (see Fig. 3.4). It is of interest to calculate the *half-life* of ^{14}C, which is defined as the time required for half of the radioactive atoms in a given sample to decay. Taking logarithms in both sides of Eq. (3.12), we obtain the following:

$$\ln\left(\frac{N}{N_0}\right) = -\lambda t \tag{3.13}$$

If $N/N_0 = 1/2$, then the half-life $t_{1/2}$ is given by the following:

$$t_{1/2} = \frac{\ln 2}{\lambda} = \frac{0.6931}{\lambda} \tag{3.14}$$

Another measure of interest is the mean life, t_{mean}, which is defined as the average time each ^{14}C atom spends in a given sample until it decays. It can be calculated as follows:

$$t_{mean} = \frac{1}{\lambda} \tag{3.15}$$

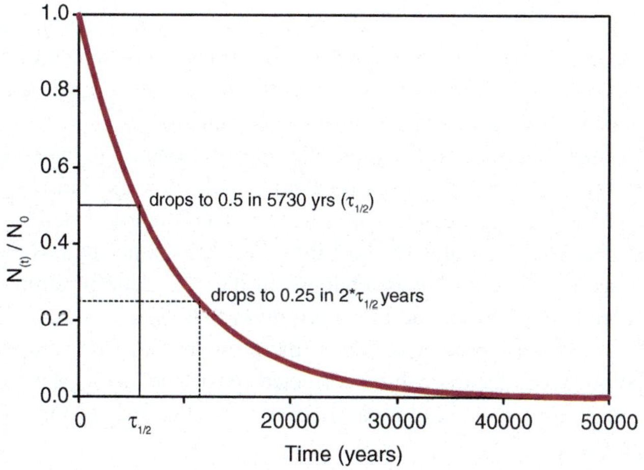

Fig. 3.4 Radioactive decay of ^{14}C showing how the proportion of the initial pool (N_t/N_0) decreases over thousands of years, with its half- and quarter-lives. The limit for ^{14}C dating is roughly 10 half-lives

In the literature, two common values are reported for the half-life and the mean life of ^{14}C. The Libby half-life $t_{1/2}$ = 5568 year and mean life t_{mean} = 8033 year, correspond to a value of $\lambda = 1.24 \times 10^{-4}$ year^{-1}. As discussed in Chap. 2, the Libby value was based on only a few measurements of the 'curve of knowns.' With more data and different methods, more recent values of the ^{14}C half-life have been proposed (Godwin 1962), which differ from the original Libby value. This value of λ (1.21×10^{-4} year^{-1}) is considered as the more correct value, equivalent to a $t_{1/2}$ = 5730 year and t_{mean} = 8267 year. Because much of the older ^{14}C literature used the Libby decay constant, it is still used by convention for reporting the ^{14}C age (Stuiver and Polach 1977). Thus, calculating a true calendar age requires a calibration curve that in part takes into account the difference in half-lives. For decay corrections associated with some of the other ways of reporting ^{14}C, the newer ('actual') half-life is used (Stuiver and Polach 1977).

Conventional ^{14}C ages, based on Libby mean life and used in archeology, are reported as follows:

$$^{14}\text{C age} = -8033\ln(F) \tag{3.16}$$

where 8033 is the mean life of ^{14}C using the Libby half-life, and F is the measured Fraction Modern. The calculation of the ^{14}C age depends on the assumptions that the initial ^{14}C content of the sample (N_0 in Eqs. 3.12 and 3.13) is known and that its value equals that of the preindustrial atmosphere (1890 wood, with ^{14}C/^{12}C ratio equivalent to 0.95 times the OX-I standard). The ^{14}C/^{12}C ratio of the sample, if it is smaller than that of the preindustrial atmosphere, gives an estimate of the time elapsed since the system was closed to exchange of ^{14}C with the atmosphere. In other words, the fraction of ^{14}C in the original sample that has been lost to radioactive decay is a measure of the time elapsed. A third assumption for ^{14}C dating is that the sample has remained a closed system—there has been no loss (other than radiodecay) or addition of ^{14}C to the sample since it ceased exchanging with the atmosphere (e.g., since death of the organism).

It is extremely important to note that conventional ^{14}C age derived from radioactive decay (Fig. 3.4) is not equivalent to calendar age and should *never* be used as such. The ^{14}C age deviates significantly from the true calendar age for two reasons: (1) because of the use of the Libby half-life (rather than the more recent one) in the calculation and (2) past changes in the ^{14}C value of atmospheric CO_2, which are large enough to violate the assumption that N_0 (the initial ^{14}C content of the sample) is constant and equals that of the preindustrial atmosphere. As will be discussed below, these factors can be corrected for by measuring the ^{14}C content of known-age samples and making a calibration curve that relates the ^{14}C age to the calendar age (Sect. 3.3.5).

We will now turn to other ways of expressing the ^{14}C measurement. These are summarized in Table 3.1. Each is derivable from the Fraction Modern (F), derived in Eq. (3.10), and each has a specific purpose or use in the literature.

3.3.4 Expressing the Absolute Amount of Radiocarbon

Because ^{14}C in the OX-I standard undergoes radioactive decay over time, its $^{14}C/^{12}C$ ratio will change depending on the year in which it is measured. For some applications, we want to report the absolute amount of ^{14}C in our sample in the year it was measured. In this case, we must apply a correction for radioactive decay undergone by the standard between 1950 (using the correct half-life of 5730 years, not the Libby half-life) and the year the sample was measured (y):

$$\Delta^{14}C = \left[\frac{\left[\frac{^{14}C}{^{12}C}\right]_{sample,-25}}{0.95 \left[\frac{^{14}C}{^{12}C}\right]_{OX1,-19} \exp^{((y-1950)/8267)}} - 1 \right] \times 1000 \quad (3.17)$$

This is the geochemical nomenclature because it reports $^{14}C/^{12}C$ relative to an absolute standard that is decay-corrected for OX-I change since 1950. It is reported in units of deviation from the standard in parts per thousand, or ‰, in a way that is analogous to the reporting of $\delta^{13}C$ data. Positive values of $\Delta^{14}C$ indicate that the sample has more ^{14}C than the preindustrial atmosphere indicating the presence of bomb ^{14}C. Figure 3.5 shows the variation in $\Delta^{14}C$ values for atmospheric CO_2 due to the production of ^{14}C by atmospheric weapons testing. A $\Delta^{14}C$ value of +1000 ‰ would be equivalent to a doubling of the ^{14}C atoms in the atmosphere, which is roughly what was done by weapons testing leading up to 1964.

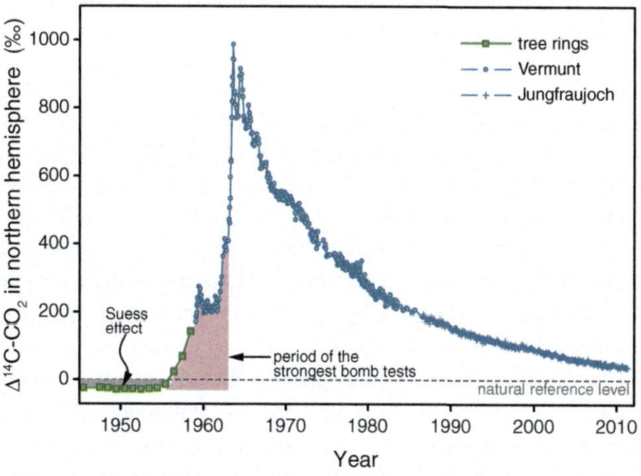

Fig. 3.5 Development of Northern Hemisphere atmospheric $\Delta^{14}CO_2$ in the last 60 years due to the production of ^{14}C by atmospheric weapons testing. Data before 1959 have been derived from tree rings (Stuiver and Quay 1981). From 1959 to 1986, measurements were performed at the Alpine site, Vermunt, and data from 1986 onward are from the Swiss High Alpine Research Station, Jungfraujoch (Levin et al. 2010)

We also define a new term, F' or the *absolute Fraction Modern*, to refer to the absolute ratio of sample to standard, corrected for radiodecay in the year of measurement (y). It is related to the Fraction Modern (F) by the following:

$$F' = \left[\frac{\left.\frac{^{14}C}{^{12}C}\right]_{\text{sample},-25}}{0.95 \left.\frac{^{14}C}{^{12}C}\right]_{\text{OX1},-19} \exp^{((y-1950)/8267)}} \right] = F \exp^{((1950-y)/8267)} \tag{3.18}$$

F', which is equivalent to $\Delta^{14}C$ /1000 + 1, is useful for tracking the absolute amount of ^{14}C in a model, for example, when one wishes to track the fate of bomb ^{14}C through a given C reservoir (see next section). Stuiver and Polach (1977) also defined the absolute percent Modern (we call this pM_{Abs}, Table 3.1), although regrettably Stuiver and Polach also call this pM (same name as 100 × F). In our nomenclature, $pM_{Abs} = 100 \times F'$.

Values of $F > 1.0$ (which lead to a negative ^{14}C age, denoted only as > Modern (since 1950)), or positive values of $\Delta^{14}C$, indicate the presence of bomb ^{14}C. Values of $F < 1$ yield meaningful ^{14}C ages and negative $\Delta^{14}C$ values, and indicate that the ^{14}C in the sample has been isolated from exchange with the atmosphere long enough for significant radioactive decay to have occurred. When $F' = 1.0$ and $\Delta^{14}C = 0$ ‰, the $^{14}C/^{12}C$ ratio of the sample (corrected, of course, for mass-dependent fractionation according to its measured $\delta^{13}C$) is the same as that for the 1890 wood standard.

3.3.5 *Known-Age Corrections and Δ: The Origin of the Calibration Curve*

The calculation of ^{14}C age using Eq. (3.18) makes the implicit assumption that the ratio of $^{14}C/^{12}C$ at the time it was formed (corrected for any mass-dependent fractionation) is the same as that in the atmosphere in 1950. We know, however, that atmospheric $^{14}CO_2$ values have changed in the past, and this affects the determination of the absolute age of an object. Causes of atmospheric ^{14}C variation (Chap. 4) include changes in the Earth's magnetic field (which determines the trajectory of cosmic ray particles hitting the upper atmosphere), solar output (variation caused by solar cycles), and the C cycle (the partitioning of C among atmosphere, ocean, and land).

Understanding past variation in atmospheric $^{14}CO_2$ is thus of great interest and the focus of a considerable amount of effort in the ^{14}C community. To determine atmospheric $^{14}CO_2$ levels in the past, it is necessary to measure ^{14}C in a sample of known age and then use the radioactive decay Eq. (3.13) to determine the initial $^{14}C/^{12}C$ ratio (R_0). Tree rings are used as known-age samples since they can be counted precisely. The tree ring record is then used to determine a *calibrated age* from a ^{14}C measurement of an unknown sample. Calibration curves are based on data from several laboratories and are updated periodically (e.g., IntCal09 and IntCal13) (Reimer et al. 2009; Reimer

2013). There are several Web-based programs (e.g., CaLIB, Oxcal) that can be used to convert a ^{14}C age to a true calendar age. As of this publication, tree ring records go back about ~14,000 years, so for older records, we use other dating methods to independently determine time (uranium–thorium in corals, dating of volcanic ash depositions, etc.). The tree ring curve necessitates another way of reporting ^{14}C in which a sample of known age is used to determine the past ^{14}C value of the atmosphere (Fig. 3.6; e.g., tree ring-based chronologies) or ocean (e.g., known-age corals or annually layered sediments). The sample is implicitly assumed to have been collected before 1950, so it is corrected for radioactive decay of the ^{14}C in the sample between the time of formation and 1950:

$$\Delta = \left[\frac{\left[\frac{^{14}C}{^{12}C}\right]_{\text{sample},-25} \exp^{((1950-x)/8267)}}{0.95\left[\frac{^{14}C}{^{12}C}\right]_{\text{OX1},-19}} - 1\right] \times 1000 \qquad (3.19)$$

where Δ, or $D^{14}C$, expresses the ^{14}C value relative to 'Modern' had the sample been measured in 1950, and *x* is the known year before 1950. The unit Δ is useful for studies attempting to show how the ^{14}C value of air (tree rings) and water (corals) changed relative to the absolute (1950) value with time and due to decay. The biggest difference between Δ and $\Delta^{14}C$ (Eq. 3.17) is that in Δ notation the standard is not corrected for radioactive decay after 1950.

As we get farther away from 1950, the corrections for radioactive decay in the standard become more and more significant, and understanding the (once subtle) differences in nomenclature become more important (Table 3.1, Fig. 3.7). In sum, the Δ or $\Delta^{14}C$ notation is used for modeling past changes in atmospheric ^{14}C, or when the absolute number of ^{14}C atoms needs to be tracked such as in efforts that trace the fate of bomb ^{14}C.

It is important to realize that the reported value of $\Delta^{14}C$ depends on the year of measurement. Either *F* and $\Delta^{14}C$ should be reported together since *F* does not change with time, or the year of measurement should be reported along with $\Delta^{14}C$

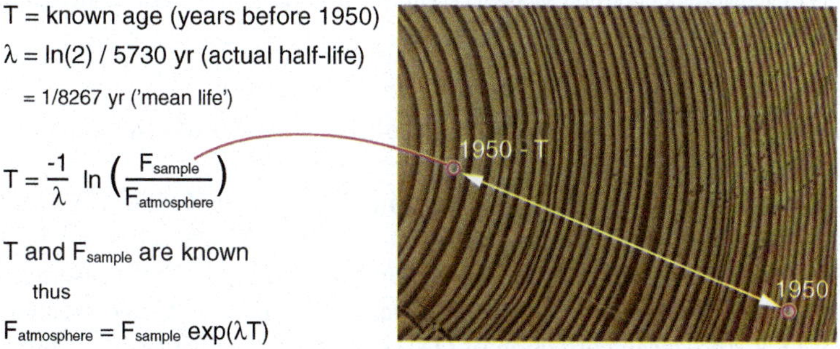

Fig. 3.6 Tree rings, each of which has a known age, can be used to determine the past ^{14}C value of the atmosphere. The Fraction Modern of the atmosphere can be calculated using the Fraction Modern of the sample (i.e., a tree ring), the ^{14}C decay constant *λ*, and the age of the sample

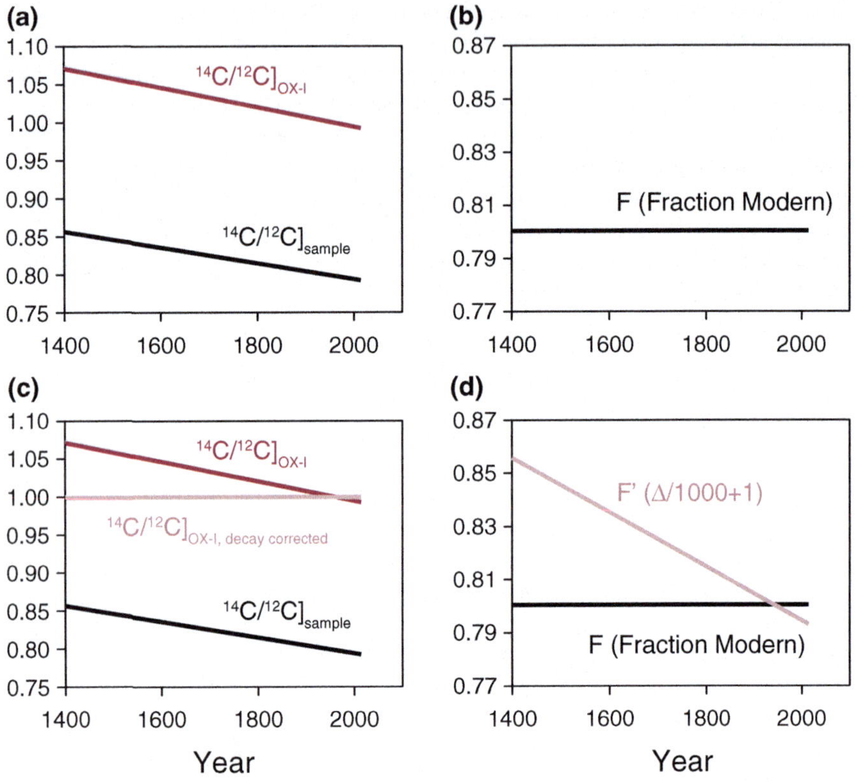

Fig. 3.7 Four different ways of reporting ^{14}C: (**a**) shows the difference between the sample and standard values using the Δ notation where the sample is corrected for decay prior to 1950, and the standard is not decay-corrected; (**b**) shows the value for Fraction Modern, which does not change over time as neither the sample nor the standard are decay-corrected, but the ^{14}C in them decays at the same rate (e.g. as in a (bolded)); (**c**) shows how the standard values differ when the standard is corrected for radioactive decay between 1950 and the year of measurement (post-1950) as in the $\Delta^{14}C$ notation (*light pink line*) versus when it is not; (**d**) shows the difference between Fraction Modern (*F*) and absolute Fraction Modern (*F'*) in which the standard is decay-corrected for decay undergone from 1950 to the year of measurement

values. In some cases, the year of sample collection may also be important to report if different from the year of measurement.

A final note is that we have avoided using the terms $d^{14}C$ and $\delta^{14}C$ defined by Stuiver and Polach (1977). These terms 'undo' the ^{13}C correction to the sample made in calculating the Fraction Modern and instead report the measured (uncorrected) ratio to those of the absolute standard. $d^{14}C$ is the ^{13}C-uncorrected equivalent of $D^{14}C$ (no decay correction of OX-I) and $\delta^{14}C$ the uncorrected equivalent of $\Delta^{14}C$ (decay-corrected OX-I). These terms are used in some models that separately account for mass-dependent fractionation of ^{13}C and ^{14}C isotopes (e.g., Wang et al. 1994), but are not commonly used in the literature.

3.4 Interpreting Radiocarbon Data

Using ^{14}C to improve our understanding of the Earth's C cycle first depends on the state of the C that is being measured. The C in the sample can be considered either an open system that is continuously exchanging with the atmosphere or a closed system that is isolated at a specific time from atmospheric exchange (Fig. 3.8). For a closed system, such as a macrofossil like a seed or a bone, it is appropriate to use the calibration curve to calculate a true age or calendar age for the C in the sample (Sect. 3.4.1). For C in an open system, such as dissolved inorganic C (DIC) in seawater, the ^{14}C value may tell you about the rate of exchange between seawater

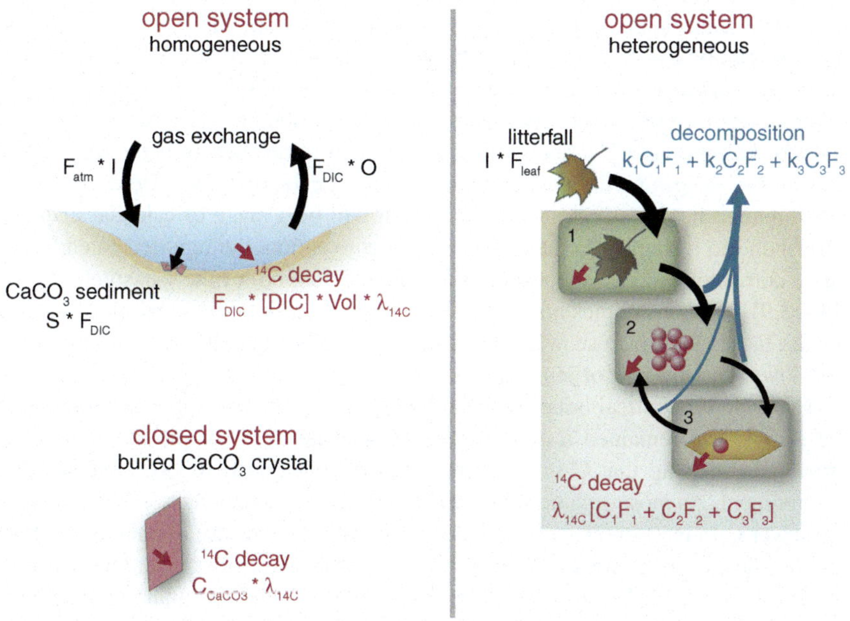

Fig. 3.8 Illustrations of open and closed systems, and homogenous and heterogeneous systems. An example of an open system that is homogenous (i.e., every ^{14}C atom has an equal probability of exchange with the atmosphere) is a shallow, closed-basin lake with high dissolved inorganic C (DIC) content. The rate of exchange can be estimated from the ^{14}C age of C in the DIC. A calcium carbonate crystal formed in the lake and subsequently buried without further C exchange with pore waters would record the age of the DIC in the lake plus the time since it was buried. An example of an open system that is not homogenous is typically found in soils. Here, we can approximate different sub-components of the total system as individual 'pools' with approximately homogenous behavior. A test of the homogeneity is to compare the ^{14}C age of the C in the system with that of the C being respired from the system

and the atmosphere, or alternatively about mixing of water masses with different ^{14}C values, or the time since deep water was isolated from the surface (Chap. 5). For open systems, we can use a model to estimate the rate of exchange of C between the reservoir and the atmosphere from its ^{14}C value (Sect. 3.4.2). For complex mixtures such as organic matter in soils and sediments, we cannot assume that all of the C in the reservoir we have measured is exchanging C with the atmosphere at the same rate; it may represent a combination of open and closed systems (Fig. 3.8). In such cases, more complex models are required, and they need to be constrained by additional information about the system if it is available.

3.4.1 Calibrated Ages

Calibration curves have been assembled from thousands of measurements of known-age samples such as tree rings (for terrestrial samples) or corals and shells (for marine samples) and represent the heroic efforts of many laboratories. The most up-to-date calibration curves (they undergo continuous inter-comparisons and updates) can be found in the journal *Radiocarbon* (the latest is INTCAL13; Reimer 2013) and are used in the various calibration programs posted on the Web by different laboratories. These community-accepted calibration curves can be used to convert the measured ^{14}C age (which, as you will remember, uses the old 'Libby' half-life) to determine the corresponding range of calendar years. Because the ^{14}C measurement has an error estimate, the result will be a range of calendar ages that correspond to a single ^{14}C measurement. By convention, the reported calendar age range corresponds to $\pm$ one standard deviation (1-sigma) of the measured ^{14}C age (Fig. 3.9). Marine samples have their own, separate calibration curve that accounts for the fact that surface seawater has lower ^{14}C values than the atmosphere due to the slower mixing rate of atmospheric $^{14}CO_2$ into the ocean (Chap. 5).

There are two critical issues with calibrated ages. The first is the assumption that the sample has remained a closed system since the death or preservation of the organism, meaning that C in the sample has not continued to exchange with the environment. As most organic materials are buried in sediment or soil, it is common for newer C to have entered the sample, particularly in the case of charcoal or bone that has large surface area and absorptive properties. Consider a sample that is so old that it no longer has any detectable ^{14}C (>70,000 years). Contamination of this sample with only 1 % Modern C will yield a ^{14}C age of 37,000 years BP. With the advent of the small sample sizes that can be measured by AMS, a major advantage has been to isolate specific compounds such as intact proteins or plant waxes that are more likely to be unchanged since deposition. It is always important to test for the presence of contamination that might affect the inferred age of a sample (Chap. 9).

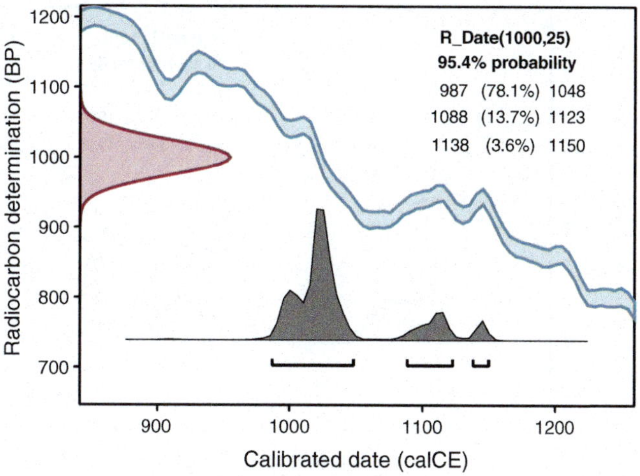

Fig. 3.9 An example of the output of a calibration program (in this case, we have based the figure on the OxCal program) that illustrates the relationship between the ^{14}C age, the calibrated age, and the changes in uncertainty that include changes in slope in the calibration curve. Data adapted from OxCal v4.1.7 Bronk Ramsey (2010), Reimer et al. (2009)

The second issue limits the usefulness of ^{14}C for determining a calendar age for most of the period from 1650 to 1950 AD. The main reason for this is the presence of a plateau, or flattening, in the calibration curve during this time period, such that ^{14}C in the atmosphere declined at a rate approximately the same as the rate of ^{14}C decay (Fig. 3.10). In combination with the inherent precision of a ^{14}C measurement, calendar ages within this time period cannot be distinguished from one another. Such plateaus in the calibration curve also occur at many other interesting times in history, and these have a strong effect on the range of calendar dates that are reported from single ^{14}C measurements.

The calculation of calendar ages using bomb ^{14}C has provided a new and different way to accurately determine the calendar date of C fixation from the atmosphere since 1956. This is done simply by comparing the $\Delta^{14}C$ value of the sample, calculated for the year the sample was taken, to the $\Delta^{14}C$ value of the atmosphere in that year, while making sure that is also expressed for the year of sampling. There is always an issue of whether the sample was fixed during the ascending or descending part of the bomb curve (Fig. 3.5). In many cases, there is the opportunity for using multiple measurements representing older and younger parts of a sample, such as two separate tree rings to most accurately assign calendar years with this method. This approach can also potentially be used in older samples to help resolve some of the plateau periods in the calibration curve.

Radiocarbon measurements not only are used to estimate the calendar ages of samples, but can also be used to estimate residence time—the average time a C molecule resides in a reservoir. The theory for these applications is outlined in the following sections.

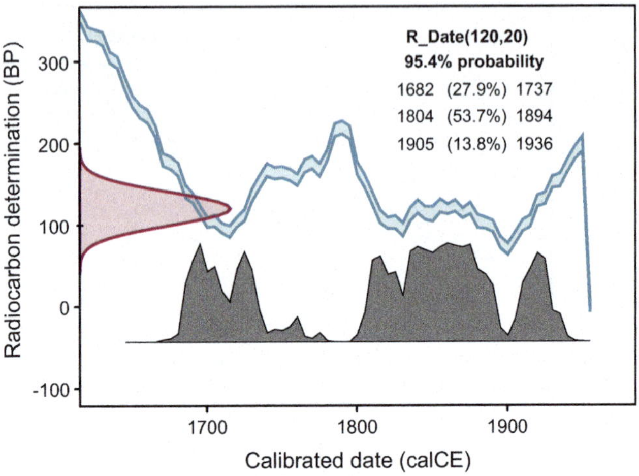

Fig. 3.10 The special problem of ^{14}C calibration for the calendar time period 1650–1950. Data adapted from OxCal v4.1.7 Bronk Ramsey (2010), Reimer et al. (2009)

3.4.2 *Single- and Multiple-Pool Turnover Models*

For open systems that continuously exchange C with their surroundings, ^{14}C measurements reflect the average time a molecule of C resides in a reservoir. This section reviews the theory for using ^{14}C to estimate the rates at which C is transferred among different reservoirs within the Earth system. This goal inevitably implies the description of these different components with a model that facilitates the estimation of the overall rate at which C is being cycled and the average time a C atom spends in a reservoir (Rodhe 2000).

Several assumptions are often made in modeling ^{14}C as it enters and leaves different reservoirs. A reservoir can be the ocean, a lake, soil, vegetation, or their components. Commonly, it is assumed that ^{14}C cycles within a single homogeneous reservoir—in other words, all ^{14}C atoms in the reservoir have an equal probability of leaving the reservoir. This assumption ignores any internal heterogeneity in dynamics of C for different components within the system. This heterogeneity, in turn, can be modeled as a set of compartments (i.e. one for each component) with connections among them, or with continuous distributions describing the arrangement of multiple parts (elements of Fig. 3.8 can be expressed as box models as seen in Fig. 3.11). In this section, these different approaches are reviewed in the context of predicting rates of C flow through the entire system being studied. But first, we will revisit the influence of radioactive decay in the context of an open system that is continually exchanging C with the environment, unlike the closed systems described in the previous sections.

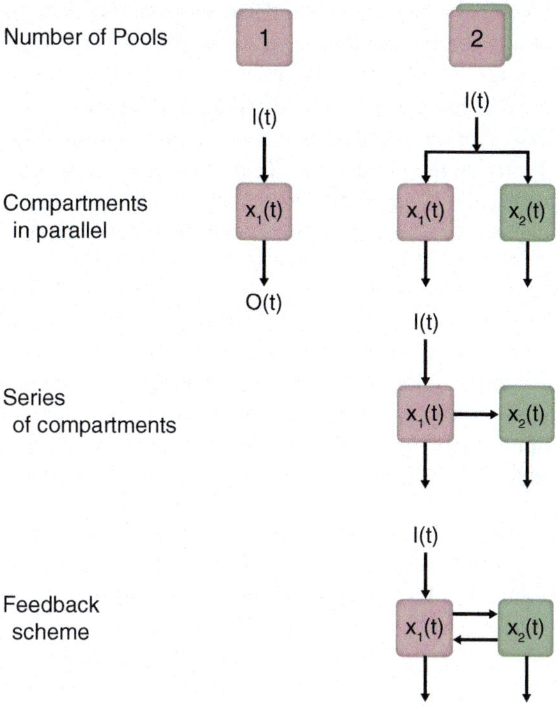

Fig. 3.11 Examples of one- and two-pool box models. The one-pool model is appropriate for a homogenous system, while the two-pool model is for a heterogeneous system (Fig. 3.7). The heterogeneity can be modeled as a set of pools with connections among them. More pools can be added to describe more heterogeneous systems

3.4.2.1 Single Homogeneous Reservoir: Pre-1950

It is generally of interest to calculate the average time that C atoms remain in a given homogeneous reservoir because this is a key parameter that influences how fast C pools in the Earth system may respond in the face of a changing environment (Rodhe 2000). Let us consider a system in which C enters and leaves a single homogeneous reservoir as follows:

$$\frac{\mathrm{d}C(t)}{\mathrm{d}t} = I(t) - O(t) \tag{3.20}$$

where $I(t)$ represents the amount of inputs and $O(t)$ the amount of outputs at a given time t. Assuming that C leaves the reservoir at a rate that is proportional to the amount of C present at any given time, the outputs can be represented as $O(t) = k\,C(t)$, where k represents the rate constant for release. In terms of ^{14}C, Eq. (3.20) can be used to represent the rate of change of ^{14}C atoms in the reservoir as follows:

$$\frac{\mathrm{d}^{14}\mathrm{C}(t)}{\mathrm{d}t} = I^{14}\mathrm{C}(t) - k^{14}\mathrm{C}(t) - \lambda^{14}\mathrm{C}(t) \tag{3.21}$$

where λ is the radioactive decay constant (1.21×10^{-4} $year^{-1}$). The number of ^{14}C atoms can be approximated as the total amount of C (mostly ^{12}C) times a measure of the $^{14}C/^{12}C$ ratio. In other words, $^{14}C(t) = F(t)\ C(t)$. Depending on the application, this $F(t)$ could be the Fraction Modern F (for models applied prior to 1950) or F' (for models that include the incorporation of bomb ^{14}C into the reservoir since 1950), so the number of ^{14}C atoms is also relative to the standard. If the reservoir is receiving C directly from the atmosphere, like a plant that is growing and depositing C into the soil, we can obtain a simpler form for the balance of ^{14}C:

$$\frac{dFC}{dt} = IF_{atm} - (k + \lambda)FC \tag{3.22}$$

A simple solution to Eq. (3.22) can be obtained assuming the system is in steady state, which is defined as when the inputs (the addition of C and ^{14}C from the atmosphere) are equal to the outputs (loss of C and ^{14}C by decomposition and radioactive decay):

$$IF_{atm} = (k + \lambda)FC \tag{3.23}$$

Equation (3.23) can be rearranged as follows:

$$FC = \frac{IF_{atm}}{k + \lambda} \tag{3.24}$$

This equation implies that when a reservoir (C) is in equilibrium, the amount of ^{14}C in the system can be calculated as the ratio between the inputs of ^{14}C from the atmosphere and the sum of the release rate k and the decay constant λ.

When analyzing systems that were in equilibrium *prior* to 1950, it is reasonable to assume that F_{atm} was constant and approximately 1.0, since the deviations from 1 represented by the calibration curve are less than 10 % for the last several thousand years. In steady state, the amount of C in the reservoir is given by $C = I/k$, which helps to further reduce Eq. (3.24) to a simpler form:

$$F = \frac{k}{k + \lambda} \tag{3.25}$$

This equation implies that the fraction of ^{14}C in homogenous reservoirs at steady state prior to 1950 can be calculated solely from the values of k and λ. This estimate is subject to uncertainty given the fact that F_{atm} in fact varied in the past, though careful analysis with the calibration curve could reduce or at least quantify this uncertainty. Equation (3.25) can be rearranged to solve for the value of k:

$$k = \frac{\lambda F}{1 - F} \tag{3.26}$$

which is a simple formula to calculate the release rate under the assumption of a homogeneous reservoir in steady state prior to 1950. The mean age of C in this case would be $1/k$. It is possible to find examples in the literature where researchers have taken the ^{14}C age (i.e. $-8033*\ln(F)$), and interpreted it directly as the mean time C resides in the reservoir (see Table 3.2). To demonstrate how misleading this can be, let us assume two example samples, sample A with Fraction Modern (F) of 0.8 and sample B with an F of 0.2. Using Eq. (3.26), we would estimate the k for samples A and B as 0.00048 year^{-1} and 0.000030 year^{-1}, or mean times C resides in the reservoirs as ($1/k$) of 2070 and 33,060 years, respectively. In contrast, the ^{14}C ages we would calculate for these samples [using Eq. (3.16)] are 1790 and 12,930 years. In both cases, the ^{14}C age underestimates the mean time C resides in the reservoir. When k is close to or smaller than the decay constant for ^{14}C, this introduces very serious error.

3.4.2.2 Homogeneous Reservoir: Incorporation of Bomb Radiocarbon Since 1950

After 1950, the amount of ^{14}C in the atmosphere varied considerably, and the assumption of $F_{atm} = 1$ is no longer a reasonable approximation. The addition of ^{14}C by weapons testing means that the amount of ^{14}C in the atmosphere is not constant, and therefore, we must track the fate of ^{14}C atoms. This requires that we use F' as defined in Eq. (3.18) to keep track of the ^{14}C within the reservoir during the bomb period. The amount of ^{14}C in the atmosphere has been estimated and reported consistently for the bomb period and reported as calibration curves, and numerical approximations can be used for modeling ^{14}C as it enters different compartments.

It is possible to track the varying ^{14}C content of a reservoir with a constant amount of ^{12}C (i.e., at steady state for the amount of C) using a numerical approximation to Eq. (3.22). In a general form, this numerical approximation is given by

$$(F'C)_{t+1} = (F'C)_t + \partial(F'C)_t \tag{3.27}$$

where $(F'C)_{t+1}$ is the amount of ^{14}C one time step forward and $(F'C)_t$ is the amount of ^{14}C present at time t, and $\partial(F'C)_t$ is a finite difference approximation to the solution of the differential Eq. (3.22) for a given time t. There are many different methods to derive finite difference approximation using numerical techniques with the aid of computers. However, the simplest way to find an approximation is using Euler's method:

$$(F'C)_{t+h} = (F'C)_t + h[(IF_{atm}) - (k+\lambda)(F'C)_t] \tag{3.28}$$

where h is the time step used in the calculation. If the time step is annual, $h = 1$, Eq. (3.28) can be solved as follows:

$$F'_{t+1} = \frac{IF_{atm} + (F'C)_t(1 - k - \lambda)}{C_{t+1}} \tag{3.29}$$

This equation has been proposed before for modeling ^{14}C in soils (Torn et al. 2009); however, it should be used carefully because it can give important approximation errors when used at the annual time step (Fig. 3.12). To reduce the approximation error, it is possible to use a finer time step and solve Eq. (3.29) as follows:

$$F'_{t+h} = \frac{h(IF_{atm})_t + (F'C)_t(1 - h(k + \lambda))}{C_{t+h}} \tag{3.30}$$

Notice that Eq. (3.29) is a particular case of Eq. (3.30) when $h = 1$. To reduce the approximation error that can be obtained when using annual time steps (Fig. 3.12), it is possible to take the annual measurements of ^{14}C in the atmosphere and interpolate the values to shorter time steps using methods such as polynomial or spline interpolation.

Another option is to use a different numerical method to solve Eq. (3.22). The method of the trapezoid provides a smaller error than Euler's method, but it is more computationally intensive. It is also possible to use special software with more accurate solutions. For soil systems, a good option is to use *SoilR* (Sierra et al. 2012), a package implemented in the *R* environment for computing that provides solutions for Eq. (3.22) using accurate methods for solving systems of differential equations.

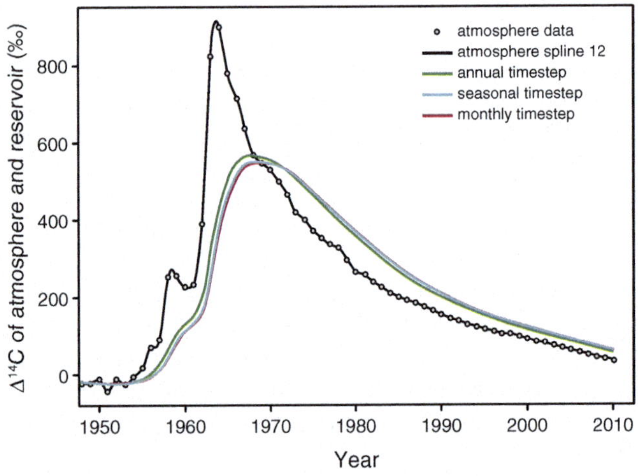

Fig. 3.12 Evolution of ^{14}C in a single homogeneous reservoir (*colored lines*) and atmospheric ^{14}C in the Northern Hemisphere (*black line* and *points*). Annual measurements of atmospheric ^{14}C are presented in points, and a 12-month interpolation using a spline function is presented as a *black line*. Radiocarbon in the reservoir was modeled using atmospheric ^{14}C at annual (h = 1, *green*), seasonal (h = 4, *blue*), and monthly (h = 12, *magenta*) time steps

In some situations, it is not correct to assume that ^{14}C enters the reservoir directly from the atmosphere. A time lag may be necessary to better describe the behavior of a system. An example is a pine needle that has been on the branch for a decade before dying and falling to the forest floor, where it takes on average 20 years to decompose. A measure of the mean age of C in the forest floor (30 years) that did not account for the 10-year 'pre-aging' of C inputs would give the erroneous result that decomposition is much slower than is actually the case. For such cases where the lag time is known, the amount of ^{14}C entering the reservoir can be calculated as follows:

$$\text{Inputs} = I_t(F_{\text{atm}})_{t+l} \tag{3.31}$$

where l is the time lag in years.

3.4.2.3 Mean Age and Turnover Time

Radiocarbon data provide a measure of the time elapsed since C in a reservoir was fixed from the atmosphere. For homogeneous reservoirs at steady state, as described above, this can be directly interpreted in terms of the dynamics of C in the reservoir. However, often reservoirs are not homogeneous, in which case additional models are required to interpret ^{14}C data.

Table 3.2 Terms used to infer C dynamics using biogeochemical box models such as the one represented in Fig. 3.13

Term	Concept
Turnover time	An inventory divided by either the input or the output flux.
Age	Time since a C atom in a system entered it (Fig. 3.13). Integrating over all atoms present in the system at a given time, one can calculate a mean age (Eq. 3.32).
Transit time	Time it takes a C atom to move through (transit) the system, which is equivalent to the age of the atoms at the time they leave it (Fig. 3.13). Integrating over all atoms in the output flux, one can calculate the mean transit time (Eq. 3.33).
Residence time	Variably defined in the literature, sometimes equivalent to age, sometimes to transit time. The use of this term (including in this book) depends on the scientific discipline and the context where it is applied.
Conventional radiocarbon age	Time since C in a system was fixed from the atmosphere as determined by the radiocarbon dating method using Libby's half life (see Eq. 3.16), and assuming a closed system for C.
Calibrated radiocarbon age	Calendar year in which the C in system was fixed from the atmosphere, assuming a closed system and using a stated calibration curve (see Sect. 3.4.1).

Note that only in single-pool, homogeneous systems in steady-state (one for which the probability of every atom leaving is equal), are *turnover time, mean age, mean transit time, and mean residence time* equal

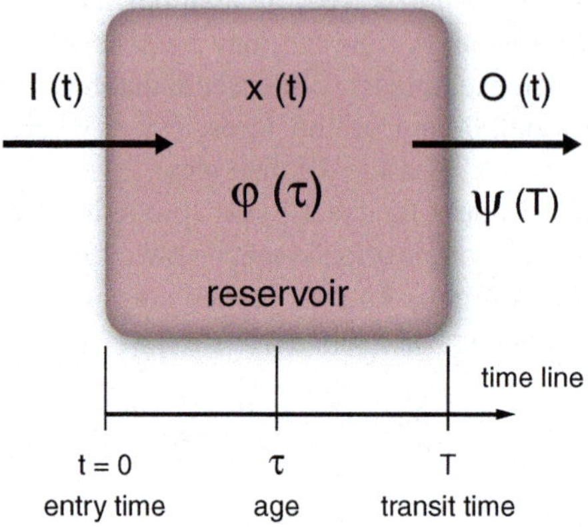

Fig. 3.13 Carbon enters a reservoir at age $\tau = 0$ and leaves at age $\tau = T$ (the transit time). The age density distribution $\varphi(\tau)$ represents the proportion of particles in the reservoir of age τ, while the transit time density distribution $\psi(T)$ represents the proportion of particles leaving the reservoir at age $\tau = T$. Figure modified from Manzoni et al. (2009)

Carbon enters a reservoir at age $\tau = 0$ and leaves at age $\tau = T$, where T is often defined as the transit time (i.e., the time a particle spends in the reservoir between entry and exit; Fig. 3.13; Eriksson 1971; Manzoni et al. 2009). In terms of what ^{14}C can measure, i.e., the age of C, there are two particular properties of interest, the **age density distribution** $\varphi(\tau)$ and the **transit time density distribution** $\psi(\tau)$. The age density distribution represents the proportion of particles in the reservoir of age τ, while the transit time density distribution represents the proportion of particles leaving the reservoir at age τ. We have defined most of these terms in Table 3.2, and caution the reader that often these terms are used interchangeably in the literature. Two additional properties of interest are often defined, the mean age (τ_r) and the mean transit time (τ_0), both formally defined as (Eriksson 1971):

$$\tau_r = \frac{\int_0^\infty \tau\varphi(\tau)\mathrm{d}\tau}{\int_0^\infty \varphi(\tau)\mathrm{d}\tau} \tag{3.32}$$

$$\tau_0 = \frac{\int_0^\infty \tau\psi(\tau)\mathrm{d}\tau}{\int_0^\infty \psi(\tau)\mathrm{d}\tau} \tag{3.33}$$

The mean transit time can also be interpreted as the mean age of matter coming out from the reservoir.

The mean age and mean transit time are only identical under a very specific set of circumstances. For a homogeneous reservoir in steady state, modeled under the assumptions of Eq. (3.20), age time and transit times are identical and equal to the inverse of the release rate k (Eriksson 1971; Manzoni et al. 2009):

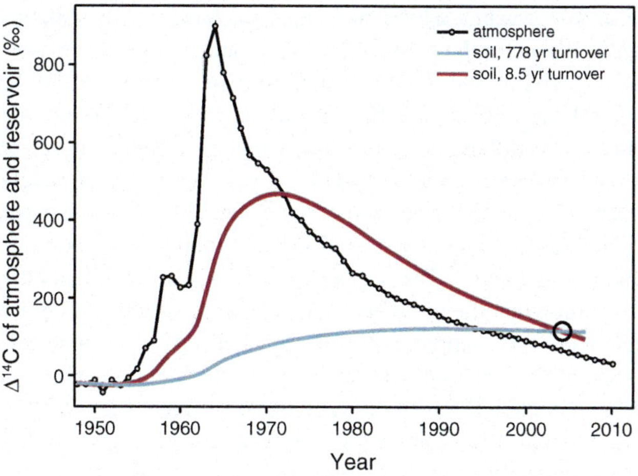

Fig. 3.14 Possible solutions of a homogenous reservoir model tested against one observation point. $\Delta^{14}C$ measured in a soil in Germany in 2004 gave a value of 115 ‰. There were two possible solutions to the model, $\tau = 8.5$ years and $\tau = 778$ years

$$\tau_r = \tau_0 = \frac{1}{k} \tag{3.34}$$

An example of this would be the ^{14}C value of DIC in a well-mixed lake. In this case, the use of ^{14}C measurements is straightforward to estimate the mean age of C in the lake, or the mean age of C leaving the lake. As noted previously, for homogenous reservoirs and ^{14}C data collected prior to the bomb period, mean ages and transit times can be calculated as follows:

$$\tau_r = \tau_0 = \frac{1 - F}{\lambda F} \tag{3.35}$$

For the bomb period, it is possible to use Eq. (3.29) or (3.30) to estimate the mean age and transit time of C in a homogenous reservoir assuming that they are the inverse of the release rate as in Eq. (3.34). In this case, it is possible to optimize Eq. (3.30) to find the solution that best fits the observed ^{14}C data (Fig. 3.14; Yu et al. 2007). However, it must be taken into account that it is always possible to find two different solutions to the optimization problem because of the shape of the bomb curve. In this case, additional information may be needed to constrain the estimation of mean ages and mean transit times (Chap. 6).

However, most real systems are not homogeneous reservoirs. One example would be a human population where the probability of death increases with age. In such a system, the age of elements leaving the reservoir (dying) is older than the average age of the living population (Rodhe 2000). A more common example in C cycling is to have a mixture of faster cycling and slower cycling components. For example, DIC in the ocean can be separated into surface water that exchanges rapidly with the atmosphere, and deep water, which exchanges much more slowly. In this case, the age of C leaving the ocean for the atmosphere will reflect the age of C in the surface ocean and be relatively young. On the other hand, the greater mass of DIC in deep water means the overall age of C in the DIC of the total ocean will be older than the C lost to the atmosphere. This difference demonstrates that the ocean must be considered as at least two reservoirs (surface and deep DIC) if we want to appropriately interpret the ^{14}C data in terms of the response time of the ocean DIC pool to a perturbation.

Systems such as sedimentary or soil organic matter often consist of faster and slower cycling components. The application of ^{14}C measurements to such systems requires recognition that the model is an approximation of the system, and testing of assumptions used—for example, testing to see whether the ^{14}C value of CO_2 being lost due to decomposition (a measure of transit time) is the same as the mean age of C in the reservoir being modeled. If transit time and mean age are not similar, the assumptions of a model with a single reservoir are not met. Alternatively, measurements of different physically or chemically separated 'fractions' may indicate clearly that the C in the reservoir is not homogeneously aged (e.g., as surface water DIC-^{14}C is distinct from deep water DIC-^{14}C). In such cases, alternative models that break down the reservoir into elements or pools that are individually approximated as homogeneous reservoirs are required for estimating the rates of C cycling appropriately from ^{14}C data.

3.4.2.4 Multiple Discrete Pools

Commonly, a reservoir is composed of different C components that cycle at different rates. In these cases, the theory of homogeneous reservoirs above is insufficient to characterize the overall behavior of heterogeneous reservoirs (Eriksson 1971; Rodhe 2000).

The total amount of C $C(t)$ at any given time in a reservoir can be divided into n different pools i:

$$C(t) = \sum_{i=1}^{n} C_i(t) \tag{3.36}$$

Each pool can have different rates of change over time and may or may not depend on the cycling rates of other pools. This can be generalized as a linear system of ordinary differential equations (ODEs) of the form:

$$\begin{aligned} \frac{dC_1(t)}{dt} &= f(t, C_1, ..., C_n) \\ \frac{dC_2(t)}{dt} &= f(t, C_1, ..., C_n) \\ &\vdots \\ \frac{dC_n(t)}{dt} &= f(t, C_1, ..., C_n) \end{aligned} \tag{3.37}$$

Table 3.3 Equations for the calculation of mean transit time ($\bar{T}$) and mean age ($\bar{\tau}$) of a one-pool model and different structures of a two-pool model

Model	$\bar{T}$	$\bar{\tau}$
Individual pool	$\frac{1}{k}$	$\frac{1}{k}$
Series of two pools	$\frac{(1-r)k_1+k_2}{k_1k_2}$	$\frac{1}{k_1}+\frac{1}{k_2}+\frac{1}{(1-r)k_1+k_2}$
Two pools in parallel	$\frac{(1-\alpha_1)k_1+\alpha_1k_2}{k_1k_2}$	$\frac{1}{k_1}+\frac{1}{k_2}-\frac{1}{(1-\alpha_1)k_1+\alpha_1k_2}$
Feedback system	$\frac{(1-r)k_1+k_2}{rk_1+k_2}$	$\frac{(1-r)k_1(k_1+2k_2)+k_2^2}{rk_1+k_2[(1-r)k_1+k_2]}$

Modified from Manzoni et al. (2009)

where $f(t, C_1, \ldots, C_n)$ is a linear function that determines the instantaneous behavior of each pool over time (Sierra et al. 2012). If the instantaneous behavior of each pool is proportional only to the amount of C in that pool ($dC_i/dt = f(t, C_i)$), the reservoir is represented as a system of parallel compartments (Fig. 3.11). If the instantaneous behavior of one or more pools also depends on the amount of C in other pools, the reservoir can behave as a series of compartments or in a feedback scheme (Bruun et al. 2004; Manzoni et al. 2009).

In reservoirs with multiple pools, such as the surface and deep ocean or organic matter reservoirs discussed above, the mean ages and transit times are not equivalent, and their mathematical formulation depends on the specific form of Eq. (3.37). In other words, depending on the number of pools and the connection among them, ages and transit times have different formulas. Some examples of the exact equations of ages and transit times for different two-pool models are presented in Table 3.3.

The fraction of ^{14}C in a reservoir with multiple pools is also represented by a system of linear ODEs of the form:

$$\begin{aligned} \frac{dF'C_1(t)}{dt} &= f(\lambda, t, F', C_1, \ldots, F', C_n) \\ \frac{dF'C_2(t)}{dt} &= f(\lambda, t, F', C_1, \ldots, F', C_n) \\ &\vdots \\ \frac{dF'C_n(t)}{dt} &= f(\lambda, t, F', C_1, \ldots, F', C_n) \end{aligned} \tag{3.38}$$

where λ is the ^{14}C decay constant and F' is the absolute Fraction Modern ^{14}C. Solving Eq. (3.38), it is possible to obtain ages and transit times for each pool in the reservoir. This can be achieved in two different ways. One is determining the size and fluxes in and out of all the pools in the reservoir. Actual ^{14}C measurements can be used to estimate the mean ages and transit times for individual pools using the concepts of homogeneous pools aforesaid (e.g., Gaudinski et al. 2000). Another option is to use the measurements of C and ^{14}C in different pools to obtain the solution of Eq. (3.38) using Bayesian methods such as Markov chain Monte Carlo (e.g., Yu et al. 2007). In this case, the mean ages and transit times for each individual pool are found by repeatedly calculating the values for their given prior knowledge of the original values. The calculations are repeated a large number of times, and the estimates that best match the data are saved. At the end of the simulation, a distribution of the possible values is obtained, which also quantifies the uncertainty in the estimation of mean ages and transit times for individual pools.

The various examples in subsequent chapters demonstrate the use of models to interpret C dynamics from ^{14}C data. In each case, it is important to remember that the results very much depend on the assumed model structure, including factors such as the 'pre-aging' of C before it enters the reservoir, or the exact pathways by which C is transferred from one reservoir to another. Thus, ^{14}C data provide the estimates of dynamics that depend critically on model structure, and it is important that models and their inherent assumptions are clearly described along with interpretations.

3.4.3 Isotope Mixing Models: Source Partitioning

In many cases, the inputs of C to a reservoir are from several sources. For example, C in the atmosphere comes from plant respiration, microbial respiration, and anthropogenic activities. Stable isotopes have been used regularly to partition reservoirs and fluxes into their component sources. For example, ^{13}C and ^{15}N have been used in food web studies to decipher what an organism has eaten and in what relative proportions (e.g., Urton and Hobson 2005), but ^{13}C has also been used to partition C fluxes into its sources (e.g., Dorrepaal et al. 2009). More recently, ^{14}C has been used by itself and with ^{13}C to partition a C mixture into its source components. Most often this source partitioning is performed on soil and ecosystem respiration fluxes (Chap. 6).

The mathematical models used to calculate the relative contribution of each source to the mixture fall into two categories. The first category includes the determined models with unique analytical solutions. In this category, there are no more than $n + 1$ sources, with n equaling the number of isotopes. Using only ^{14}C, a mixture can be partitioned into two sources whose contributions have a unique solution, and while using ^{14}C and $\delta^{13}C$ together, a mixture can be partitioned into three sources with unique solutions (assuming the two isotopes give independent information). In the second category are the undetermined models without unique solutions, where there are $>n + 1$ sources, and each source has a range of possible contributions to the mixture that satisfy the mathematical model. The mathematical equations and computer programs one uses to solve the mixing models and the certainty of the resulting source contributions depend on whether unique solutions are possible.

3.4.3.1 Visualizing Mixing Model Data

Before applying any of the models, however, plotting the isotopic values of sources and mixtures is recommended. The plot will be a single line, if using only ^{14}C, or a two-axis graph with the source values connected creating a polygon (Fig. 3.15), if

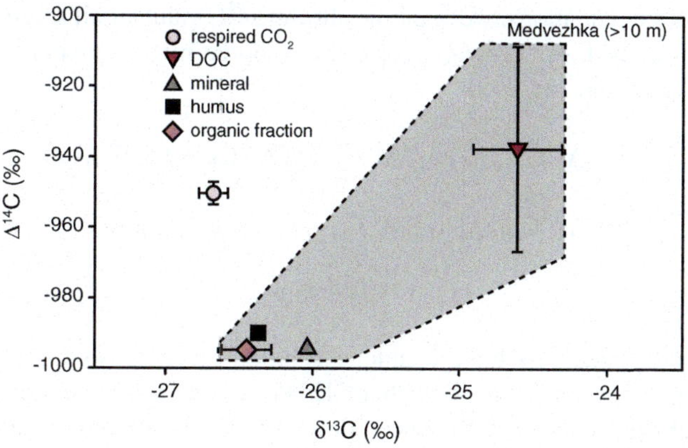

Fig. 3.15 A mixing polygon of $\Delta^{14}C$ and $\delta^{13}C$ values from CO_2 respired from a laboratory soil incubation (the mixture) and from bulk (not respired) soil C fractions (the sources). Polygon plots are useful because they show where a mixture falls relative to its sources and can point to potential problems. In this case, the mixture isotopic value for respired CO_2 falls outside of the polygon created by the source values, indicating there is a missing source that was not measured. Figure modified from Dutta et al. (2006)

using ^{14}C and $\delta^{13}C$. These plots will demonstrate where the mixture falls in relation to the sources and its geometry can flag potential problems. The plots can indicate whether there is a missing source when the mixture's value falls outside the line or polygon connecting the source values. The plot can also indicate when there will be a wide range of solutions such as when the mixture's value falls in the middle of the polygon between all the sources or when a source's value falls within the polygon bounded by other sources, which indicates that source is not needed to solve the model (Phillips and Gregg 2003).

3.4.3.2 Models with Unique Solutions

Isotope mixing models use a system of mass-balance equations (Phillips 2001). In the mass-balance equations, the isotopic values of the mixture and sources are known, and the proportions of the sources that make up mixture's isotopic value are the unknowns. For a ^{14}C only, two-source mixing model, the following two equations are used:

$$\Delta^{14}C_{\text{mixture}} = f_A \Delta^{14}C_A + f_B \Delta^{14}C_B \tag{3.39}$$

$$1 = f_A + f_B \tag{3.40}$$

where $\Delta^{14}C_{\text{mixture}}$, $\Delta^{14}C_A$, and $\Delta^{14}C_B$ are the ^{14}C values of the mixture and sources of A and B, respectively, and f_A and f_B are the proportions of the sources that make up the mixture. We use the $\Delta^{14}C$ notation here because, for consistency when using source values in equations, we want the absolute ^{14}C value of all sources in the year they were sampled. For a two-isotope, three-source mixing model using ^{14}C and ^{13}C, three equations are used:

$$\Delta^{14}\mathrm{C}_{\mathrm{mixture}} = f_A \Delta^{14}\mathrm{C}_A + f_B \Delta^{14}\mathrm{C}_B + f_C \Delta^{14}\mathrm{C}_C \tag{3.41}$$

$$\delta^{13}\mathrm{C}_{\mathrm{mixture}} = f_A \delta^{13}\mathrm{C}_A + f_B \delta^{13}\mathrm{C}_B + f_C \delta^{13}\mathrm{C}_C \tag{3.42}$$

$$1 = f_A + f_B + f_C \tag{3.43}$$

Because these equations have a unique solution for each source proportion, no special software is needed to solve them. However, one can use computer programs such as IsoError to calculate the uncertainty associated with proportion estimates—variances, standard errors, and confidence intervals—using the measured variability of the mixture and source isotopic values (Phillips and Gregg 2001).

3.4.3.3 Models Without Unique Solutions

There are several approaches and software programs that can be used with undetermined models (those with $>n + 1$ sources) in order to find a range of feasible relative source contributions. One is IsoSource (Phillips and Gregg 2003) that can calculate the ranges of contribution from any number of sources using one or more isotopes. IsoSource iteratively (using steps of 1–2 %) calculates every possible combination of sources that sum to 100 %. The corresponding isotopic value of the mixture is compared to the measured isotopic value, and if this difference falls within a stated error range, the source combination is accepted. The result of such a model is a range of potential solutions wherein all solutions are equally feasible.

Generally, badly constrained ranges of solutions occur when there are numerous sources in the model, when the source values are not sufficiently distinct from one another, or when the mixture's isotopic value is centralized in relation to source values (Phillips and Gregg 2003). Some researchers have developed an iterative approach similar to IsoSource such as the one developed by Schuur et al. (2003), where the source contributions varied from 0 to 100 % in 12.5 % intervals and results were kept if the predicted mixture value matched the actual mixture's value $\pm$ its standard error.

Several methods have been used to constrain solutions in these undetermined mixing models. One approach is to combine sources with similar values so that the model can have $n + 1$ sources and therefore a unique solution (Phillips et al. 2005). Another is to combine ranges of sources that are related to each other after the model is run, which generally produces a narrower range of solutions for the aggregated source (Phillips et al. 2005). Lastly, one can constrain solutions a posteriori based on ecological observations. When partitioning soil respiration from incubations, Dioumaeva et al. (2002) discarded solutions where wood and microbially modified organic matter had a greater contribution than roots, because roots are known to decompose faster. When partitioning CO_2 released from a forest fire, Schuur et al. (2003) only accepted solutions that followed a series of rules based on the observations of source flammability (i.e., leaf litter must burn before deeper organic layers).

Drawbacks of iterative methods to solve undetermined mixing models are the inability to include error and variation explicitly and inability to report central tendency, as all solutions in a range are equally feasible. Recently, several Bayesian methods, such as MixSIR (Moore and Semmens 2008) and SIAR (Parnell et al. 2010), have been developed for solving undetermined mixing models. These methods can explicitly include errors. Because the solutions also are true probability distributions, they indicate which source proportions are most likely, although developers of both methods advocate reporting results as ranges because there is still no unique solution. The two Bayesian approaches to mixing models both allow other information about the system to be included as priors in the model as long as that information can be distilled down to proportional contributions of each source. Both use the Dirichlet prior (Semmens et al. 2009), which is purposely vague so the prior distributions act to constrain the posterior distributions without influencing them significantly. If there is no prior information, then an uninformative prior can be used, meaning that no additional constraint is imposed on the system other than the isotope observations.

Both these Bayesian approaches can incorporate the added model complexity if necessary. For instance, the source isotopic values themselves, previously treated as fixed, can become estimated parameters in the Bayesian model, each with their own underlying distributions (Ward et al. 2010). In this case, the measured isotopic value means and variances would be used in the prior distributions of the model. This approach explicitly accounts for uncertainties in the mean and variances of source values, which is important to consider when working with small sample sizes (Ward et al. 2010). Tests of this 'fully Bayesian' approach demonstrated that it increased the ranges of source proportions but reduced bias of estimates. Bayesian mixing models can also be made hierarchal and include random effects. Semmens et al. (2009) adjusted the MixSIR model to allow the source proportions to change among individual mixtures at different scales. Methods that include further complexity are in development, to include the additional random model effects such as repeated measures.

3.5 Conclusions and Future Directions

The reporting and interpretation of ^{14}C data differ depending on the type of C reservoir being studied (open versus closed) and the questions posed. This chapter has reviewed the nomenclature and approaches used for the three main uses of ^{14}C data: (1) the estimation of the ages of C in closed systems; (2) the estimation of mean ages and transit times of C in different reservoirs of the Earth system (open systems); and (3) the identification of the relative contributions of sources in a mixture. It is evident from this chapter that reporting and modeling ^{14}C are not trivial issues. Special care must be taken when reporting ^{14}C measurements following the different standard notations. Similarly, assumptions used in the mathematical formulation of models used to estimate ages and transit times and source partitioning must be stated clearly. The assumptions behind each model may have significant effects on the calculated values of different parameters of interest and therefore in the conclusions of an associated scientific investigation.

Other chapters in this book will provide specific applications of the different notations and models introduced in this chapter. In terms of notation, it is possible that the current terminology will not change dramatically within the next few decades, although the development of new optical methods to measure ^{14}C (Galli et al. 2011; Zare 2012) may bring additional reporting notations. In terms of modeling, new tools will likely be developed to improve parameter estimation and to use more efficient algorithms. However, the basic mathematical models likely will remain the same.

References

Bronk Ramsey, C., Dee, M., Lee, S., Nakagawa, T., and Staff, R. (2010). Developments in the calibration and modelling of radiocarbon dates. *Radiocarbon* 52(3): 953–961.

Bruun, S., J. Six, and L.S. Jensen. 2004. Estimating vital statistics and age distributions of measurable soil organic carbon fractions based on their pathway of formation and radiocarbon content. *Journal of Theoretical Biology* 230: 241–250.

Coplen, T.B., W.A. Brand, M. Gehre, M. Gröning, H.A. Meijer, B. Toman, and R.M. Verkouteren. 2006. New guidelines for δ 13C measurements. *Analytical Chemistry* 78: 2439–2441.

Dioumaeva, I., S. Trumbore, E.A.G. Schuur, M.L. Goulden, M. Litvak, and A.I. Hirsch. 2002. Decomposition of peat from upland boreal forest: temperature dependence and sources of respired carbon. Journal of Geophysical Research-Atmospheres **108**.

Donahue, D.J., T.W. Linick, and A.J.T. Jull. 1990. Isotope ratio and background corrections for accelerator mass spectrometry radiocarbon measurements. *Radiocarbon* 32: 135–142.

Dorrepaal, E., S. Toet, R.S.P. van Logtestijn, E. Swart, M.J. van de Weg, T.V. Callaghan, and R. Aerts. 2009. Carbon respiration from subsurface peat accelerated by climate warming in the subarctic. *Nature* 460: 616-U679.

Dutta, K., E.A.G. Schuur, J.C. Neff, and S.A. Zimov. 2006. Potential carbon release from permafrost soils of northeastern Siberia. *Global Change Biology* 12: 2336–2351.

Eriksson, E. 1971. Compartment models and reservoir theory. *Annual Review of Ecology and Systematics* 2: 67–84.

Faure, G. 1986. *Principles of Isotope Geology*, 2nd ed. New York: Wiley.

Galli, I., S. Bartalini, S. Borri, P. Cancio, D. Mazzotti, P. De Natale, and G. Giusfredi. 2011. Molecular gas sensing below parts per trillion: radiocarbon-dioxide optical detection. *Physical Review Letters* 107: 270802.

Gaudinski, J.B., S.E. Trumbore, E.A. Davidson, and S. Zheng. 2000. Soil carbon cycling in a temperate forest: radiocarbon-based estimates of residence times, sequestration rates and partitioning fluxes. *Biogeochemestry* 51: 33–69.

Godwin, H. 1962. Half-life of radiocarbon. *Nature* 195: 984.

Hoefs, J. 2009. Stable Isotope Geochemistry. 6th edition. Springer.

IAEA 2001. Environmental Isotopes in the Hydrological Cycle: Principles and Applications, vol. 1, p. 97, Fig. 7.5.

Kendall, C., and J.J. McDonnell (eds.). 1998. *Isotope tracers in catchment hydrology*. Amsterdam: Elsevier Science.
Levin, I., T. Naegler, B. Kromer, M. Diehl, R.J. Francey, A.J. Gomez-Pelaez, L.P. Steele, D. Wagenbach, R. Weller, and D.E. Worthy. 2010. Observations and modelling of the global distribution and long-term trend of atmospheric $14CO_2$. *Tellus B* 62: 26–46.
Manzoni, S., G.G. Katul, and A. Porporato. 2009. Analysis of soil carbon transit times and age distributions using network theories. Journal of Geophysical Research 114.
Mook, W.G., Jc Bommerso, and Wh Staverma. 1974. Carbon isotope fractionation between dissolved bicarbonate and gaseous carbon dioxide. *Earth and Planetary Science Letters* 22: 169–176.
Mook, W.G.E. 2000. Environmental Isotopes in the Hydrological Cycle: Principles and Applications, Vol I, Introduction, Theory, and Methods. Page 291 *in* I. H. Programme, editor. UNESCO/IAEA, Paris.
Moore, J.W., and B.X. Semmens. 2008. Incorporating uncertainty and prior information into stable isotope mixing models. *Ecology Letters* 11: 470–480.
Olsson, I. 1970. The use of oxalic acid as a standard. Page 17 *in* Radiocarbon Variations and Absolute Chronology, Nobel Symposium.
Parnell, A.C., R. Inger, S. Bearhop, and A.L. Jackson. 2010. Source partitioning using stable isotopes: coping with too much variation. Plos One **5**.
Phillips, D.L. 2001. Mixing models in analyses of diet using multiple stable isotopes: a critique. *Oecologia* 127: 166–170.
Phillips, D.L., and J.W. Gregg. 2001. Uncertainty in source partitioning using stable isotopes. *Oecologia* 127: 171–179.
Phillips, D.L., and J.W. Gregg. 2003. Source partitioning using stable isotopes: coping with too many sources. *Oecologia* 136: 261–269.
Phillips, D.L., S.D. Newsome, and J.W. Gregg. 2005. Combining sources in stable isotope mixing models: alternative methods. *Oecologia* 144: 520–527.
Reimer, P.J. 2013. IntCal13 and Marine13 radiocarbon age calibration curves 0-50,000 years CAL BP. *Radiocarbon* 55: 1869–1887.
Reimer, P.J., M.G.L. Baillie, E. Bard, A. Bayliss, J.W. Beck, P.G. Blackwell, C.B. Ramsey, C.E. Buck, G.S. Burr, R.L. Edwards, M. Friedrich, P.M. Grootes, T.P. Guilderson, I. Hajdas, T. J. Heaton, A.G. Hogg, K.A. Hughen, K.F. Kaiser, B. Kromer, F.G. McCormac, S.W. Manning, R.W. Reimer, D.A. Richards, J.R. Southon, S. Talamo, C.S.M. Turney, J. van der Plicht, and C.E. Weyhenmeye. 2009. IntCal09 and Marine09 radiocarbon age calibration curves, 0-50,000 years cal bp. *Radiocarbon* 51: 1111–1150.
Rodhe, H. 2000. Modeling Biogeochemical Cycles. In *International geophysics*, eds. R.J.C.H.R. Michael C. Jacobson and H.O. Gordon. Academic Press.
Rundel, P.W., J.R. Ehleringer, and K.A. Nagy (eds.). 1989. *Stable isotopes in ecological research*. New York: Springer.
Schuur, E.A.G., S.E. Trumbore, M.C. Mack, and J.W. Harden. 2003. Isotopic composition of carbon dioxide from a boreal forest fire: inferring carbon loss from measurements and modeling. Global Biogeochemical Cycles **17**.
Semmens, B.X., J.W. Moore, and E.J. Ward. 2009. Improving Bayesian isotope mixing models: a response to Jackson et al. (2009). Ecology Letters **12**:E6–E8.
Sierra, C.A., M. Muller, and S.E. Trumbore. 2012. Models of soil organic matter decomposition: the SoilR package, version 1.0. *Geoscientific Model Development* 5: 1045–1060.
Stuiver, M., and H.A. Polach. 1977. Reporting of C-14 data—discussion. *Radiocarbon* 19: 355–363.
Stuiver, M., and P.D. Quay. 1981. Atmospheric C-14 changes resulting from fossil fuel CO_2 release and cosmic ray flux variability. *Earth and Planetary Science Letters* 53: 349–362.
Torn, M.S., C.W. Swanston, C. Castanha, and S.E. Trumbore. 2009. Storage and turnover of organic matter in soil. In *Biophysico-chemical processes involving natural nonliving organic matter in environmental systems*, 219–272. Wiley.
Urton, E.J.M., and K.A. Hobson. 2005. Intrapopulation variation in gray wolf isotope (delta N-15 and delta C-13) profiles: implications for the ecology of individuals. *Oecologia* 145: 317–326.

Wang, Y., R. Amundson, and S. Trumbore. 1994. A model for soil (CO_2)-C14 and its implications for using C-14 to date pedogenic carbonate. *Geochimica et Cosmochimica Acta* 58: 393–399.
Wanninkhof, R. 1985. Kinetic fractionation of the carbon isotopes C-13 and C-12 during transfer of CO_2 from air to seawater. *Tellus Series B-Chemical and Physical Meteorology* 37: 128–135.
Ward, E.J., B.X. Semmens, and D.E. Schindler. 2010. Including source uncertainty and prior information in the analysis of stable isotope mixing models. *Environmental Science and Technology* 44: 4645–4650.
Yu, S.Y., J. Shen, and S.M. Colman. 2007. Modeling the radiocarbon reservoir effect in lacustrine systems. *Radiocarbon* 49: 1241–1254.
Zare, R.N. 2012. Analytical chemistry: ultrasensitive radiocarbon detection. *Nature* 482: 312–313.

Chapter 4
Radiocarbon in the Atmosphere

J.C. Turnbull, H. Graven and N.Y. Krakauer

4.1 Introduction

Carbon dioxide (CO_2) is one of the most abundant and important atmospheric trace gases. It is found naturally in the atmosphere with a preindustrial mole fraction ("concentration") of about 280 parts per million (ppm). It is stable in the atmosphere, but exchanges readily with the surface reservoirs: the oceans (Chap. 5) and terrestrial biosphere or land (Chap. 6). Both land and oceans naturally exchange large amounts of C with the atmosphere. On long timescales, CO_2 is also removed from the atmosphere by rock weathering, but this very slow process will not be discussed further here.

The land absorbs CO_2 by photosynthetic uptake in plants and re-emits CO_2 back into the atmosphere by respiration from plants, animals, and soils (Fig. 4.1). This process produces large daily and seasonal cycles in the atmospheric CO_2 mole fraction, but on an annual to decadal timescale, the natural land exchange is nearly balanced. The gross one-way fluxes into and out of the land are about 120 petagrams of C per year (Pg C $year^{-1}$) (Ciais et al. 2013), but the natural net annual land flux historically was approximately zero.

J.C. Turnbull (✉)
National Isotope Centre, GNS Science, Lower Hutt, New Zealand

J.C. Turnbull
Cooperative Institute for Research in Environmental Sciences, University of Colorado, Boulder, CO, USA

H. Graven
Department of Physics and Grantham Institute for Climate Change, Imperial College London, London, UK

N.Y. Krakauer
Department of Civil Engineering, The City College of New York, New York, NY, USA

E.A.G. Schuur et al. (eds.), *Radiocarbon and Climate Change*,
DOI 10.1007/978-3-319-25643-6_4

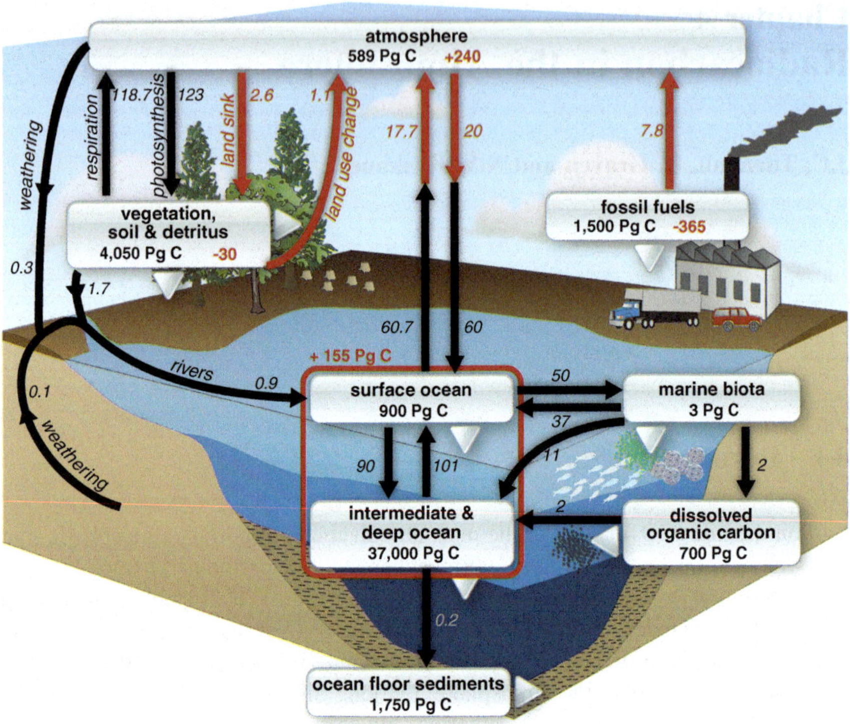

Fig. 4.1 Global C cycle showing natural C fluxes (*black*) and anthropogenic perturbations to the C cycle (*red*). Units are Pg C for pools (*boxes*) and Pg C $year^{-1}$ for fluxes (*arrows*). Numbers shown here represent the midpoints of ranges published in Fig. 6.1 of Ciais (2013) with the exception of the soil organic matter pool that harmonizes new numbers for permafrost with global soil organic carbon inventories (Schuur et al. 2015)

Ocean C uptake and release is via gas exchange with surface waters. When upwelling water is low in dissolved CO_2, atmospheric CO_2 is absorbed and the ocean acts as a C sink. Conversely, when upwelling water is CO_2-rich, CO_2 is released and the ocean acts as a C source to the atmosphere. Many factors influence the rate of exchange, and the gross one-way fluxes into and out of the ocean are about 80 Pg C $year^{-1}$ (Fig. 4.1), but on annual to decadal timescales, the net ocean flux was, like the land flux, roughly neutral.

Over the past 150 years, the atmosphere has been perturbed by the introduction of additional CO_2 from anthropogenic activities, primarily the combustion of fossil fuels and land-use change. The rate at which fossil fuel and land-use change C has been added to the atmosphere is small compared to the large natural fluxes, but it represents an additional one-way flux into the atmosphere that is not in balance. This results in an accelerating buildup of CO_2 in the atmosphere to the current mole fraction of over 400 ppm (in 2015). However, this increase in atmospheric CO_2 represents only about half of the ~500 Pg C emitted by humans since 1850

(Houghton 2008; Boden et al. 2012), and therefore, the other half must have moved into the ocean and land C reservoirs. Understanding exactly how much CO_2 is emitted from anthropogenic sources such as fossil fuels and land-use change, and the details of where, when, and how CO_2 is absorbed into the ocean and land is vital to predicting future atmospheric CO_2 levels and hence future climate.

Measurements of CO_2 in the atmosphere can potentially be used, in conjunction with atmospheric transport models, to infer the locations and magnitudes of the sources and sinks (release and uptake) of CO_2. Yet measurements of CO_2 mole fraction in themselves are not always sufficient to pinpoint the locations and magnitude of different sources and sinks with useful precision. Difficulties in doing so arise from the sparseness of the atmospheric measurement network, uncertainties and biases in the atmospheric transport models, and the inability to distinguish between different sources, particularly when they are located within the same region. A particular challenge is that the large gross fluxes of C into and out of the biosphere are usually the dominant source of CO_2 variability at any given time, even though on the annual timescale, the net biospheric flux is roughly in balance resulting in little or no net change in atmospheric CO_2. Radiocarbon (^{14}C) can provide a window into the atmospheric CO_2 budget, as different sources of CO_2 vary in ^{14}C content. This difference in source isotopic values allows partitioning of atmospheric observations into their component sources using standard partitioning models (Chap. 3).

In this chapter, we examine the controls on the ^{14}C content of CO_2 ($^{14}CO_2$) in the atmosphere over time in relation to CO_2 itself, and the specific C cycle questions that can therefore be addressed with atmospheric ^{14}C measurements. We discuss not only atmospheric observations, but also how these observations are interpreted using models of atmospheric transport, which describe the physical mixing of the atmosphere (similarly, models of the ocean and biosphere C reservoirs are used to describe ^{14}C movement through them). This discussion spans the simplest conceptual model of addition of a gas into a single well-mixed (homogeneous) box of air, to multibox (heterogeneous) models where the atmosphere is partitioned into a few latitudinal divisions, to full-fledged three-dimensional global or regional atmospheric transport models.

We start by examining the terms of the CO_2 and $^{14}CO_2$ budgets (Sect. 4.2). Then, we divide the atmospheric ^{14}C history into five different time periods, defined by times when different factors dominated atmospheric $^{14}CO_2$. First, we consider the preindustrial (before 1850) near steady state of $^{14}CO_2$, investigating its sources and sinks and how they differ from those of CO_2 itself (Sect. 4.3). Next, we examine the period from 1890 to 1945, when fossil fuel CO_2 emissions started to impact atmospheric $^{14}CO_2$ (the Suess effect, Sect. 4.4). The bomb period from 1945 to 1985 follows (Sect. 4.5), when the production and redistribution of ^{14}C from aboveground nuclear weapons testing produced a major perturbation in atmospheric $^{14}CO_2$ that propagated throughout the C cycle. The post-bomb period from 1985 until present is characterized by the strong dominance of fossil fuel emissions on $^{14}CO_2$ (Sect. 4.6). We also discuss how other sources are impacted, and how this can be used to examine C cycle processes. In Sect. 4.7, we suggest how atmospheric $^{14}CO_2$ might continue to evolve in the future. In Sect. 4.8, we delve into using $^{14}CO_2$ as a tracer for recently added fossil fuel CO_2, currently the most widely used application of atmospheric $^{14}CO_2$ measurements. This extended discussion of $^{14}CO_2$ is followed by related approaches using ^{14}C of other atmospheric species, including CH_4, carbon monoxide, and aerosols (Sect. 4.9). These have been less extensively studied than CO_2, but have potential for bringing new understanding to a number of atmospheric processes. Atmospheric ^{14}C is already used in a diverse suite of applications, yet there remain outstanding questions and future directions to be explored, which we address in Sect. 4.10.

4.2 Carbon Dioxide and Radiocarbon in the Atmosphere: Overview

4.2.1 The Global Carbon Dioxide and Radiocarbon Budgets

The atmospheric CO_2 budget can be described by:

$$\frac{dC_a}{dt} = F_p + F_r + F_{ao} + F_{oa} + F_{ff} \tag{4.1}$$

where C_a is the atmospheric CO_2 burden. F_p and F_r are the CO_2 fluxes into (photosynthesis) and out of (respiration, biomass burning, etc.) the terrestrial biosphere. From an atmospheric perspective, C uptake or release by vegetation due to land-use change is indistinguishable from natural terrestrial C exchange and is therefore implicitly included in these terms. F_{ao} and F_{oa} are the fluxes from the atmosphere into the ocean and from the ocean into the atmosphere, respectively. F_{ff} is the flux into the atmosphere from fossil fuel emissions. In this notation, fluxes into the atmosphere (F_r, F_{oa}, F_{ff}) are positive, and fluxes out of the atmosphere (F_p, F_{ao}) are negative.

The change in ^{14}C can be described in a similar way as follows:

$$\frac{dC_a\Delta_a}{dt} = F_p\Delta_p + F_r\Delta_r + F_{ao}\Delta_{ao} + F_{oa}\Delta_{oa} + F_{ff}\Delta_{ff} + F_c\Delta_c + F_n\Delta_n + F_d\Delta_d \tag{4.2}$$

where Δ is the ^{14}C value of the denoted pool or flux (Tans et al. 1979). The ^{14}C budget has three extra terms. All of these terms are pure ^{14}C fluxes, with, by definition, ^{14}C of 8.5×10^{14} ‰, so although the fluxes are very small, the impact on $^{14}CO_2$ can be large. F_c is the flux due to natural cosmogenic production of ^{14}C in the upper atmosphere. F_n is the flux due to (1) atmospheric nuclear weapons testing (bomb ^{14}C) and (2) the nuclear industry—primarily nuclear reactors, but also including spent nuclear fuel reprocessing and other ^{14}C production, such as ^{14}C created for isotopic labeling experiments and subsequently released to the atmosphere. This term comprises ^{14}C atoms created by humans. F_d is the negative flux due to radioactive decay, which is too small to influence decadal to century-scale changes in atmosphere ^{14}C and is ignored hereafter. This equation describes the entire atmosphere as a single pool, but since the stratosphere, as the primary site of cosmogenic production, is enriched in ^{14}C relative to the troposphere, it can be treated as a separate pool (Levin et al. 2010). Combining Eqs. 4.1 and 4.2, we obtain

$$C_a \frac{d\Delta_a}{dt} = F_r(\Delta_r - \Delta_a) + F_{oa}(\Delta_{oa} - \Delta_a) + F_{ff}(\Delta_{ff} - \Delta_a) + F_c(\Delta_c - \Delta_a) + F_n(\Delta_n - \Delta_a) + F_p(\Delta_p - \Delta_a) + F_{ao}(\Delta_{ao} - \Delta_a) \quad (4.3)$$

Each term on the right-hand side of Eq. 4.3 is an *isoflux*, the product of the one-way gross CO_2 flux and the isotopic difference between the source (or sink) and the atmosphere. This difference is known as the isotopic disequilibrium. A small CO_2 flux can have a large isoflux if the isotopic disequilibrium between the atmosphere and the reservoir is large, and conversely, a large CO_2 flux can have a small or zero isoflux if there is little or no isotopic disequilibrium between the atmosphere and the reservoir. Isofluxes may be positive or negative, depending on the sign of the isotopic disequilibrium, and the isoflux may be of a different sign than the CO_2 flux. When C is dissolved into the surface ocean or taken up by plants (fixed by photosynthesis), the only change in isotopic composition is due to mass-dependent fractionation during uptake, which is corrected for in the ^{14}C nomenclature (Chap. 3). Therefore, p and ao are by definition equal to a, so the isotopic disequilibrium in uptake is zero. As a result, the last two terms in Eq. 4.3 that represent the isofluxes into the biosphere and oceans are equal to zero, and

$$C_a \frac{d\Delta_a}{dt} = F_r(\Delta_r - \Delta_a) + F_{oa}(\Delta_{oa} - \Delta_a) + F_{ff}(\Delta_{ff} - \Delta_a) + F_c(\Delta_c - \Delta_a) + F_n(\Delta_n - \Delta_a) \quad (4.4)$$

The remaining terms in Eq. 4.4 are all fluxes of CO_2 into the atmosphere. So measurements of $^{14}CO_2$ detect the one-way gross fluxes into the atmosphere, rather than the net fluxes that are detected by CO_2 concentration measurements.

Four factors drive the isotopic disequilibrium in ^{14}C between the sources and the atmosphere: (1) natural production of ^{14}C by interaction of nitrogen with cosmic rays in the upper atmosphere; (2) anthropogenic production of ^{14}C from nuclear weapons testing and the nuclear industry; (3) fossil fuel emissions; and (4) the *reservoir age*, which determines the ^{14}C content in each C reservoir or source. Radiocarbon decays in reservoirs that are out of contact with the atmosphere, resulting in lower ^{14}C, which can also be presented in terms of the ^{14}C age or reservoir age in units of years by relating ^{14}C to the decay rate of ^{14}C. The reservoir age can be determined from the ^{14}C content of the atmosphere at the time the C was taken up into the reservoir, together with the residence time of C in that reservoir. It can also be determined empirically from measurements of the reservoir's ^{14}C value.

4.2.2 *Atmospheric Transport Modeling of Radiocarbon*

Interpretation of atmospheric observations of $^{14}CO_2$ and CO_2 requires an understanding of the movement and mixing of air, as well as the sources and sinks of CO_2. Carbon cycle and atmospheric transport models are used to build up detailed descriptions of these interactions, and these descriptions can be tested and adjusted by comparison with observations. The global budgets of both CO_2 and $^{14}CO_2$, described in Eqs. 4.1 and 4.2, can be simulated by models with varying degrees of resolution and complexity. Models of the terrestrial biosphere and ocean provide information on the flux of C and the isotopic disequilibrium from each reservoir, taking changes in atmospheric ^{14}C into account. One example of this type of modeling exercise is the global budgeting of CO_2 and $^{14}CO_2$ isofluxes for different time periods shown in Tables 4.1, 4.2, and 4.3. Simulated isofluxes include uncertainties that are the subject of ongoing research. These uncertainties relate to the transport rates and pathways of C on land and in the ocean and the difficulty in representing ecosystem and oceanic dynamics in models (Naegler and Levin 2009b; Levin et al. 2010; Graven et al. 2012b, c). Uncertainty in the cosmogenic ^{14}C production rate is also significant, with various estimates differing by ~25 % (Lal 1988; Masarik and Beer 1999). The source of CO_2 from fossil fuel combustion is

Table 4.1 Idealized steady-state preindustrial ^{14}C isofluxes to the atmosphere

Reservoir	Estimated flux F into the atmosphere (Pg C $year^{-1}$)	$\Delta^{14}C$ difference between reservoir and atmosphere (‰)	Isoflux (Pg C ‰ $year^{-1}$)	Impact on $\Delta^{14}CO_2$ (‰ $year^{-1}$)
Oceans	85 ± 21[a]	−65[b]	−5500 ± 1400	−9 ± 2
Terrestrial biosphere	52 ± 11[c]	−4.5[c]	−235 ± 55	−0.4 ± 0.1
Cosmogenic production	$6.5 \pm 0.8 \times 10^{-12}$ [d]	8.5×10^{14}	5500 ± 700	9 ± 1
Annual mean change			0	0

The isoflux is the product of the C flux and the isotopic difference between the reservoir and the preindustrial atmosphere ($a = 0$ ‰). The impact of each isoflux on atmospheric $^{14}CO_2$ is calculated assuming a preindustrial CO_2 mixing ratio of 280 ppm. Cosmogenic production generates pure ^{14}C, which equates to a $\Delta^{14}C$ value of 8.5×10^{14} ‰

[a]Heimann and Maier-Reimer (1996)

[b]Key et al. (2004)

[c]Naegler and Levin (2009b)

[d]Lal (1988), Masarik and Beer (1999)

Table 4.2 Estimated ^{14}C isofluxes to the troposphere in 1963 near the start of the bomb period

Reservoir	Estimated flux F into the atmosphere (Pg C year^{-1})	$\Delta^{14}C$ difference between reservoir and atmosphere (‰)	Isoflux (Pg C ‰ year^{-1})	Impact on $\Delta^{14}CO_2$ (‰ year^{-1})
Fossil fuels	2.8 ± 0.3[a]	−1570[f]	−4400 ± 400	−7 ± 1
Ocean	85 ± 21[b]	−635[g]	−54,000 ± 13,000	−80 ± 20
Terrestrial biosphere	52 ± 11[c]	−575[c]	−30,000 ± 6000	−44 ± 9
Nuclear weapons testing	2.6×10^{-10} [d]	8.5×10^{14}	220,000	320
Cosmogenic production	$6.5 \pm 0.8 \times 10^{-12}$ [e]	8.5×10^{14}	5500 ± 700	8 ± 1
Annual mean change			+140,000	+200

The isoflux is the product of the C flux and the isotopic difference between the reservoir and the atmosphere (tropospheric $\Delta^{14}CO_2$ = 570 ‰ for 1963; Levin et al. 2010). The impact of each isoflux on atmospheric $^{14}CO_2$ is calculated assuming an atmospheric CO_2 mixing ratio of 319 ppm for 1963 (Keeling and Whorf 2005). Note that this budget omits ^{14}C accumulation in the stratosphere

[a]Marland et al. (2006), including 10 % uncertainty in emissions
[b]Heimann and Maier-Reimer (1996)
[c]Naegler and Levin (2009b), assuming negligible change in ^{14}C from preindustrial state
[d]Levin et al. (2010)
[e]Lal (1988), Masarik and Beer (1999)
[f]Zero ^{14}C content, or −1000 ‰ in fossil C
[g]Key et al. (2004), assuming negligible change in ^{14}C from preindustrial state

estimated using economic data on fuel use and carries an uncertainty of 5–10 % for global emissions (Marland 2010).

Thus far, we have used Eq. 4.4 to describe the global ^{14}C budget, yet we often want to understand the C budget in smaller regions. The biosphere, fossil fuel, and nuclear isofluxes are all predominantly over the Northern Hemisphere land. In contrast, the ocean exchange is concentrated in the Southern Ocean, due to upwelling of old water and fast wind-driven air–sea exchange there. These geographic differences result in a north–south interhemispheric gradient in $^{14}CO_2$. The cosmogenic production source is concentrated in the upper atmosphere at higher latitudes because of the structure of the Earth's magnetic field, but is roughly symmetric between the two hemispheres (e.g., Lal 1988). There is also considerable heterogeneity in the isofluxes at regional scales and through time. The mixing timescale of the troposphere between hemispheres is about one year; it involves barriers associated with the intertropical convergence zone near the equator, as well as divergent subtropical and convergent subpolar regions (Jacob 1999). Mixing occurs more rapidly in the west–east direction over the mid-latitudes of each hemisphere due to predominantly zonal winds, with a timescale of a few weeks.

Table 4.3 Estimated ^{14}C isofluxes to the atmosphere in 1995 (post-bomb period)

Reservoir	Estimated flux F into the atmosphere (Pg C $year^{-1}$)	$\Delta^{14}C$ difference between reservoir and atmosphere (‰)	Isoflux (Pg C ‰ $year^{-1}$)	Impact on $\Delta^{14}CO_2$ (‰ $year^{-1}$)
Fossil fuels	6.4 ± 0.6[a]	−1115[f]	−7100 ± 700	−9 ± 1
Ocean	85 ± 21[b]	−60[g]	−5100 ± 1200	−7 ± 2
Terrestrial biosphere	52 ± 11[c]	25 ± 25[c]	1300 ± 1500	2 ± 2
Nuclear weapons testing	8×10^{-13} [d]	8.5×10^{14}	700	1
Cosmogenic production	$6.5 \pm 0.8 \times 10^{-12}$ [e]	8.5×10^{14}	5500 ± 700	7 ± 1
Annual mean change			−4700	−6

The isoflux is the product of the C flux and the isotopic difference between the reservoir and the atmosphere (using a atmospheric $\Delta^{14}CO_2$ = 115 ‰ for 1995; Levin et al. 2010). The impact of each isoflux on atmospheric $^{14}CO_2$ is calculated assuming an atmospheric CO_2 mixing ratio of 361 ppm for 1995 (Keeling and Whorf 2005). Note that if mixing is assumed to occur only in the troposphere, then these values increase by about 25 %
[a]Marland et al. (2006), including 10 % uncertainty in emissions
[b]Heimann and Maier-Reimer (1996)
[c]Naegler and Levin (2009b)
[d]UNSCEAR (2000)
[e]Lal (1988), Masarik and Beer (1999)
[f]Zero ^{14}C content, or −1000 ‰ in fossil C
[g]Key et al. (2004)

Mixing with the stratosphere is slower, with a timescale of several years (Jacob 1999). This variability in the isofluxes and in atmospheric mixing results in spatial and temporal variability in ^{14}C and in CO_2 mole fraction in the atmosphere.

Atmospheric observations are made at specific locations, and the atmospheric variability needs to be understood to interpret these measurements at global, regional, and local scales. Atmospheric transport models are commonly used to address this. These models use a combination of theoretical formulations and observations of meteorological parameters to describe atmospheric transport. The isofluxes are then input to the model and convolved with the transport to obtain simulated atmospheric ^{14}C and CO_2 mole fractions.

The simplest atmospheric transport models are box models that divide the atmosphere into a few compartments or *boxes*, such as the GRACE model (Levin et al. 2010). Typically, the global atmosphere is divided into 2–8 zonal bands and/or separated vertically into troposphere and stratosphere boxes. A separate stratosphere is particularly important for describing vertical gradients of ^{14}C, since this is the initial location of both cosmogenic ^{14}C production and bomb-produced ^{14}C. These simple box models typically work on time steps of one year. At each time step, an estimated isoflux from each source (biosphere, ocean, fossil fuels, nuclear industry, and cosmogenic production) is entered into each box. The isoflux estimates are best guesses that are usually determined offline from other models or information. The isofluxes are mixed between the atmospheric boxes using simple parameterizations of the mixing rates between the various boxes. These box models can approximate the overall latitudinal patterns observed in both CO_2 concentration and in ^{14}C isotope ratios, and can be used to test isoflux estimates. Simple models are computationally cheap, so they are often used to examine variability over long time periods of hundreds to thousands of years. However, their simplicity means that they cannot adequately describe more detailed regional and temporal variability.

More complex global atmospheric transport models divide the world into many more boxes. Typically, the Earth's surface is gridded into boxes of up to a few degrees (100–500 km) on a side, and between 15 and 100 vertical levels. As computing time and storage become cheaper, these models are moving to higher resolution (more gridboxes). These models usually ingest meteorological observations to *force* the transport, but can also be driven by their own physics-based simulation of atmospheric transport, or some combination of the two. The largest difficulty comes in dealing with phenomena that occur at a smaller scale than the gridboxes and must be *parameterized*, convective mixing being one example. These models are immensely helpful in understanding observed distributions. They are widely used in *inversions*, a method of adjusting the a priori flux estimates to optimize the agreement between observations and model simulations (Gurney et al. 2002). Typically, a set of a priori flux estimates are obtained from bottom-up inventory information and/or process models such as terrestrial and ocean C cycle models and gridded to the same spatial and temporal scale as the atmospheric transport model. These fluxes are ingested into the model to provide simulated ^{14}C and CO_2 mole fractions. The simulated and observed values are compared, and adjustments (typically using least squares minimization) are made to the fluxes to improve the agreement between the two, obtaining improved flux estimates, known as a posteriori fluxes. Thus far, $^{14}CO_2$ inversions have been done in only a simplified way, whereby model simulations are performed with several different sets of a priori fluxes to test which fluxes agree best with the observations (e.g., Turnbull et al. 2011).

However, more work is needed to improve the accuracy of the atmospheric transport models. Most atmospheric transport models do not accurately predict the vertical transport of tracers such as CO_2, leading to too much or not enough accumulation in the surface boundary layer (Stephens et al. 2007). As yet, only a handful of simulations with three-dimensional atmospheric transport models have been performed with a complete depiction of the ^{14}C isofluxes (Randerson et al. 2002; Turnbull et al. 2009b; Miller et al. 2012). These studies have shown that over the continents in the post-bomb period (since 1985), the ^{14}C variability is usually dominated by the fossil fuel isoflux (Sect. 4.6). Thus, the models can be used to simulate continental $^{14}CO_2$ gradients resulting from fossil fuels and compare them to observed $^{14}CO_2$ gradients (Sect. 4.8).

Regional models are similar to global atmospheric transport models, but allow increased spatial and temporal resolution by modeling only a small portion of the atmosphere. They typically cover regions with spatial extent of a few hundreds to thousands of kilometers and often only examine the lower regions of the atmosphere. The higher resolution is beneficial for regional studies, but these models must address "boundary conditions"—the mixing ratios and isotopic values of air entering the edges of the model domain (Hsueh et al. 2007; Palstra et al. 2008; Riley et al. 2008). At much smaller scales of tens to hundreds of meters, plume models provide an alternative modeling method. These describe the dispersion of a plume of gas within the atmosphere and can be useful for describing atmospheric enrichment near a source, such as that caused by ^{14}C emissions from nuclear power plants (Levin et al. 2003) or emissions from point sources (Turnbull et al. 2014). At distances of more than a few kilometers from a source, their assumptions usually break down.

4.3 Preindustrial Distribution of Radiocarbon

In the preindustrial atmosphere before 1850, the fossil fuel and nuclear isofluxes were zero, so only cosmogenic production, radioactive decay, and land and ocean exchanges influenced atmospheric $^{14}CO_2$. Cosmogenic production was the only positive isoflux, increasing ^{14}C in the atmosphere by about 9 ‰ year^{-1} (Table 4.1). Both the terrestrial biosphere and oceans had negative isofluxes, since C resides in each of these reservoirs for some time and becomes depleted in ^{14}C by radioactive decay before returning to the atmosphere. The biospheric residence time is typically years or decades (Chap. 6), whereas the ocean reservoir age is typically hundreds or thousands of years (Chap. 5). Therefore, in the preindustrial atmosphere, the ocean isoflux by far dominated over the land isoflux (Table 4.1) in its influence on $^{14}CO_2$ variability.

4.3.1 Mean Isotope Ratio

The ^{14}C content of CO_2 in the past atmosphere can be measured from C samples with known ages, such as tree rings. These can be analyzed for ^{14}C content today, which can be converted to past ^{14}C (Chap. 3) by counting the exact age of each tree ring and correcting for radioactive decay that occurred since the time of growth. Considerable effort has been made to construct tree-ring chronologies for paleoclimate applications (Chap. 7), and the tree-ring record presently extends ~14,000 years before 1950 ("before present" or BP) (Friedrich et al. 2004; Hua et al. 2009). Reconstruction of atmospheric $^{14}CO_2$ content goes beyond this, back to the detection limit of ^{14}C measurements at ~50,000 years BP, based on ^{14}C in cross-dated coral carbonate and foraminifera (Reimer 2013). Neither corals nor foraminifera reflect the atmospheric $^{14}CO_2$ content as precisely as tree rings because their ^{14}C is different from atmospheric $^{14}CO_2$ due to the ocean reservoir age (Reimer 2013). Uncertainty in the reservoir age and its variability through time make the reconstruction of atmospheric $^{14}CO_2$ content in the distant past potentially less accurate than the measurements from tree rings.

In a steady-state Earth system, $^{14}CO_2$ will be constant, with cosmogenic ^{14}C being produced in the atmosphere, then mixing throughout the C reservoirs, where it radioactively decays. Table 4.1 shows the globally integrated $^{14}CO_2$ isofluxes for an idealized preindustrial steady state, where cosmogenic ^{14}C production in the atmosphere is balanced almost entirely by C exchange with the oceans, which have lower ^{14}C due to radioactive decay within the oceans themselves. Biospheric exchange makes a smaller contribution to the preindustrial $^{14}CO_2$ balance since the land biosphere has a much shorter mean C residence time and smaller C pool than the oceans.

Yet the Earth System is dynamic, and records show that $^{14}CO_2$ has varied through time, due to variability in both cosmogenic production and in the global C cycle. The various records show that $^{14}CO_2$ increases when the cosmogenic ^{14}C production rate increases (F_c in Eq. 4.4), while it decreases when the atmospheric CO_2 concentration increases and dilutes the ^{14}C content (in Eq. 4.4, increasing C_a results in decreased a). During the last ice age, when the atmospheric CO_2 concentration was up to 35 % lower than during interglacial periods, $\Delta^{14}CO_2$ was as high as +600 ‰ (Reimer 2013) (Fig. 4.2). Over the last 10,000 years, $\Delta^{14}CO_2$ has stayed within a 100 ‰ range, mostly decreasing gradually, and showed quasi-cyclic variations on decade to century timescales associated with solar activity, which drives changes in cosmogenic ^{14}C production (Stuiver and Braziunas 1993). Millennial and longer scale fluctuations in $^{14}CO_2$ are largely correlated with the Earth's magnetic field strength, which also affects the ^{14}C production rate (Mazaud et al. 1991). By assuming that the C cycle (i.e., plant photosynthesis and respiration and ocean CO_2 exchange) was constant over the Holocene, the $^{14}CO_2$ record can be inverted to yield a time series of smoothed fluctuations in cosmogenic ^{14}C production (Usoskin and Kromer 2005). However, part of

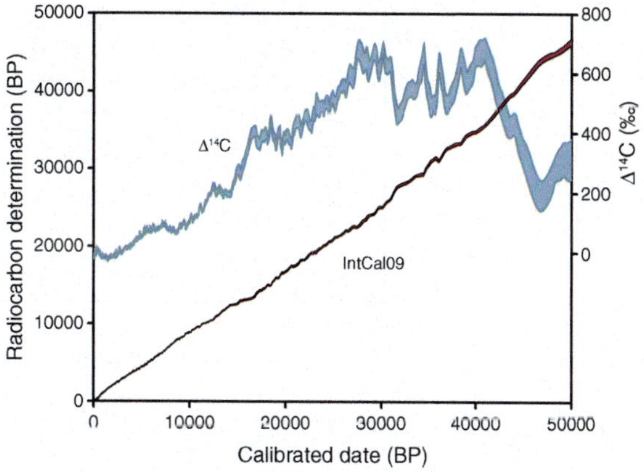

Fig. 4.2 The IntCal09 calibration curve and age-corrected $\Delta^{14}C$ (‰) with one standard deviation envelopes. Figure modified from Reimer et al. (2009)

the $^{14}CO_2$ variability is likely due to changes in the C cycle, driven for example by ocean upwelling variability (Muscheler et al. 2004; Kohler et al. 2006). A simple box model of the C cycle performed better in reproducing the higher $^{14}CO_2$ in the last ice age if exchange between surface and deep ocean water is decreased by 50 % and the shell carbonate sedimentation rate is decreased by 10 % compared with Holocene values (Hughen et al. 2004). In the preindustrial atmosphere, the biosphere likely did not have a large impact on $^{14}CO_2$ gradients, even if the biospheric CO_2 flux varied, since the biospheric turnover time is on the order of decades, and the ^{14}C of respired CO_2 was therefore only slightly lower than atmospheric $^{14}CO_2$ (Braziunas et al. 1995). This topic is also covered in more detail in Chap. 7.

4.3.2 Spatial Gradients

The reconstructions of $^{14}CO_2$ show hemispheric and regional variability (Hua and Barbetti 2004; Hua et al. 2013) due to the spatial distribution of the isofluxes, with the ocean isoflux concentrated in the Southern Hemisphere, the biospheric isoflux mostly in the Northern Hemisphere, and the cosmogenic isoflux at high altitudes near the poles (Sect. 4.2.2). The mixing timescale of the troposphere is about one year between hemispheres, but only a few weeks in the west–east direction at the mid-latitudes of each hemisphere. Due to the faster mixing time in the longitudinal direction, along with the primarily latitudinal pattern of spatial variability in the fluxes, latitudinal variability in preindustrial $^{14}CO_2$ was characteristically much greater than longitudinal variability. Models have shown that preindustrial longitudinal gradients are likely below the detection limit of present-day measurement techniques because of the fast mixing in the west–east direction (Fig. 4.3, Braziunas

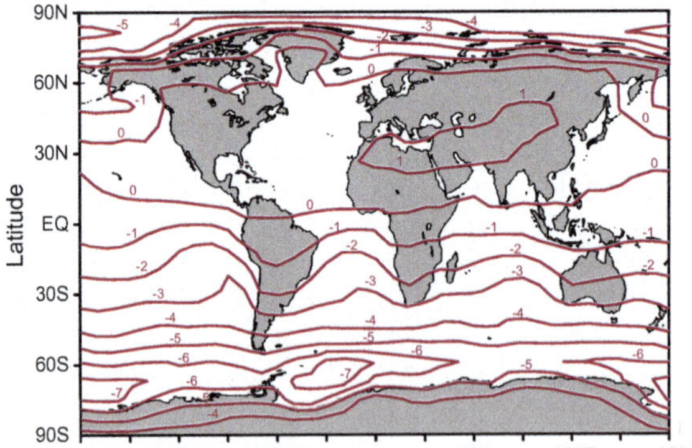

Fig. 4.3 Geographic distribution of simulated preindustrial atmospheric $\Delta^{14}C$. Contours are in units of permil and represent deviations from a fixed value of 0 ‰ for the Olympic Peninsula, Washington, USA. Figure modified from Braziunas et al. (1995)

et al. 1995). Most studies have therefore focused on the larger preindustrial latitudinal gradients, particularly the interhemispheric difference.

The first accurate measurements of the interhemispheric gradient in preindustrial $^{14}CO_2$ showed that Southern Hemisphere tree-ring samples from the 1830s had $\Delta^{14}CO_2$ 4.5 ± 1 ‰ lower than Northern Hemisphere samples (Lerman et al. 1970). Subsequent studies found similar north–south offsets from mid-latitude sites in the late eighteenth to the late nineteenth centuries (Vogel et al. 1993; McCormac et al. 1998; Stuiver and Braziunas 1998). This preindustrial interhemispheric difference is primarily due to the predominance of the ocean isoflux in the Southern Hemisphere. In addition to the large ocean surface area there, upwelling of old water and fast wind-driven air–sea exchange increase the Southern Ocean isoflux. Two studies (Hogg et al. 2002; McCormac et al. 2002) analyzed decadal-average tree-ring samples from Great Britain and New Zealand going back to 950 AD, with replicate measurements conducted in the Belfast and Waikato laboratories to control for interlaboratory calibration differences. They found an average difference of 5.0 ± 1.6 ‰ over 950–1850, but with detectable centennial-scale variability of up to ±4 ‰ (Fig. 4.4). The variability has been attributed to changing wind patterns over the Southern Ocean during the Medieval warm period and the Little Ice Age (Rodgers et al. 2011). Such analyses are at the edge of the accuracy of current measurement technology and require stringent calibration and quality control procedures. Recent work, making use of newly developed, longer New Zealand tree chronologies, finds interhemispheric offsets for the first millennium AD that are similar in magnitude and variability to those of the preindustrial second millennium (Hogg et al. 2009b). With a lower degree of precision, *wiggle matching* (matching the timescales by lining up the patterns of variability in ^{14}C) with floating Southern Hemisphere tree-ring chronologies extends this conclusion through previous millennia in the Holocene (Hogg et al. 2009a).

Less attention has been paid to reconstructing preindustrial $^{14}CO_2$ in regions outside the mid-latitudes. The complexity of atmospheric transport in the tropics is illustrated by work with Thai trees (19°N) from the seventeenth and eighteenth

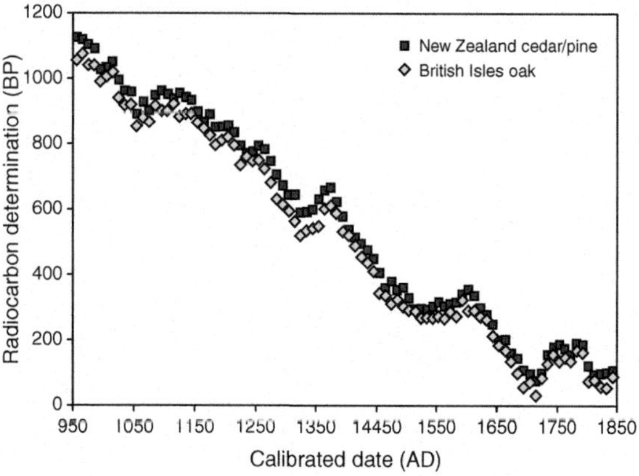

Fig. 4.4 Radiocarbon ages of independently dated Southern and Northern Hemisphere wood. Figure modified from Hogg et al. (2002)

centuries, which showed that $^{14}CO_2$ levels were intermediate between those of northern and southern mid-latitudes despite their location in the Northern Hemisphere (Hua et al. 2004). This is attributed to northward entrainment of Southern Hemisphere air, with lower $^{14}CO_2$, during the Asian summer monsoon.

4.4 The "Suess" Period: 1890–1945

The first observation of anthropogenic influence on atmospheric $^{14}CO_2$ was made by Suess (1955), showing decreasing $^{14}CO_2$ in the early part of the 1900s in tree-ring records from North America. The decline in atmospheric $^{14}CO_2$ content since the industrial revolution demonstrated the addition of ^{14}C-free fossil fuel CO_2 to the atmosphere (Fig. 4.5a). Suess' tree-ring measurements showed a decrease of 25 ‰ in atmospheric $\Delta^{14}CO_2$ between 1890 and 1950 (Suess 1955). This decrease, now called the *Suess effect*, is not determined solely by the rate of fossil fuel

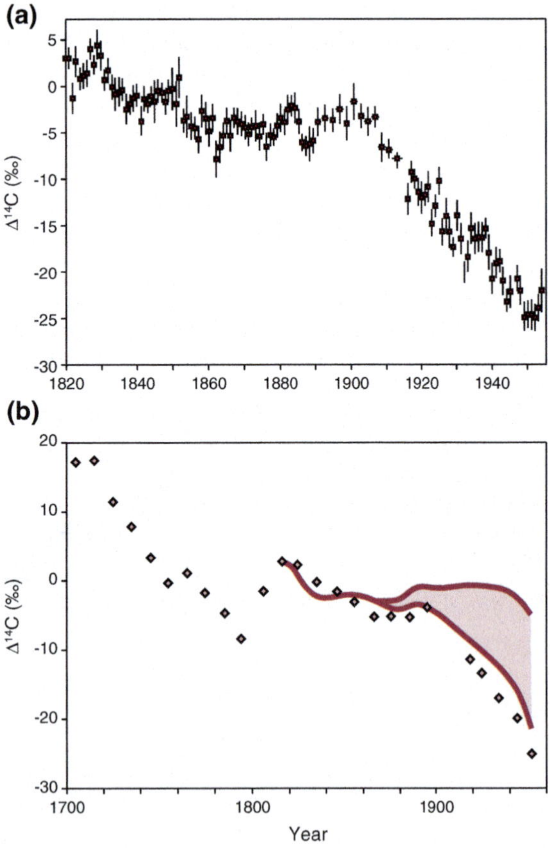

Fig. 4.5 (**a**) Observations of Δ^{14}C in tree rings from the Northwestern US. (**b**) Carbon cycle model of $\Delta^{14}CO_2$ calculated for the natural atmospheric $^{14}CO_2$ level (top line) and taking into account the release of fossil CO_2 (bottom line). Symbols are observed $^{14}CO_2$ values. Figures modified from Stuiver and Quay (1981)

emissions, however. Gross fluxes of C between the atmosphere and the oceans and terrestrial biosphere moderate the dilution of atmospheric $^{14}CO_2$ by effectively increasing the reservoir of C into which the ^{14}C-free fossil fuel CO_2 is mixed. Via gross exchanges, some fossil-derived C enters the oceanic and terrestrial reservoirs and is replaced in the atmosphere by C with the ^{14}C of the oceanic and terrestrial reservoirs.

Representation of the Suess effect in models of the global C cycle is sensitive to the exchange rates associated with air–land and air–ocean fluxes (Revelle and Suess 1957; Oeschger et al. 1975; Stuiver and Quay 1981). In one of the first applications of a C cycle model, the observed Suess effect was used in a very simple box model that included one atmospheric box and one oceanic box to estimate the global air–sea exchange rate (Revelle and Suess 1957). Since then, the Suess effect has been simulated in more advanced C cycle models, including one that also considered how the observed Suess effect might have been influenced by variations in cosmogenic ^{14}C production (Fig. 4.5b, Stuiver and Quay 1981). Utilizing a model with various possible relationships between ^{14}C production and the observed number of sunspots between 1735 and 1952, this study demonstrated that the observed decrease in $^{14}CO_2$ could not have been caused by natural fluctuations in cosmogenic production.

Further compilation of tree-ring records from different locations provided an early indication that the magnitude of the Suess effect varied regionally. The decrease in $^{14}CO_2$ until the 1930s was 5–10 ‰ greater in European trees than in North American trees (Tans et al. 1979; De Jong and Mook 1982). These observations demonstrated that the Suess effect was enhanced in regions with strong local combustion sources, a finding that would later be exploited to estimate regional fossil fuel CO_2 emission rates (Sect. 4.8.1).

4.5 The "Bomb" Period: 1945–1985

The first direct measurements of $^{14}CO_2$ in the atmosphere were made in 1954 in New Zealand (Rafter and Fergusson 1957; Currie et al. 2011). Globally distributed networks of ground-based atmospheric sampling stations followed shortly after (Nydal 1963; Manning et al. 1990; Levin et al. 1992). These networks were initially deployed to observe the radioactive fallout caused by intensive nuclear weapons testing in the 1950s and 1960s. Production of ^{14}C by nuclear weapons testing nearly doubled the tropospheric burden of $^{14}CO_2$ in the Northern Hemisphere (Fig. 4.6). In the Northern Hemisphere, $^{14}CO_2$ peaked in 1963 with the maximum of large aboveground thermonuclear (hydrogen fusion) bomb tests immediately before the atmospheric nuclear test ban treaty came into effect. In the Southern Hemisphere, $^{14}CO_2$ peaked a few years later. $^{14}CO_2$ then began to decrease rapidly as the negative isoflux from the ocean and terrestrial biosphere became larger than the positive isoflux from the stratosphere (Figs. 4.5a and 4.6). The observed trend in $^{14}CO_2$ resembled an exponential curve, initially falling rapidly (by more than 40 ‰ $year^{-1}$)

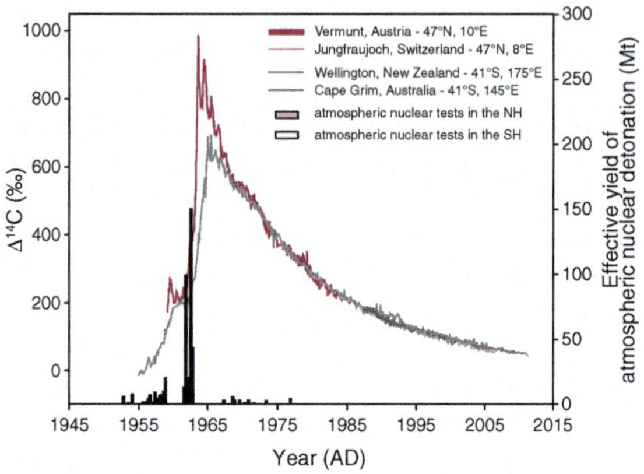

Fig. 4.6 Observations of $\Delta^{14}CO_2$ and the magnitude of nuclear explosions occurring in the Northern Hemisphere (*black lines* and *bars*) and Southern Hemisphere (*gray lines* and *white bars*). Figure modified from Hua and Barbetti (2007), with observations from Levin and Kromer (2004), Currie et al. (2011) and Levin et al. (2010)

then slowing with time (Fig. 4.6). Estimated isofluxes for the start of the bomb period are shown in Table 4.2.

Due to the long half-life of ^{14}C, excess ^{14}C produced by weapons testing will remain for many thousands of years. As discussed for fossil fuel CO_2 and the Suess effect, the natural C cycle exchange acts to redistribute this bomb ^{14}C throughout the atmosphere, biosphere, and oceans. The rate of redistribution of bomb ^{14}C provides a method for tracing natural C exchanges that has been used in many aspects of C cycle studies. In this section, we will describe the input of ^{14}C to the atmosphere from weapons testing and the initial transfers of bomb ^{14}C between C reservoirs, focusing on the influence of these processes on atmospheric $^{14}CO_2$ and how these observations have been used to study atmospheric C exchange.

4.5.1 Global Bomb Radiocarbon Budget

Most of the nuclear weapons tests occurred in the Northern Hemisphere (UNSCEAR 2000) (Fig. 4.6), and the explosive force injected ^{14}C into the stratosphere roughly 10–17 km above sea level. The large initial concentration of bomb ^{14}C in the northern stratosphere was observed using high-altitude aircraft, revealing $\Delta^{14}CO_2$ values as large as 5000–20,000 ‰ in the stratosphere (Telegadas 1971; Hesshaimer and Levin 2000). Bomb ^{14}C gradually entered the troposphere via stratosphere–troposphere exchange processes in the mid- to high latitudes. Investigation of the time-evolving budget of bomb ^{14}C in C cycle studies and assessment of the human exposure to radioactive fallout requires estimates of the total amount of ^{14}C produced by weapons testing (Fig. 4.7). These estimates use emission factors that assume ^{14}C produced in each test was roughly proportional to the explosive force of the bomb (e.g., UNSCEAR 2000). There are significant uncertainties in these estimates of bomb ^{14}C production since the production of neutrons may vary with different types of fuel or bomb designs and not all neutrons react to produce ^{14}C. Several studies have augmented estimates from scaling factors with observations of bomb-derived ^{14}C and C cycle models (e.g., Hesshaimer et al. 1994; Naegler and Levin 2006). The current best estimate of total bomb ^{14}C is 598–632 × 10^{26} atoms (Naegler and Levin 2006). Though a few tests were performed after the nuclear test ban in 1963, almost all bomb ^{14}C was produced by thermonuclear bomb tests between 1961 and 1963 (Fig. 4.6). For comparison, it would take 250 years to produce this amount of ^{14}C cosmogenically (Masarik and Beer 1999). Besides the strong pulse of ^{14}C from bomb testing, a small amount of anthropogenic ^{14}C production continues to occur through nuclear industrial and research applications (UNSCEAR 2000). Annual anthropogenic ^{14}C production comprised approximately 10 % of natural cosmogenic production in recent decades (Graven and Gruber 2011).

Since all of the bomb ^{14}C produced must be allocated into atmospheric, oceanic, or terrestrial reservoirs, accounting for bomb ^{14}C in a realistic C cycle model constrained with observations of bomb ^{14}C in the troposphere, stratosphere, and ocean can help to refine the estimates of the total bomb ^{14}C produced. By creating ^{14}C budgets that incorporate information from C cycle models, bomb ^{14}C observations, and bomb detonation histories, researchers have also argued for a lower estimate of the air–ocean gas exchange rate (Hesshaimer et al. 1994) and provided an estimate of the total biospheric inventory of bomb ^{14}C (Naegler and Levin 2009a) (Fig. 4.7). The total biospheric inventory of bomb ^{14}C can provide a useful measure of the average residence time and ^{14}C disequilibrium of terrestrial C, which

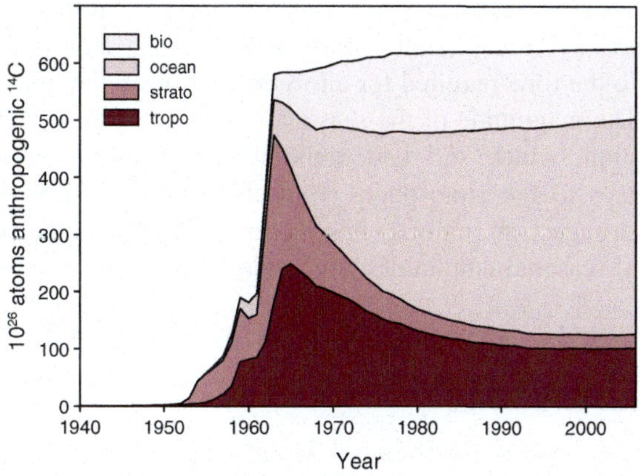

Fig. 4.7 The time-evolving global inventory of anthropogenic ^{14}C, derived primarily from bomb testing but also from ongoing ^{14}C production by the nuclear power industry, in tropospheric, stratospheric, terrestrial, and oceanic reservoirs. Figure modified from Naegler and Levin (2009b)

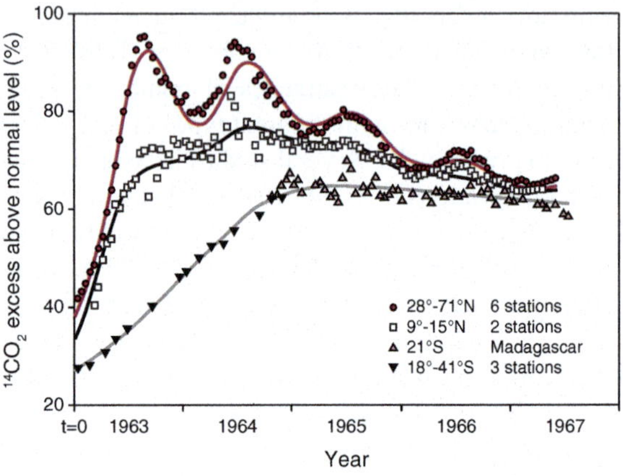

Fig. 4.8 Observations of $\Delta^{14}CO_2$ during the early bomb period, from stations grouped by latitude. Figure modified from Nydal and Lovseth (1983)

is difficult to observe directly because of the heterogeneity in biomass and ecosystem types (Chap. 6).

4.5.2 Atmospheric Radiocarbon Seasonality During the Bomb Period

The seasonal cycles of $^{14}CO_2$ had amplitudes of more than 100 ‰ at Northern Hemisphere ground-level stations in the mid-1960s (Fig. 4.8). These observations clearly demonstrated the strong seasonality in the descent of stratospheric, bomb ^{14}C-enriched air across the tropopause (Nydal and Lovseth 1965), showing that the strength of stratosphere-to-troposphere exchange has a maximum in late spring in each hemisphere. The seasonal maximum in $^{14}CO_2$ at ground level occurred in summer, delayed by a few months from the strongest stratosphere–troposphere exchange due to the time required for air to descend from the upper troposphere to ground level. The magnitude of the seasonal amplitude was influenced by heterotrophic respiration, which still had prebomb ^{14}C values, resulting in strongly negative isofluxes to the atmosphere (Randerson et al. 2002). As heterotrophic respiration is strongest in summer, this isoflux offset the stratospheric input and slightly reduced seasonal amplitudes from what they otherwise would have been.

4.5.3 Spatial Gradients

By tracing the movement of bomb ^{14}C through the atmosphere, early $^{14}CO_2$ observations helped to establish the rates and seasonality of stratosphere-to-troposphere exchange and of latitudinal and cross-equatorial mixing of the troposphere. Strong latitudinal gradients in the troposphere were created by the concentrated bomb ^{14}C input to the northern latitudes where nuclear weapons were tested and subsequent transport of that bomb ^{14}C through the troposphere. Observed $\Delta^{14}CO_2$ was 100–500 ‰ higher at northern stations than at southern stations between 1961 and 1965 (Figs. 4.6 and 4.7). Strong latitudinal gradients were also present within the Northern Hemisphere, where observed $\Delta^{14}CO_2$ values were 20–200 ‰ higher at mid- to high latitudes than low latitudes (Fig. 4.8).

Observed $\Delta^{14}CO_2$ values during the bomb period were used to estimate that the timescale of interhemispheric mixing across the equator is approximately one year (Lal and Rama 1966; Nydal 1968). This mixing timescale is an important determinant of latitudinal gradients of CO_2 and several industrial gases, since their emissions and surface exchanges occur primarily in the Northern Hemisphere. Therefore, accurate representation of interhemispheric mixing is essential for model-based estimates of regional fluxes of CO_2 and other gases (Denning et al. 1999). The early $^{14}CO_2$ observations have also been used to constrain cross-equatorial and cross-tropopause exchange in atmospheric box models (Lal and Rama 1966; Nydal 1968; Johnston 1994; Naegler and Levin 2006) and to test three-dimensional atmospheric general circulation models (Kjellstrom et al. 2000; Land et al. 2002). In addition to providing a chemical tracer to estimate transport rates, observations of $^{14}CO_2$ and other radioactive fallout species have also helped in developing theories on internal stratospheric dynamics and stratosphere–troposphere exchange (Holton et al. 1995).

Latitudinal gradients observed in the bomb period were also influenced by the spatial distribution of terrestrial, oceanic, and fossil fuel isofluxes (globally integrated isofluxes for 1963 are summarized in Table 4.2). During the period of testing and the subsequent few years, these influences were minor in comparison with the effect of the bombs and stratospheric ^{14}C input to the Northern Hemisphere. However, after 1968, stratospheric ^{14}C input decreased substantially and latitudinal gradients became much smaller. In the late 1960s and early 1970s, stratospheric and oceanic isofluxes made roughly equal contributions to the interhemispheric $^{14}CO_2$ gradient, each increasing Northern Hemisphere $\Delta^{14}CO_2$ by 10–20 ‰, compared to the Southern Hemisphere (Randerson et al. 2002; Levin et al. 2010; Fig. 4.9c). Even though negative isofluxes were occurring over the entire ocean, the isofluxes were the largest over the Southern Ocean due to stronger air–sea ^{14}C gradients and high winds (see also Chap. 5). Stronger negative isofluxes over the Southern Ocean decreased Southern Hemisphere $^{14}CO_2$ values relative to Northern Hemisphere $^{14}CO_2$ values, reinforcing the gradient caused by stratospheric ^{14}C input in the north. Small negative isofluxes from the biosphere and fossil fuel emissions, focused over the tropics and northern continents, counteracted the stratospheric and oceanic influences on latitudinal gradients (Randerson et al. 2002; Levin et al. 2010; Fig. 4.9c).

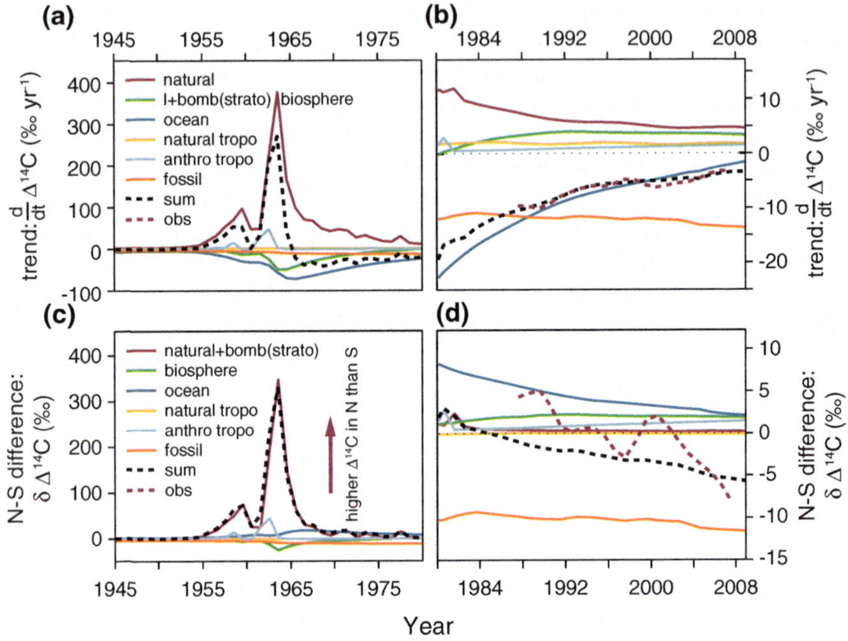

Fig. 4.9 Simulated components of the global tropospheric $^{14}CO_2$ trend (**a**, **b**) and North–South interhemispheric gradient (**c**, **d**) for 1945–1980 (**a**, **c**) and 1980–2008 (**b**, **d**). The *black dashed line* shows the sum of all simulated components, and the *purple dashed line* in the *right panels* shows the observed global trend (**b**) and the observed gradient between Jungfraujoch and Cape Grim, 1987–2007 (**d**). Figure modified from Levin et al. (2010)

4.5.4 Evolving Isofluxes During the Bomb Period

As bomb ^{14}C dispersed through the atmosphere and entered oceanic and terrestrial reservoirs, changes to $^{14}CO_2$ and the subsequent changes in the ^{14}C of oceanic and terrestrial C caused ^{14}C isofluxes to evolve. For example, in 1963, respiration of short-term C pools in the terrestrial biosphere, such as leaves and twigs, added CO_2 that was fixed before the peak in $^{14}CO_2$. Therefore, the ^{14}C of leaves and twigs was much lower than $^{14}CO_2$ and the respiration of leaves and twigs caused a large negative isoflux (Fig. 4.9a; Table 4.2). By 1975, a large fraction of the C in leaves and twigs had been fixed after the peak in $^{14}CO_2$ and the respiration of leaves and twigs contributed much less to the decreasing trend in $^{14}CO_2$. Since the amount and average residence time of C in the terrestrial biosphere is smaller than in the ocean, the negative isoflux from the terrestrial biosphere was smaller and decreased more rapidly than the oceanic isoflux (Randerson et al. 2002; Naegler and Levin 2009a; Levin et al. 2010). Negative isofluxes from fossil fuel emissions were much smaller than the negative isoflux from the ocean and terrestrial biosphere before 1970, but after 1970 fossil fuel emissions became a principal influence on the decreasing trend in $^{14}CO_2$.

4.6 The Post-bomb Period: 1985–Present

We consider the post-bomb period as the period commencing in the mid-1980s, after a few decades had passed since the peak $^{14}CO_2$ level in the troposphere and the influences on atmospheric $^{14}CO_2$ had changed significantly. The rapid uptake of bomb ^{14}C during the bomb period that decreased the tropospheric $^{14}CO_2$ content simultaneously increased ^{14}C in land biosphere and ocean reservoirs. In the post-bomb period, rapidly overturning C reservoirs came closer to equilibrium with tropospheric $^{14}CO_2$, while at the same time, uptake of bomb ^{14}C into longer term reservoirs continued and fossil fuel emissions increased. These processes changed the disequilibrium in terrestrial, oceanic, and stratospheric C reservoirs, thereby altering the influence of those exchanges on tropospheric $^{14}CO_2$ (Fig. 4.9b, d). The start of this period is somewhat arbitrary, since the processes are continuous, but we consider this period to start when the global biosphere isoflux changed sign in about the mid-1980s (Randerson et al. 2002; Naegler and Levin 2009b). This section will describe the influences on the long-term trend, interhemispheric gradient, and seasonal cycles in tropospheric $^{14}CO_2$ over the post-bomb period, including how each process changed since the bomb period.

The history of $^{14}CO_2$ through the post-bomb period has been directly measured by continued measurements from a few locations. The longest records, both of which continue today, are from New Zealand (Currie et al. 2011) and the European Alps (Levin and Kromer 2004), shown in Fig. 4.6. Observations from these and several other shorter records from around the globe are available from the Carbon Dioxide Information Analysis Center (http://cdiac.ornl.gov/). In recent years and decades, additional observations of $^{14}CO_2$ at both clean-air and polluted measurement sites have begun at laboratories in Europe (Levin et al. 2010; Van Der Laan et al. 2010), the US (Turnbull et al. 2007; Graven et al. 2012b, c) and Japan (Kitigawa et al. 2004).

4.6.1 Fossil Fuel Carbon Dioxide

CO_2 emissions from fossil fuel burning have grown substantially over the post-bomb period, increasing by ~50 % between 1985 and 2005 (Marland et al. 2006). Dilution of $^{14}CO_2$ by fossil-derived CO_2 is now the strongest contribution to the long-term trend and interhemispheric gradient of $^{14}CO_2$ in unpolluted background air (Fig. 4.9) and is one of the main influences on seasonal cycles of $^{14}CO_2$ in the Northern Hemisphere (Levin et al. 2010; Graven et al. 2012b, c). Fossil fuel CO_2 emissions decreased tropospheric $\Delta^{14}CO_2$ by about 10 ‰ $year^{-1}$ (Table 4.3).

Despite the large growth in fossil fuel emissions, their influence on the $^{14}CO_2$ trend and interhemispheric gradient has remained rather steady over the post-bomb period (Levin et al. 2010; Graven et al. 2012b, c; Fig. 4.9b). This can be attributed to a decrease in the sensitivity of $^{14}CO_2$ to fossil fuel emissions between the 1980s and 2000s: The troposphere–fossil fuel isotopic disequilibrium grew smaller as $\Delta^{14}CO_2$ dropped from 250 to 50 ‰ and the fractional change in the CO_2 mixing ratio caused by an added increment of fossil fuel CO_2 grew smaller as tropospheric CO_2 increased from 350 ppm to the 2014 level of about 395 ppm (Levin et al. 2010; Graven et al. 2012c).

4.6.2 *Nuclear Industry*

The growth of nuclear power over the post-bomb period also has an impact on $^{14}CO_2$ because ^{14}C is released from nuclear power plants and fuel reprocessing sites, causing small, continual increases to the global inventory of ^{14}C (UNSCEAR 2000; Levin et al. 2010; Graven et al. 2012b). Globally, this term is small, at about 0.5–1 ‰ $year^{-1}$ increase in $\Delta^{14}CO_2$ (Table 4.3), but it can be important at the local and regional scale (Graven and Gruber 2011).

4.6.3 *Terrestrial Carbon*

By the 1980s, the terrestrial biosphere had assimilated ~20 % of the total bomb ^{14}C, and mean ^{14}C of respired C was close to that of tropospheric $^{14}CO_2$ (Randerson et al. 2002; Naegler and Levin 2009b). In essence, the terrestrial biosphere had nearly reached equilibrium with tropospheric $^{14}CO_2$ after roughly 20 years. Yet, as tropospheric $^{14}CO_2$ continued to decline due to air–ocean exchange and fossil fuel emissions, complete equilibrium could not be reached. Atmospheric $^{14}CO_2$ became, on average, lower than the ^{14}C of C respired by terrestrial ecosystems. Thus, in the post-bomb period, the net effect of terrestrial respiration was to return bomb ^{14}C back to the troposphere (Fig. 4.7), and the biospheric isoflux reversed sign (Randerson et al. 2002; Naegler and Levin 2009b; Levin et al. 2010; Fig. 4.10). Regionally, terrestrial ecosystems release C of various ages during heterotrophic respiration, and ecosystems in different locations can release C with different mean ages (Chap. 6). Therefore, although the global average near-equilibrium in ^{14}C distribution was reached in the 1980s, this near-equilibrium was reached sooner in the tropics where C cycling is rapid, and later in boreal and arctic systems where C cycling is slower (Randerson et al. 2002).

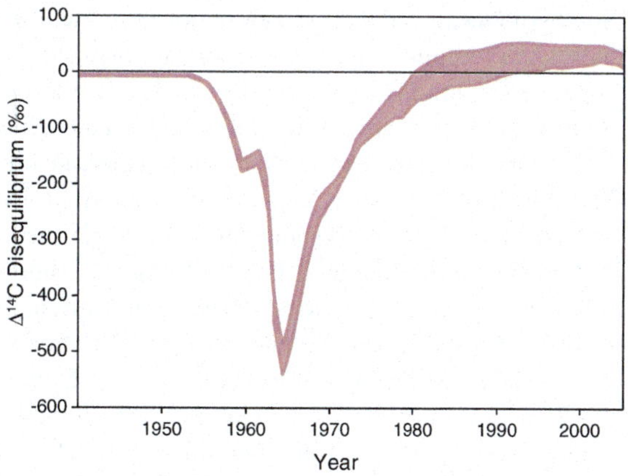

Fig. 4.10 Disequilibrium between $\Delta^{14}C$ in terrestrial respiration and tropospheric CO_2 over 1940–2005 as simulated by a two-box biosphere model. Figure modified from Naegler and Levin (2009b)

4.6.4 Oceans

Water at the ocean surface is composed of varying mixtures of young water that has recently been in contact with the air and older water that has been sequestered in the deep ocean for centuries (Chap. 5). Because of this, the ocean surface has even stronger regional variability in the mean age of C exchanged with the air than does the terrestrial biosphere. Water in subtropical gyres does not mix readily with deeper water and maintains close contact with the atmosphere, while in the Southern Ocean and North Pacific Ocean, there is a significant fraction of old water that has been upwelled from depth after being out of contact with the atmosphere for hundreds to thousands of years.

In the post-bomb period, water in the subtropical gyres contains a large amount of bomb ^{14}C that has not been dispersed by mixing with deeper water, and thus, isofluxes here resemble those from the terrestrial biosphere. They do incorporate some upwelled water that is depleted in ^{14}C; however, the source of upwelling is shallow, relatively young water that has already assimilated some bomb ^{14}C. Thus, subtropical areas of the ocean had reached near-equilibrium with tropospheric $^{14}CO_2$ by 1995, when large-scale hydrographic surveys showed that ^{14}C values in surface waters of the subtropics were nearly the same as tropospheric $^{14}CO_2$ values (Key et al. 2004; Graven et al. 2012a) (Fig. 4.10). Since 1995, ^{14}C values in these ocean regions has become even higher than tropospheric $^{14}CO_2$ values (Graven et al. 2012a). Like the terrestrial biosphere, these subtropical surface waters have started to release bomb ^{14}C back to the atmosphere where $^{14}CO_2$ content continues to decrease; thus, the isoflux has changed sign.

In contrast, in both the Southern Ocean and North Pacific Ocean, significant uptake of bomb ^{14}C is still occurring. Since surface waters in these regions mix with deep water, into which bomb ^{14}C has not yet intruded, ^{14}C there is still substantially depleted in comparison with the troposphere and has not come to this near-equilibrium state (Key et al. 2004; Graven et al. 2012a; Fig. 4.10). The

Southern Ocean is the dominant region of oceanic influence on atmospheric $^{14}CO_2$, with large regions of negative ocean–atmosphere gradients stronger than 100 ‰ as compared to gradients of 50–100 ‰ over smaller areas in the North Pacific. In the Southern Ocean, strong circumpolar winds drive strong upwelling of deep, ^{14}C-depleted water and enhance gas transfer, leading to strong negative isofluxes. These isofluxes cause $^{14}CO_2$ to be reduced in air above the Southern Ocean (Levin and Hesshaimer 2000; Randerson et al. 2002; Fig. 4.11).

Globally, the integrated flux of bomb ^{14}C into the ocean over the bomb and post-bomb periods provides a strong constraint on the average gas exchange velocity (Chap. 5; Revelle and Suess 1957; Broecker et al. 1985; Sweeney et al. 2007). Other estimates of the average gas exchange velocity must extrapolate from point measurements of ocean–atmosphere fluxes, which results in large uncertainties (Frost and Upstill-Goddard 1999). Both oceanic and atmospheric ^{14}C measurements have been crucial in estimating and refining the global mean gas exchange velocity (Hesshaimer et al. 1994; Krakauer et al. 2006; Naegler and Levin 2006), which is needed to calculate the magnitude and locations of uptake of anthropogenic CO_2 by the ocean using observations of ocean–atmosphere gradients in pCO_2 (Takahashi et al. 2009).

Over the post-bomb period, the integrated flux of bomb ^{14}C into the oceans has diminished and global ocean bomb ^{14}C inventories in the 2000s are nearly stagnant (Graven et al. 2012a), due to near balance between the high-latitude uptake and

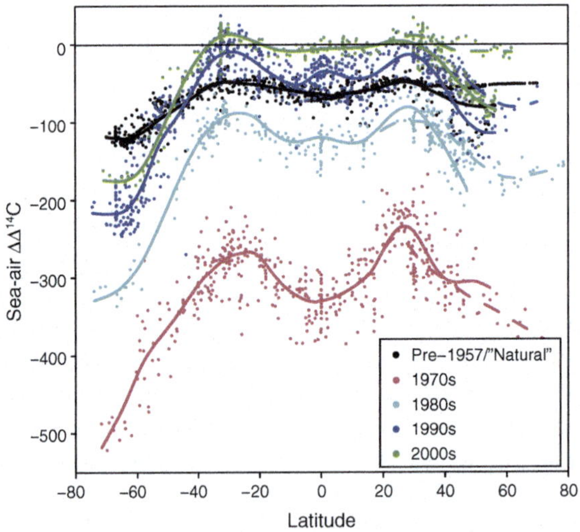

Fig. 4.11 Ocean–atmosphere gradients in $\Delta^{14}C$ versus latitude by decade in the bomb and post-bomb periods. Estimates of preindustrial surface water $\Delta^{14}C$ (Key et al. 2004) or observations made prior to 1957 are compared to the preindustrial atmospheric ^{14}C of 0 ‰ and shown in *black*. *Solid lines* show smoothed curves for data in all regions south of 30°N and in the North Pacific, north of 30°N. *Dashed lines* show smoothed curves for data in all regions south of 30°N and in the North Atlantic, north of 30°N. Figure modified from Graven et al. (2012a)

low-latitude release of bomb ^{14}C. As the low-latitude release surpasses high-latitude uptake, the oceans are now beginning an overall transfer of bomb ^{14}C back to the atmosphere.

4.6.5 Stratosphere

In the post-bomb period, the stratosphere continues to be enriched in ^{14}C relative to the troposphere, even though the bomb ^{14}C that was initially injected into the stratosphere has dispersed. Observations of $\Delta^{14}CO_2$ in stratospheric air indicate that it was 50–150 ‰ above that measured in the troposphere in the early 1990s (Turnbull et al. 2009b; Fig. 4.12). One reason for higher ^{14}C content in the stratosphere during the post-bomb period is that most cosmogenic production of ^{14}C occurs in the stratosphere (O'Brien 1979; Jöckel 1999). While the natural stratosphere–troposphere gradient was amplified by the input of bomb ^{14}C, the stratosphere–troposphere gradient has been positive throughout history due to this effect. Another reason for higher ^{14}C content in the stratosphere during the post-bomb period is the residence time of air in the stratosphere, on the order of five-to-ten years, which creates a lag in stratospheric response to the surface exchanges that are reducing tropospheric $^{14}CO_2$.

This stratosphere–troposphere $^{14}CO_2$ gradient continues to be exploited in studies of cross-tropopause transport of air and stratospheric residence times during the post-bomb period. Observed stratosphere–troposphere gradients of $^{14}CO_2$ in 1989–1990 were used to estimate an average turnover time of the stratosphere of

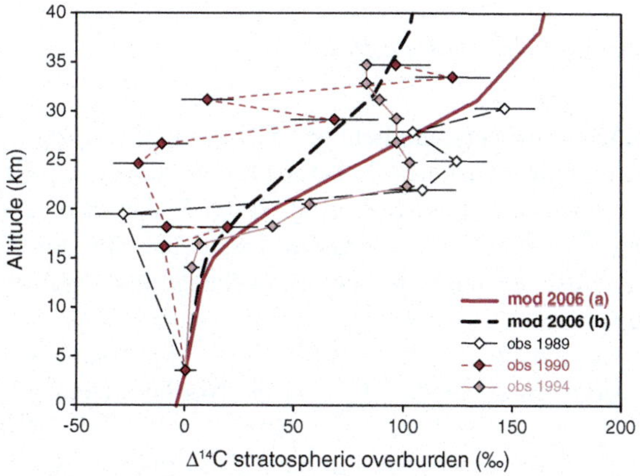

Fig. 4.12 Vertical profiles of stratospheric $\Delta^{14}CO_2$ above Japan (40°N) in 1989–94, relative to $\Delta^{14}CO_2$ in the free troposphere, from stratospheric and high tropospheric samples taken in 1989, 1990, and 1994. Two model simulations for 2006 are shown, model (**a**) uses cosmogenic production from Lal (1988), model (**b**) moves the production to lower altitudes within the stratosphere. Figure modified from Turnbull et al. (2009b)

~9 years (Nakamura et al. 1992, 1994). In addition, very strong vertical gradients observed within the stratosphere indicate vertical mixing of stratospheric air is slow. Investigation of stratosphere–tropospheric dynamics can also utilize ^{14}CO, the dominant form of newly produced ^{14}C atoms which oxidizes to $^{14}CO_2$ over a timescale of weeks to months (Jöckel and Brenninkmeijer 2002; Sect. 4.9.1).

4.6.6 *Long-Term Trend*

In the post-bomb period, there is a complex interplay between positive influences on tropospheric $^{14}CO_2$ from the terrestrial biosphere, low-latitude oceans, stratosphere and nuclear industry, and negative influences from the high-latitude oceans and fossil fuel combustion (Levin et al. 2010; Graven et al. 2012b, c; Fig. 4.9b; Table 4.3). The long-term trend of tropospheric $^{14}CO_2$ continues to be negative (Figs. 4.6 and 4.9b), although the rate of decline has slowed from 10 ‰ year^{-1} in the 1980s to 5 ‰ year^{-1} since 2000 (Meijer et al. 2006; Levin et al. 2010). The primary cause of this slowing is the weakening effect of ocean–atmosphere exchange, which has resulted in the emergence of fossil fuel emissions as the dominant influence on tropospheric $^{14}CO_2$. Since 2000, all influences on $^{14}CO_2$ each contribute less than ±5 ‰ year^{-1} to the trend, except for fossil fuel emissions, which contribute −10 to −15 ‰ year^{-1} (the exact magnitude varies slightly through time and depends on whether CO_2 from fossil fuel emissions is assumed to mix into the entire atmosphere or only into the troposphere) (Levin et al. 2010; Graven et al. 2012b; Fig. 4.9b).

4.6.7 *Interhemispheric Gradient*

The observed interhemispheric gradient of $^{14}CO_2$ was close to zero over most of the post-bomb period, reflecting a near balance between the effect of fossil fuel emissions and other processes (Randerson et al. 2002; Meijer et al. 2006; Levin et al. 2010; Fig. 4.9d). The dilution of $^{14}CO_2$ by fossil fuel emissions in the Northern Hemisphere was offset primarily by air–sea exchange that decreased $^{14}CO_2$ in the Southern Hemisphere. ^{14}C release by terrestrial ecosystems and nuclear power plants, concentrated in the Northern Hemisphere, also counteracted the effect of fossil fuel emissions. Since 2000, $^{14}CO_2$ in the Northern Hemisphere has become lower than $^{14}CO_2$ in the Southern Hemisphere (Levin et al. 2010; Graven et al. 2012c), apparently due to a reduction in the ^{14}C isoflux from the Southern Ocean. However, the total interhemispheric gradient simulated by models has not matched the observed gradient over the post-bomb period, and the changing interhemispheric gradient has not yet been fully explained (Randerson et al. 2002; Levin et al. 2010).

4.6.8 Seasonal Cycles

Seasonal cycles in $^{14}CO_2$ were much smaller during the post-bomb period than during the bomb period (Randerson et al. 2002; Levin et al. 2010; Fig. 4.13). The largest cycles now occur at northern mid- and high-latitude sites, which show amplitudes of several permil with maxima in late summer or fall (Meijer et al. 2006; Turnbull et al. 2007; Levin et al. 2010; Graven et al. 2012c). Seasonality in $^{14}CO_2$ primarily results from seasonal variations in atmospheric transport that regulates the stratospheric and fossil fuel isofluxes, rather than seasonality in the isofluxes themselves. As during the bomb period, the seasonal input of stratospheric air increases $^{14}CO_2$ in summer and fall at mid- to high latitudes, although the effect is much smaller in the post-bomb period (Randerson et al. 2002; Levin et al. 2010). Although fossil fuel CO_2 emissions themselves have only a small seasonal cycle, more vigorous atmospheric mixing during summer and fall allows fossil fuel CO_2 emissions to be transported away from the surface and mixed into a larger volume of air, increasing local tropospheric $^{14}CO_2$ relative to the winter months (Randerson et al. 2002; Turnbull et al. 2009a; Levin et al. 2010; Graven et al. 2012c). Smaller effects result from terrestrial and oceanic exchanges. Stronger summertime respiration slightly enriches $^{14}CO_2$ in summer months in the Northern Hemisphere, and stronger ocean–atmosphere exchange slightly depletes $^{14}CO_2$ in winter months over the Southern Ocean (Randerson et al. 2002; Turnbull et al. 2009a; Levin et al. 2010).

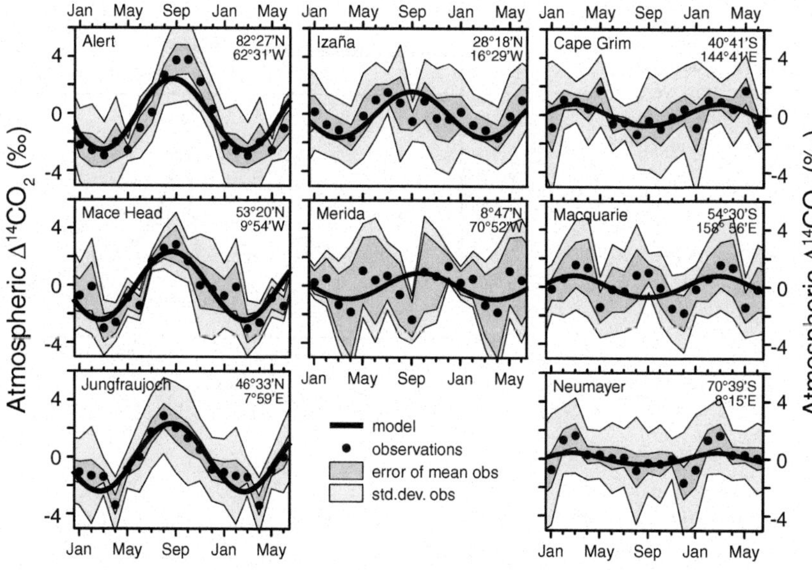

Fig. 4.13 Observed and simulated average seasonal cycles of $\Delta^{14}CO_2$ for sites in the Heidelberg University observation network over 1995–2005. Figure modified from Levin et al. (2010)

4.7 The Future Trajectory of Atmospheric Radiocarbon

Tropospheric $\Delta^{14}CO_2$ is now (in 2014) about 30 ‰, that is, 3 % above the preindustrial level. If fossil fuel emissions continue to increase in a "business-as-usual" scenario (Ciais et al. 2013), atmospheric $\Delta^{14}CO_2$ values will likely drop below the preindustrial level (0 ‰) within the next decade. The continued decrease in $^{14}CO_2$ will cause the C reservoirs in the land and surface ocean that are presently enriched in ^{14}C to sustain their positive isofluxes to the atmosphere, while negative isofluxes from ^{14}C-depleted reservoirs such as the Southern Ocean will become weaker as their disequilibrium with $^{14}CO_2$ is reduced. Atmospheric $\Delta^{14}CO_2$ values may reach −150 ‰ by the year 2100 (Caldeira et al. 1998; Fig. 4.14). At this point, even the deep waters upwelling in the Southern Ocean will return CO_2 enriched in ^{14}C to the atmosphere and become a positive isoflux. Observation of such changes in the coming decades will continue to illuminate the C cycle exchange processes that determine how ^{14}C is redistributed in response to anthropogenic perturbations.

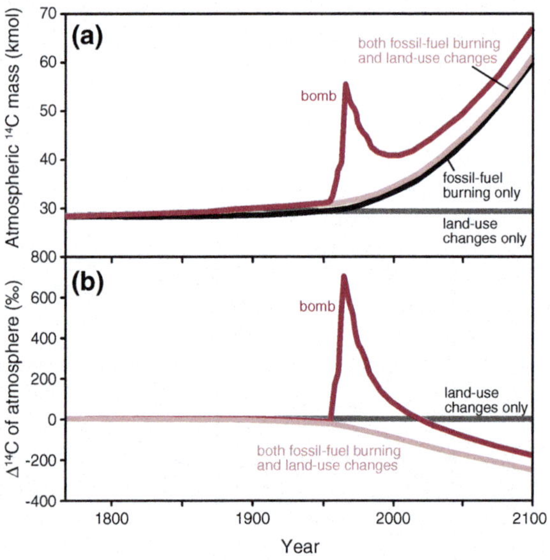

Fig. 4.14 Model simulations of the atmospheric ^{14}C inventory and $^{14}CO_2$ over 1765–2100 for scenarios in which perturbations from land-use change only, land-use change and fossil fuel burning or land-use change, fossil fuel burning and bomb testing were included. Figure modified from Caldeira et al. (1998)

4.8 Atmospheric Monitoring of Fossil Fuel Carbon Dioxide Emissions

4.8.1 Determination of Fossil Fuel Mole Fraction from Radiocarbon Observations

In recent years, atmospheric $^{14}CO_2$ measurements have been recognized as the most unambiguous tracer method to quantify fossil fuel CO_2 (CO_{2ff}) emissions. Due to the influence of the fossil fuel isoflux on atmospheric $^{14}CO_2$ gradients, $^{14}CO_2$ can provide an objective method of evaluating the CO_{2ff} emissions reported by governments and industry. Equally importantly, the separation of fossil fuel and biological CO_2 sources by $^{14}CO_2$ allows examination of biological CO_2 exchange processes.

In order to calculate the amount of fossil fuel CO_2 present in an atmospheric sample in which $^{14}CO_2$ has been measured, the mass balances for C and ^{14}C from Eqs. 4.1 and 4.2 are integrated to:

$$CO_{2obs} = CO_{2bg} + CO_{2ff} + CO_{2other} \tag{4.5}$$

and

$$\Delta_{obs}CO_{2obs} = \Delta_{bg}CO_{2bg} + \Delta_{ff}CO_{2ff} + \Delta_{other}CO_{2other}, \tag{4.6}$$

then manipulated to solve for CO_{2ff}:

$$CO_{2ff} = \frac{CO_{2obs}(\Delta_{obs} - \Delta_{bg})}{\Delta_{ff} - \Delta_{bg}} - \frac{CO_{2other}(\Delta_{other} - \Delta_{bg})}{\Delta_{ff} - \Delta_{bg}}. \tag{4.7}$$

To emphasize the influence of fossil fuel combustion, the other terms have been condensed into a single CO_{2other} term, which is essentially a correction for non-fossil influences on $^{14}CO_2$. This represents CO_2 added or removed by non-fossil sources or sinks (respiration and biomass burning, nuclear industry, and to a lesser extent the oceans and cosmogenic production) and Δ_{other} represents the weighted mean ^{14}C of the other CO_2 sources. The integration adds two other terms, CO_{2bg} and Δ_{bg}, which represent the background air CO_2 mixing ratio and its $\Delta^{14}CO_2$ value, respectively. In this approach, the "background" is chosen to represent the initial composition of a parcel of air, which then moves across a region of interest, modifying its CO_2 mole fraction and $^{14}CO_2$ by the addition of CO_{2ff} and CO_{2other}. An upwind clean-air reference site can be assumed to represent the background; this is especially useful when examining large-scale gradients in $^{14}CO_2$ across continents.

In Eq. 4.7, ^{14}C of photosynthetic uptake (gross primary productivity) is implicitly assumed to be equal to Δ_{bg}. This is strictly true in the limit that the time (and space) between background and observation goes to zero. Some authors (e.g.,

Kuc et al. 2007; Riley et al. 2008) have instead assumed that ^{14}C of photosynthesis is equal to Δ_{obs}, and in this case, Eq. 4.7 can be rewritten as follows:

$$CO_{2ff} = \frac{CO_{2bg}(\Delta_{obs} - \Delta_{bg})}{\Delta_{ff} - \Delta_{obs}} - \frac{CO_{2other}(\Delta_{other} - \Delta_{obs})}{\Delta_{ff} - \Delta_{obs}} \tag{4.8}$$

When integrated $^{14}CO_2$ sampling (e.g., absorption onto NaOH over a period of days or weeks) is used, this formulation may be advantageous, as CO_{2obs} is not required, but the mean CO_{2bg} must be estimated. In the case of flask sampling, CO_{2obs} will usually be directly measured from the same flask, so Eq. 4.7 is more convenient. If CO_{2other} is zero, then Eqs. 4.7 and 4.8 are exactly equivalent, but slight differences in the calculated CO_{2ff} of up to 0.1 ppm can occur when both photosynthetic uptake of CO_2 and the $\Delta_{obs} - \Delta_{bg}$ differences are large (Turnbull et al. 2009a). These differences are small relative to the current CO_{2ff} detection capability.

In the current atmosphere, with about 395 ppm CO_2 mole fraction and $\Delta^{14}CO_2$ of about 30 ‰ (in 2014), the addition of 1 ppm of CO_{2ff} results in a decrease in $\Delta^{14}CO_2$ of $\sim$2.7 ‰ (e.g., Turnbull et al. 2006). Current ^{14}C measurement uncertainties are, at best, just under 2 ‰ in ^{14}C, from both accelerator mass spectrometry (AMS) and gas counting methods (Graven et al. 2007; Turnbull et al. 2007; Levin et al. 2010). Since both Δ_{obs} and Δ_{bg} are used in calculating CO_{2ff}, the resulting uncertainty in CO_{2ff} due to measurement uncertainty alone is about $\pm$1 ppm. Ongoing development of measurement techniques may bring the uncertainties down to 0.6–0.7 ppm in the near future.

Confounding influences, which are wrapped up in the second term in Eq. 4.7, can be important in some regions. This term can be regarded as the potential bias in CO_{2ff} if the non-fossil influences are not accounted for. Over the continents, heterotrophic respiration and biomass burning are the most important of these influences. In the tropics, and possibly at high northern latitudes, biospheric sources may have particularly substantial impacts on $^{14}CO_2$. In the tropics, the large magnitude of the respiration CO_2 flux means that even if the ^{14}C disequilibrium is small, the impact on $^{14}CO_2$ can be significant. At high latitudes, biospheric C may have resided in the soil for long periods, particularly if warming unlocks C that has been frozen for centuries or millennia (Chap. 6). This may result in a large and variable isotopic disequilibrium. The bias from biospheric sources on $^{14}CO_2$-derived estimates of CO_{2ff} has been estimated at 0.2–0.5 ppm over the Northern Hemisphere mid-latitudes and may be larger in the tropics and at high latitudes (Turnbull et al. 2009a). Since the bias is much smaller than typical ^{14}C measurement precision, some researchers have assumed it to be zero (e.g., Meijer et al. 1996; Levin et al. 2003).

Ethanol and other biofuels mixed with or replacing fossil gasoline may also need to be accounted for. In the ^{14}C method, these are not distinguishable from biomass burning or heterotrophic respiration fluxes. In contrast, other methods, particularly bottom-up inventory-based methods, may include biofuels in the gasoline budget. Some researchers have included biofuels in their ^{14}C-based CO_{2ff} estimates by measuring the ^{14}C of vehicle exhaust (Djuricin et al. 2010). Where biofuel content is known (e.g. mandated by government), adjusted ^{14}C of gasoline can also be calculated.

Another influence over continental regions arises from nuclear activities which produce ^{14}C from nuclear power generation and combustion or reprocessing of radioactive waste. The total ^{14}C isoflux from these activities is small relative to other sources, but the effect can be significant close to nuclear sites. For example, a study of fossil fuel emissions in Heidelberg, Germany, required an explicit correction of about 4.8 ‰ in $\Delta^{14}CO_2$ to account for the effects of a nearby nuclear reactor (Levin et al. 2003). A dataset of estimated $^{14}CO_2$ emissions from all known nuclear reactors is available (Graven and Gruber 2011) that takes into account varying ^{14}C emissions from different reactor types. For example, heavy water reactors used in Canada and gas-cooled reactors used in the UK emit much more ^{14}C for the same amount of electrical power generation than other reactor types. Most reactor types produce ^{14}C predominantly as $^{14}CO_2$, with the exception of pressurized water reactors, which produce ^{14}C predominantly as methane ($^{14}CH_4$). There is a lag time of several years before $^{14}CH_4$ is oxidized to $^{14}CO_2$, which affects where and when that $^{14}CO_2$ would be detected in atmospheric measurements. The magnitude and timing of nuclear reactor ^{14}C releases are imperfectly known, and there appears to be substantial variability in the amount of ^{14}C released even amongst reactors of the same type. Incineration of biomedical and other radioactive waste that contains ^{14}C in high concentrations is permitted to occur, but is poorly documented even though it can have a strong effect on local $^{14}CO_2$ measurements (e.g., Trumbore et al. 2002).

Over marine or coastal regions, a potential bias from ocean exchange may also be present. This is likely to be largest in areas close to upwelling regions in the Southern Ocean and the northwest Pacific Ocean where the isotopic disequilibrium between atmosphere and ocean is greatest (Randerson et al. 2002; Key et al. 2004; Turnbull et al. 2009a). For most continental regions of interest, this effect can be eliminated by careful choice of background, using a measurement station that reflects the same ocean influence as the region of interest.

Different sampling methods can be used to make the ^{14}C measurements. Flask samples are collected over a few minutes or up to one hour (Meijer et al. 1996; Turnbull et al. 2007; Graven et al. 2009; Djuricin et al. 2010; Turnbull et al. 2012) and can therefore be examined in the context of specific wind conditions and emission plumes. The same flask samples can be measured for other species to provide additional information (Turnbull et al. 2006; Miller et al. 2012). Integrated samples (either CO_2 absorbed into alkali, or plant material proxies) represent a longer term mixture of air masses and can be used to easily and cost-effectively examine broadscale gradients and long-term changes (Levin et al. 2008).

4.8.2 *Large-Scale Spatial Fossil Fuel Carbon Dioxide Variability and Plant Material Proxies*

Global $^{14}CO_2$ modeling studies (Randerson et al. 2002; Turnbull et al. 2009a; Miller et al. 2012) have been used to examine the spatial variability of $^{14}CO_2$ in the recent atmosphere. These models use best estimates of the ^{14}C isoflux magnitudes and spatial and temporal patterns to predict the atmospheric distribution of $^{14}CO_2$. They produce reasonable representations of the observed spatial distribution, although they do not resolve all the finer details, particularly at small scales. These global models have shown that $^{14}CO_2$ variability over the Northern Hemisphere land is dominated, at the continental scale (i.e., across a single continent), by CO_{2ff} emissions (Fig. 4.15).

On the observation side, plant materials have proven to be an excellent proxy for $^{14}CO_2$ at the continental, annual scale, allowing them to be used to understand the broad spatial distribution of $^{14}CO_2$ and to compare with model simulations. Since photosynthetic uptake does not alter ^{14}C, short-lived annual plants reflect atmospheric $^{14}CO_2$ averaged over their growing season, with the caveats that most plants

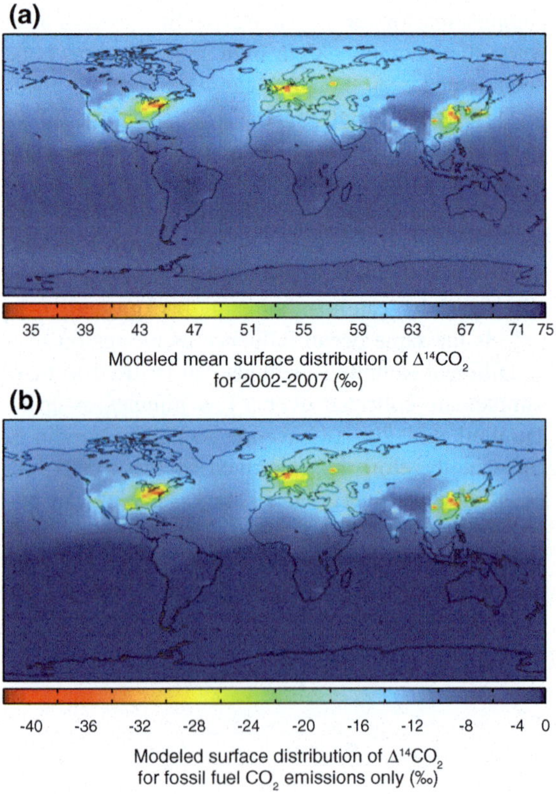

Fig. 4.15 (**a**) Modeled mean surface $\Delta^{14}CO_2$ distribution from the LMDZ model for 2002–2007. All known ^{14}C fluxes are included. (**b**) Modeled mean surface $\Delta^{14}CO_2$ distribution for 2002–2007 if fossil fuel CO_2 is the only source that alters $\Delta^{14}CO_2$ (i.e., no other sources are considered). The range of the two scales is the same (40 ‰). Figure modified from Turnbull et al. (2009b)

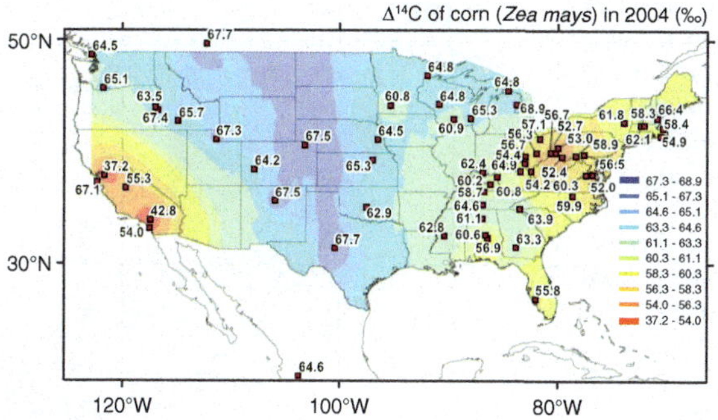

Fig. 4.16 Radiocarbon measurements of corn (*Zea mays*) across North America during the summer of 2004 (in units of ‰). *Squares* show actual measured values, and colors are the interpolated spatial distribution. During this time period, a decrease of 2.8 ‰ corresponded to ~1 ppm of added fossil fuel CO_2. Figure modified from Hsueh et al. (2007)

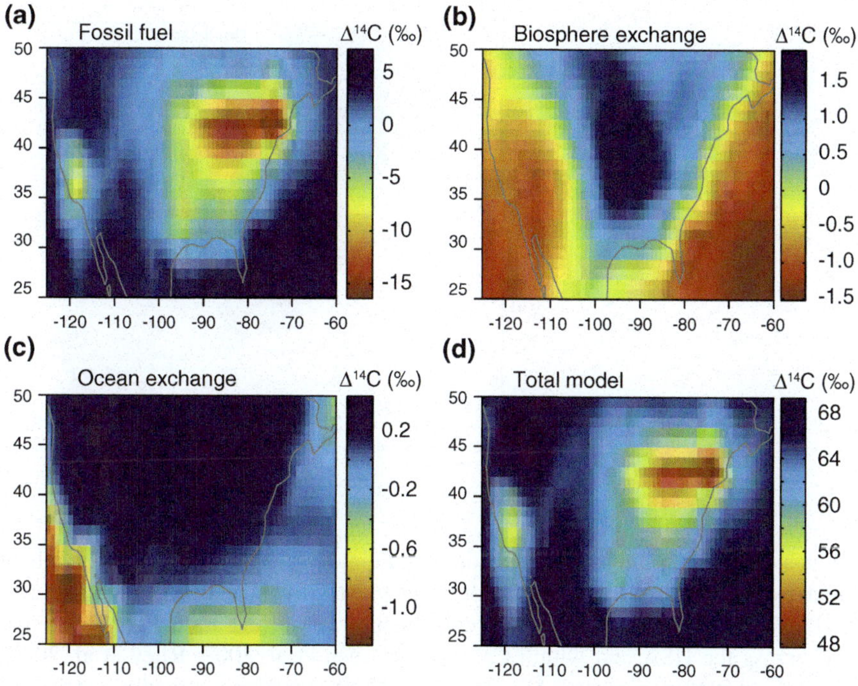

Fig. 4.17 Modeled contributions to surface atmospheric Δ^{14}C anomalies (with units of ‰) caused by (**a**) fossil fuel emissions, (**b**) terrestrial biosphere exchange, and (**c**) ocean exchange as derived from a global model during May–July of 2004 and (**d**) the combined influence of these processes, along with the effects of stratosphere–troposphere mixing. The background ^{14}C in (**d**) was adjusted by a single uniform scalar so that the mean of the model was the same as the mean of corn observations from western US mountains. Figure modified from Hsueh et al. (2007)

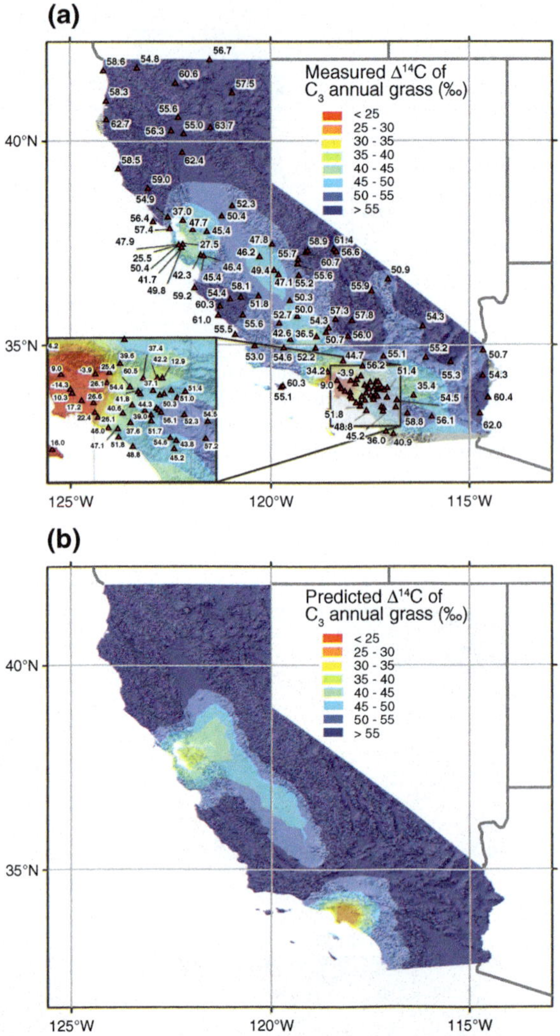

Fig. 4.18 Spatial pattern of $\Delta^{14}C$ from C_3 annual grasses in California, USA, from (**a**) observations in 2005 and (**b**) model simulation for annual grasses. In (**a**), *triangles* are actual observation sites, and *color scale* shows the interpolated spatial pattern. An expanded view of the Los Angeles area observations is shown on the *bottom left* of panel (**a**). Figures modified from Riley et al. (2008)

photosynthesize C only during the day, and that the magnitude of photosynthetic uptake varies with weather conditions and plant growth phase (Bozhinova et al. 2013). Two of the first studies used corn (maize) leaves and annual grasses to map ^{14}C across the USA and within California (Hsueh et al. 2007; Riley et al. 2008). The spatial pattern measured in corn across North America is broadly matched by models, with the lowest $\Delta^{14}CO_2$ values, indicative of strong CO_{2ff} sources, in the northeastern USA and on the California coast, and higher $\Delta^{14}CO_2$ values over the central region and Rocky Mountains (Figs. 4.16 and 4.17). In California, variations in the spatial distribution compared to model simulations show where model transport could be improved at the regional scale (Riley et al. 2008; Fig. 4.18). For example, the model used by Riley et al. (2008) predicts an area of low $\Delta^{14}CO_2$ values over Los Angeles. The observations reflect the same broad pattern, but show even lower values in Los Angeles, and also indicate an area of low $\Delta^{14}CO_2$ to the east of Los Angeles which is not predicted by the model.

Wine ethanol has also been used to map $^{14}CO_2$, and since wine vintages are rigorously documented, both the spatial distribution and the recent history of $^{14}CO_2$ can be obtained (Palstra et al. 2008). In Japan, ubiquitous rice grains provide similar spatial and historical $^{14}CO_2$ information (Shibata et al. 2005). In a study of rice, samples taken near the large number of nuclear power plants in Japan exhibited higher $\Delta^{14}CO_2$ values and were screened out to obtain the CO_{2ff} influence. In another study of large-scale ^{14}C patterns, flask samples of air were used to map $^{14}CO_2$ from a train-based platform across Eurasia (Turnbull et al. 2009b). Several of these samples were influenced by nearby nuclear power plant emissions. The nature of the near-instantaneous flask samples made it possible to screen out these samples to reveal the continental-scale $^{14}CO_2$ gradient, which was generally consistent with a modeled result.

4.8.3 Quantification of Fossil Fuel Carbon Dioxide Emissions at the Urban and Regional Scale

At the smaller scale, $^{14}CO_2$ is commonly used in urban regions to quantify fossil fuel emissions and to examine emissions of other anthropogenic trace gases. The method began to be used in a quantitative way in the late 1980s, with several studies calculating CO_{2ff} in flask samples of air from Europe (Levin et al. 1989; Meijer et al. 1996; Zondervan and Meijer 1996). Longer time series using CO_2 absorbed into alkali from German air followed (Levin et al. 2003, 2008). These measurements demonstrated large CO_{2ff} enhancements in the city of Heidelberg relative to clean-air sites (Fig. 4.19).

The $^{14}CO_2$ method has been shown to be less prone to biases than other tracer methods of quantifying CO_{2ff} (Turnbull et al. 2006; Djuricin et al. 2010; Fig. 4.20). However, the utility of the method is limited by two major factors. First, the accuracy of calculated CO_{2ff} is limited by the ^{14}C measurement precision. Second, the complex measurement techniques mean that samples must be transported and analyzed in a laboratory, limiting the temporal and spatial resolution of measurements. Therefore, ^{14}C measurements are often combined with other, more easily measured tracers to increase the resolution of CO_{2ff} estimates. The most widely used of the other tracers is CO, which is co-emitted with CO_2 during incomplete combustion. If the CO/CO_{2ff} emission ratio is known, CO_{2ff} can be calculated from CO measurements, but the emission ratio is variable depending on combustion conditions, and CO has other sources and sinks (Turnbull et al. 2006). These confounding problems can often be addressed by obtaining $^{14}CO_2$ and CO measurements from flasks or time-integrated samples (e.g., air collected in bags) to

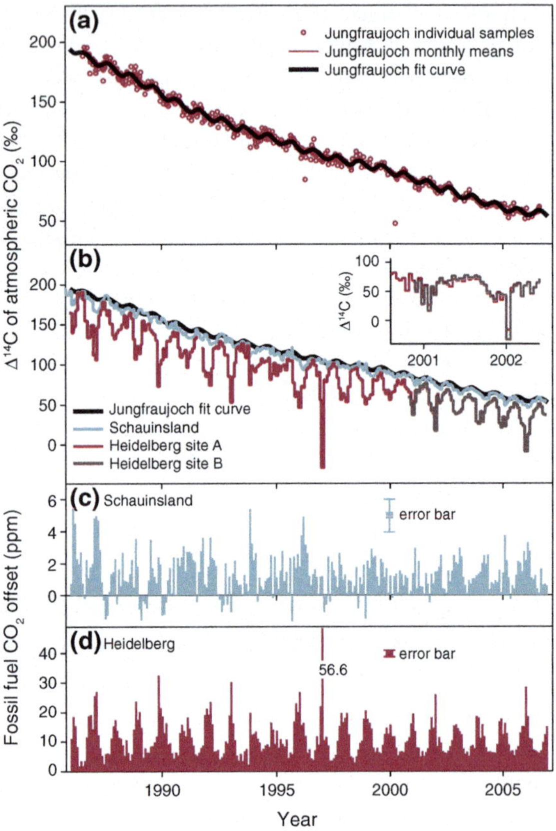

Fig. 4.19 (**a**) Long-term $\Delta^{14}CO_2$ observations at Jungfraujoch; the smooth solid line is a harmonic fit to the monthly mean observations. (**b**) Monthly mean $\Delta^{14}CO_2$ at Schauinsland and Heidelberg. The *inlay* shows the comparison of observations at two different Heidelberg sites about 500 m apart. (**c**) Monthly mean fossil fuel CO_2 at Schauinsland calculated using Jungfraujoch as background. (**d**) Monthly mean fossil fuel CO_2 at Heidelberg using Jungfraujoch as background. Figure modified from Levin et al. (2008)

quantify the CO/CO_{2ff} ratio at a given location and time period. The derived emission ratio can then be used to obtain high-resolution CO_{2ff} from continuous CO measurements at the same location (Levin and Karstens 2007; Vogel et al. 2010; Turnbull et al. 2011).

These high-resolution CO_{2ff} estimates from combined CO and $^{14}CO_2$ measurements have been used in combination with estimates of atmospheric transport to determine the CO_{2ff} emission flux and compared with reported emission fluxes from bottom-up inventories. In these studies, the estimated emission fluxes have been reasonably consistent with reported inventory data, but have large uncertainties that are dominated not by the ^{14}C measurement uncertainty, but by uncertainties in the atmospheric transport of emissions. Turnbull et al. (2011) used a simple Gaussian plume model to describe atmospheric transport at the urban scale, but this type of model requires explicit knowledge of the wind speed and the planetary boundary layer height (emissions are assumed to mix within the planetary boundary layer, but not to escape into the free troposphere). The planetary boundary layer height in particular is difficult to represent in models and is the subject of

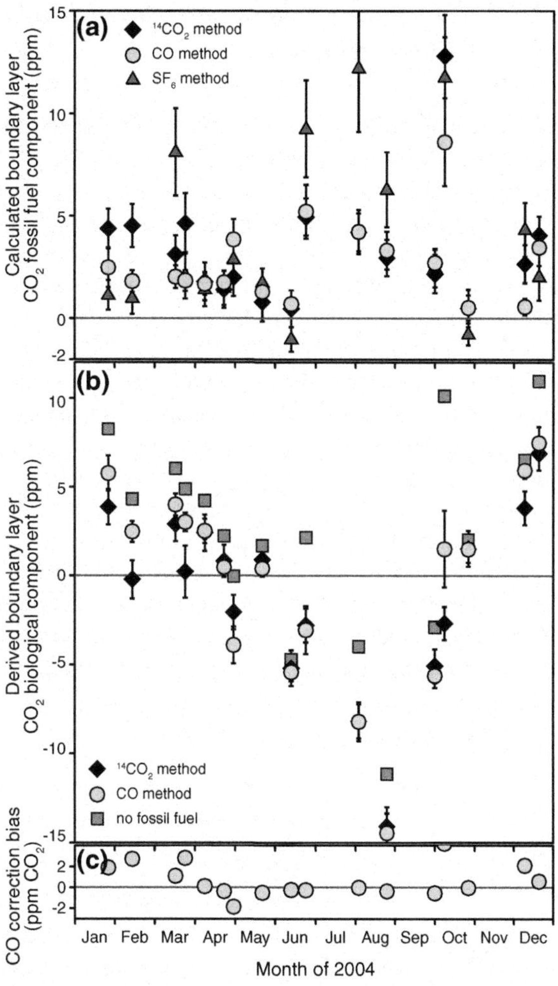

Fig. 4.20 (**a**) Calculated boundary layer fossil fuel CO_2 at Harvard Forest, USA, using $^{14}CO_2$ and the CO and SF_6 correlate tracer methods. (**b**) Derived boundary layer biological CO_2 component from total CO_2 and CO_{2ff} derived from each method. (**c**) Bias in the CO-derived CO_{2ff} relative to $^{14}CO_2$-dervied CO_{2ff}. Figure modified from Turnbull et al. (2006)

ongoing research. At even smaller scales such as single point source monitoring, averaging over long time periods may be needed to reduce uncertainties due to turbulent mixing and obtain reasonable agreement between model and observations (Turnbull et al. 2014). Van Der Laan et al. (2010) used a different approach, avoiding the need for explicit knowledge of the meteorological parameters by correlating their CO_{2ff} measurements with ^{222}Rn. ^{222}Rn is a gas produced by radioactive decay in the soil and released to the atmosphere. Its short half-life of a few days and relatively consistent emission across all land regions means that it is an excellent tracer for boundary layer stability and mixing. By assuming the ^{222}Rn emissions, and correlating CO_{2ff} with ^{222}Rn, it is possible to back-calculate the CO_{2ff} emissions. In this method, uncertainties in the ^{222}Rn emissions dominate.

4.8.4 Other Applications of Fossil Fuel Carbon Dioxide Measurements

Radiocarbon-derived CO_{2ff} estimates are also useful in quantifying CO_2 exchange with the terrestrial biosphere. Once the CO_{2ff} mole fraction is known, any remaining CO_2 variability over a land region must be ascribed to biological CO_2 exchange (including photosynthesis, respiration, biomass burning, and biofuel use). Contrary to initial expectations, CO_{2ff} is not always the dominant source of CO_2 variability even in highly urbanized regions, and respiration can be an important component, even in winter. A study in Los Angeles, California, USA, in 2007 and 2008, partitioned CO_2 sources using ^{14}C, ^{13}C, and ^{18}O. At a highly urbanized site, in a semi-arid environment, Djuricin and co-workers found that fossil fuel emissions contributed 30–50 % of the total CO_2 enhancement relative to background CO_2 values at Trinidad Head in Northern California and background $^{14}CO_2$ values at Point Barrow, Alaska (Djuricin et al. 2010; Fig. 4.21). In another study, the biosphere was shown to contribute up to 30 % of the total CO_2 enhancement in samples taken over Sacramento, California, in February and March 2009 (Turnbull et al. 2011). This study also found that the biosphere contribution could be different in sign between the urban area and the surrounding rural region. A third study over Denver, Colorado, USA, showed that respiration contributed 30–50 % of the total CO_2 enhancement in summer during the morning (Graven et al. 2009). All of these studies made measurements relative to continental clean-air or free troposphere background values. When more local background sites are used, it is likely that the

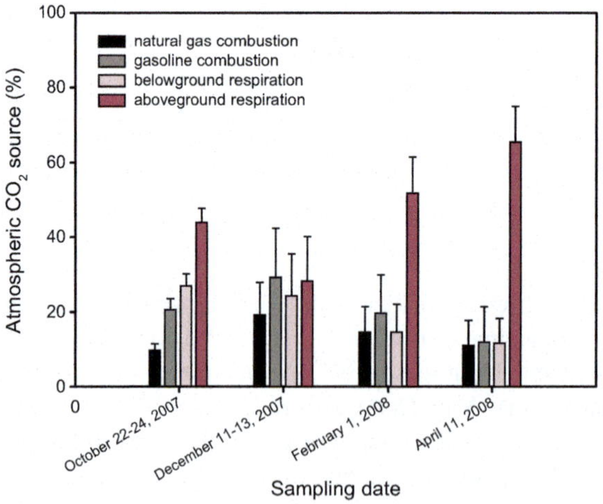

Fig. 4.21 Partitioning of CO_2 sources in Los Angeles, USA, for four episodes in 2007 and 2008. Sources were partitioned using $^{14}CO_2$ to obtain the fossil fraction, $^{13}CO_2$ to separate the different types of fossil fuel sources, and ^{18}O of CO_2 to identify the different types of respiration. Note also that bioethanol additive in gasoline was accounted for in the fossil fraction. Figure modified from Djuricin et al. (2010)

biosphere contribution in an urban area will appear much smaller, as much of the biosphere influence is likely coming from the surrounding countryside rather than the urban area itself.

The CO_{2ff} emission flux, while somewhat uncertain, is better known than fluxes of most other anthropogenic trace gases. The CO_{2ff} emission flux is known within 3–40 % at the annual, national scale, and within 20–50 % at regional and urban scales (Gurney et al. 2009; Marland 2010; Peylin et al. 2011). Fluxes of other species (e.g., hydrocarbons, halocarbons, criteria air pollutants) may be known to within only 50–100 % at the annual, national scale, and even more poorly at the urban and regional scales. The emission ratio of CO_{2ff} to another trace gas can therefore be used to improve the constraint on the flux of any trace gas with sources that are colocated with fossil fuel combustion. This ratio method avoids the need for atmospheric transport to be explicitly known or modeled. Meijer et al. (1996) pioneered this approach to examine CO emissions in the Netherlands, and other studies have since shown that bottom-up emission inventories of CO for much of the USA were too high (Turnbull et al. 2006; Graven et al. 2009; Miller et al. 2012), whereas observed CO/CO_{2ff} ratios were consistent with emission inventories for Germany and California (Vogel et al. 2010; Turnbull et al. 2011). The same method has also been used to evaluate hydrocarbon and halocarbon emissions in California (Turnbull et al. 2011) and the eastern USA (Miller et al. 2012).

4.8.5 Volcanic Carbon Dioxide Emissions

The volcanic CO_2 isoflux is very small compared to the other isofluxes and hence is ignored at the global and regional scales. Yet offsets in local atmospheric $^{14}CO_2$ are measurable in volcanic calderas and within 1 km of springs and seeps where there is outgassing of ^{14}C-free magmatic CO_2; this can be thought of as a natural analogue to the $^{14}CO_2$ depletion near cities and power plants where fossil fuel burning is concentrated. One of the first measurements of local $^{14}CO_2$ depletion attributable to volcanic activity was made on reeds and tree leaves at the shores of a lake in the volcanic Eifel region of western Germany with extensive CO_2 bubbling. Plants growing next to the lake showed $\Delta^{14}C$ decreases of ~100 ‰, but the depletion was undetectable ~100 m from the lake (Bruns et al. 1980). Other studies have shown that at the very local scale, $^{14}CO_2$ can be used as a sensitive tracer of CO_2 outgassing, just as it is a sensitive tracer of fossil fuel combustion at larger scales (Pasquier-Cardin et al. 1999; Saurer et al. 2003).

4.9 Atmospheric Radiocarbon in Other Compounds

4.9.1 ^{14}CO as a Detector for Hydroxyl Radicals

The isotope ^{14}CO tells a very different story about the atmosphere than $^{14}CO_2$. Produced naturally in the upper atmosphere, ^{14}C oxidizes extremely rapidly to ^{14}CO. The oxidation of ^{14}CO by hydroxyl radical (OH) to $^{14}CO_2$ occurs with a timescale of about two months (Weinstock and Niki 1972). Hydroxyl radicals provide the oxidizing power of the atmosphere, removing many important trace gases, including CH_4, CO, volatile organic compounds, and hydrochlorofluorocarbons. OH is produced naturally in the atmosphere during sunlit hours and its high reactivity means that it has a fleeting atmospheric lifetime of only a fraction of a second, making it extremely difficult to directly measure and study (Krol and Lelieveld 2003). OH might be expected to vary due to changes in its production and changes in the abundance of the trace gases which react with it. This variability in OH can have significant consequences for the global atmosphere. For example, lower OH production would result in a longer atmospheric lifetime for CH_4, a strong greenhouse gas (Kirschke et al. 2013). If the ^{14}C production rate is known, then measurements of ^{14}CO concentration provide a method to indirectly estimate OH abundance and variability via its effect on CO lifetime.

This method has been used to estimate short-term variations in OH of 10 % in the Southern Hemisphere high latitudes (Manning et al. 2005; Fig. 4.22), which are attributed to volcanic and biomass burning emissions. Hemispheric differences in OH abundance can also be tested with ^{14}CO (Jöckel and Brenninkmeijer 2002). The short lifetime and high-altitude source of ^{14}CO makes it more sensitive to OH concentrations at high altitudes and high latitudes than to OH concentrations at low altitudes and low latitudes (Krol et al. 2008). The distribution of ^{14}CO can additionally be used to test stratosphere–troposphere exchange in atmospheric models (Jöckel and Brenninkmeijer 2002).

4.9.2 Fossil Methane Source Identification

Methane, like CO_2, is a greenhouse gas with both natural and anthropogenic sources. Methane levels have increased from ~800 parts per billion (ppb) in the preindustrial era to 1800 ppb at present. The CH_4 budget is more complex than that of CO_2, with a plethora of sources, and unlike CO_2 it is oxidized in the atmosphere with a lifetime of about 10 years (Kirschke et al. 2013). The ^{14}C content of CH_4 is one way of elucidating the CH_4 budget. Fossil CH_4 sources are ^{14}C-free, whereas CH_4 produced from modern biomass by microorganisms (e.g., in rice paddies, termites, and cows) and fires has a ^{14}C content similar to that of the atmosphere. The fossil fraction can, in principle, be obtained in an analogous manner to CO_{2ff} (Townsend-Small et al. 2012). However, $^{14}CH_4$ from nuclear power generation

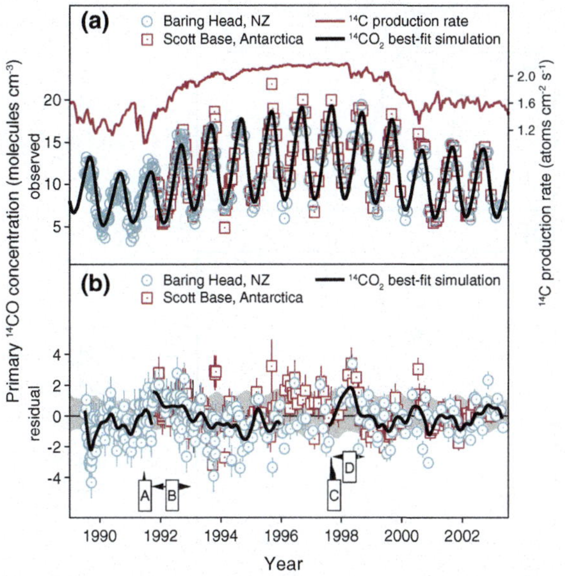

Fig. 4.22 (**a**) ^{14}CO concentrations for individual samples from Baring Head, New Zealand (*blue circles*) and Scott Base, Antarctica (*purple squares*) from 1989 to 2003. The *purple* and *black curves* show ^{14}C production rates (*right-hand scale*) and a best-fit simulation of ^{14}CO. (**b**) Residuals (observed minus simulated) for each measurement with uncertainties (1-sigma) and a smooth curve fit to the residuals (*black line*). The *gray shaded* regions show the deviations from the best fit that would be caused by a 10 % change in OH concentrations. *A*—time of eruption of Mt Pinatubo; *B*—period of high Southern Hemisphere CH_4 concentrations; *C*—time profile of Indonesian fires; *D*—period of high Southern Hemisphere CO concentrations. Figure modified from Manning et al. (2005)

comprises perhaps 20–40 % of the $^{14}CH_4$ budget (Quay et al. 1999), complicating the interpretation of $^{14}CH_4$ measurements. Nevertheless, observations and modeling have estimated the fossil CH_4 fraction from $^{14}CH_4$ measurements at ~30 % versus 20 % from bottom-up CH_4 emission inventories (Lassey et al. 2007).

A novel and challenging use of $^{14}CH_4$ measurements has been to identify the cause of the abrupt CH_4 increase from about 500 to over 700 parts per billion (ppb) during the Younger Dryas–Preboreal transition about 11,600 years ago (Petrenko et al. 2009; Fig. 4.23). Several mechanisms could have driven the CH_4 increase, with increased wetland CH_4 emissions and destabilization of marine CH_4 hydrate (solid CH_4 and water mixtures quite commonly found on the ocean floor) being the most likely. Tiny amounts of CH_4 were extracted from air bubbles in ~1000 kg of Greenland ice and measured for $^{14}CH_4$. Wetland $^{14}CH_4$ should be similar to that of the contemporaneous atmosphere, whereas CH_4 hydrate CH_4 was ancient and contained no ^{14}C. The measurements were confounded by in situ cosmogenic production of $^{14}CH_4$ molecules in the Arctic ice. After modeling cosmogenic production to remove this effect (which increased $\Delta^{14}CH_4$ by about

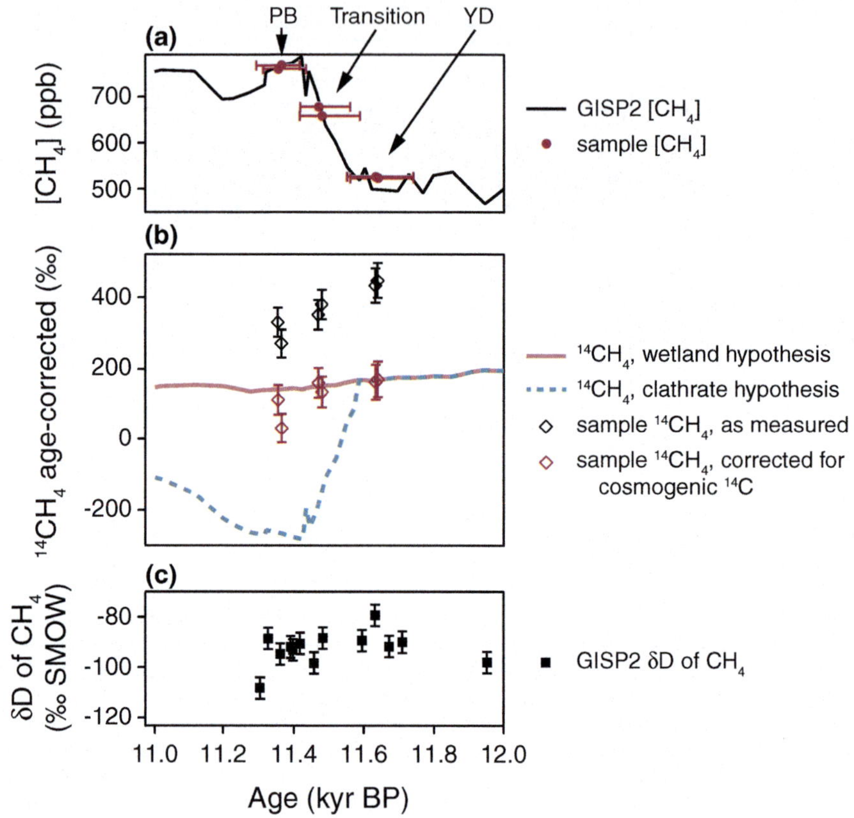

Fig. 4.23 $^{14}CH_4$ during the Younger Dryas (YD)—Preboreal (PB) transition. For the wetland hypothesis, 100 % biospheric CH_4 emissions are assumed for all times, with $^{14}CH_4$ equal to contemporaneous $^{14}CO_2$. For the CH_4 hydrate hypothesis, all of the CH_4 rise during the transition is assumed to be due to CH_4 hydrate emissions which decrease linearly to 0 over 1000 years, and have $^{14}CH_4$ of −1000 ‰. The transient $^{14}CH_4$ increase at 11.44 thousand years B.P. in the CH_4 hydrate model line corresponds to the simultaneous transient drop in CH_4 mole fraction. Horizontal error bars represent the maximum possible range of air ages. Figure modified from Petrenko et al. (2009)

200 ‰), the researchers found that $\Delta^{14}CH_4$ during the Younger Dryas—Preboreal transition stayed roughly constant at about 180 ‰. The CH_4 hydrate hypothesis predicts a decrease of $\Delta^{14}CH_4$ to about −200 ‰ during this period, so the CH_4 rise must have been due primarily to wetland sources, and not due to large-scale CH_4 hydrate destabilization.

4.9.3 *Aerosol Source Attribution*

Aerosols are another important area of atmospheric research, having both a climate effect and a negative impact on human health. As for other atmospheric trace species, understanding the sources is a critical part of the puzzle, and aerosol sources can be partitioned into fossil and biogenic components using ^{14}C measurements of the C in the aerosols.

Radiocarbon measurement can determine the fossil fuel contribution to total aerosol (or a collected size range) versus biogenic sources. Yet this misses an important part of the story, as anthropogenic biomass burning is also a substantial contributor to aerosols which cannot be distinguished from natural biogenic sources of aerosol. The elemental C (soot) fraction of aerosols is unique in being produced only as primary particles from fossil fuel combustion and biomass burning, whereas the organic C aerosol component is a mixture of primary sources including fossil fuel, biomass burning, and natural biogenic materials such as plant debris and pollen as well as secondary aerosol formed from gaseous precursors (Slater et al. 2002).

A study in Zurich, Switzerland, in 2002 and 2003 separated and measured both the soot and organic aerosol fractions (Szidat et al. 2006). The $\Delta^{14}C$ of the soot fraction ranged from less than −900 ‰ in summer to −750 ‰ in winter. They assigned a biomass burning $\Delta^{14}C$ of 240 ‰ in Zurich, assuming residential wood burning of 30- to 50-year-old wood, and fossil fuel $\Delta^{14}C$ of −1000 ‰. Using a simple two-member mixing model akin to the fossil fuel CO_2 method (Sect. 4.8), they determined that soot aerosol is produced solely by fossil fuel combustion in summer, but in winter, biomass burning contributes up to 25 %, even though biomass burning is only a tiny portion of local energy use. In contrast, organic aerosols had $\Delta^{14}C$ values of about −200 ‰ in all seasons of the year. This implies that the fossil fuel contribution was about 30 % of organic aerosol throughout the year. By assuming a consistent emission ratio of soot and organic biomass burning aerosols, they could further show that of the remaining non-fossil 70 % of organic aerosol, anthropogenic biomass burning was the main contributor in winter, but that natural biogenic sources became dominant in summer.

4.10 Conclusions and Future Directions

4.10.1 *Improved Radiocarbon Measurement Techniques and Comparability*

Atmospheric ^{14}C measurements are expanding to measure new trace gas species and to answer ever more detailed questions about our Earth System. A major limitation is the sample size requirement and difficulty of both AMS and decay counting measurement techniques. Currently, single sample AMS precision in

$\Delta^{14}CO_2$ is ±2 to ±5 ‰, with only a few laboratories able to achieve precision of ±2 ‰ or better. In comparison with other materials that are measured for ^{14}C, the analysis of atmospheric samples of CO_2 requires less sample handling, while the required level of precision is higher. Therefore, laboratories measuring $^{14}CO_2$ have initiated intercomparison activities specific to $^{14}CO_2$ (Miller et al. 2010, 2013; Graven et al. 2013; Turnbull et al. 2013), in addition to the routine ^{14}C intercomparison activities that use solid materials (e.g., Scott et al. 2007). The $^{14}CO_2$ intercomparison will identify any laboratory offsets specific to $^{14}CO_2$ with high precision and enable measurements from different laboratories to be merged. A further issue that needs to be addressed is the comparability of measurements of CO_2 from flask samples conducted by AMS to those of integrated CO_2 samples conducted by counting techniques. Addressing this issue will require measurement and interpretation of atmospheric variability that is captured in flask sampling but averaged over by continuous sampling.

As global $^{14}CO_2$ decreases and CO_2 mixing ratio increases through time, the sensitivity of $^{14}CO_2$ to CO_{2ff} is gradually decreasing, so that the same measurement precision will result in larger CO_{2ff} uncertainties. Incremental improvements in AMS precision can drive down this uncertainty and expand the CO_{2ff} detection capability. Measurement precision drops markedly with small sample sizes, particularly limiting the usefulness of ^{14}C measurements for the less abundant trace gases. Developments in measuring smaller sample sizes to better precision will open up many more possibilities for the use of ^{14}C in atmospheric studies.

Optical ^{14}C measurement techniques currently under development have the potential to revolutionize this field, particularly for $^{14}CO_2$, which could then be measured in situ at high resolution. Advances in laser technology mean that lasers can now be created with a very fine range of output wavelengths. The general principle is that a laser can be tuned to a specific wavelength where $^{14}CO_2$ absorption occurs. The amount of absorption of that laser light as it passes through air or concentrated CO_2 would determine the $^{14}CO_2$ content. The particular challenges for optical measurement of $^{14}CO_2$ arise from its very low abundance. First, broadening of nearby absorption lines of other gases, particularly $^{12}CO_2$ and $^{13}CO_2$, likely obscures many of the $^{14}CO_2$ absorption lines. Second, the low abundance will also mean that detection of absorption is difficult. Preconcentration of air into pure or at least much higher CO_2 concentration is needed, and the path length through the cell will need to be extremely long Galli et al. (2011). The initial results are promising, but much work remains to demonstrate reliable high precision on reasonably sized samples. It is likely that other similar methods will also be investigated over the next few years.

4.10.2 Developing Radiocarbon Observations into Policy-Relevant Fossil Fuel Carbon Dioxide Emission Estimates

The use of $^{14}CO_2$ has gradually been recognized as the best way to quantify CO_{2ff} from atmospheric observations. It may provide the only method for objectively verifying reported CO_{2ff} emissions. This will likely remain the most important, and expanding, field of atmospheric ^{14}C research for quite some time. To this end, observation networks will need to be expanded, the techniques for inferring CO_{2ff} emissions from the atmospheric observations will need to evolve, and a larger capacity for $^{14}CO_2$ measurements will be essential.

In order to provide relevant estimates of CO_{2ff} emissions, a reasonable target for $^{14}CO_2$ observation and modeling programs is to aim toward achieving uncertainties of roughly ±20 % in CO_{2ff} emissions. This level of uncertainty is similar to the differences in bottom-up inventories estimated by different groups for regions encompassing whole countries, groups of neighboring countries, or individual cities or states/provinces (Marland et al. 2006; Rayner et al. 2010; Peylin et al. 2011). Of course, this may underestimate the actual uncertainty because of shared biases due to the use by different groups of the same economic data and similar extrapolation techniques. In any case, this level of uncertainty is larger than the agreed-upon emission reductions for most nations that ratified the Kyoto Protocol (less than 10 %). Therefore, observational techniques using $^{14}CO_2$ are likely to be employed as a method for validating bottom-up inventories, rather than providing a precise top-down estimate of fossil fuel emissions that verifies that emission reductions have been achieved by an individual country.

Even accuracies of ±20 % in CO_{2ff} emissions will be challenging to achieve from atmospheric observations. Pacala et al. (2010) estimated the CO_{2ff} enrichment for several large urban regions to be 3–15 ppm, depending on the total emissions, population density, and meteorological characteristics of each city. This equates to a $\Delta^{14}CO_2$ decrease of −6 to −40 ‰. Continental-scale signals in $\Delta^{14}CO_2$ caused by regionally concentrated CO_{2ff} emissions are likely to be −5 to −20 ‰, as shown in Sect. 4.8.2 (e.g., Hsueh et al. 2007). Attaining ±20 % uncertainty in CO_{2ff} emissions will therefore require reducing the combined uncertainties in the measurement of $^{14}CO_2$, in the specification of non-fossil influences to $^{14}CO_2$, and in modeled transport from their present levels.

Understanding atmospheric transport is perhaps the most difficult problem in interpreting atmospheric $^{14}CO_2$ measurements. The methodology for quantifying CO_{2ff} mole fraction in atmospheric samples is now well established, but translating those atmospheric observations to emission fluxes is still challenging. Improvements in atmospheric transport models can potentially solve this problem by more accurately describing the transport of emissions to the observation site. Other approaches can reduce the reliance on atmospheric transport models. Levin et al. (2008) showed how long-term $^{14}CO_2$ measurements at a single site, paired with long-term observations at a mountaintop reference site, could be used to detect changes in emissions through time. This technique reduces the reliance on knowledge of atmospheric transport, assuming that interannual variations and trends in winds or boundary layer ventilation do not change the influential source regions or the dilution of surface emissions. In Sect. 4.8.3, we described how Van Der Laan et al. (2010) used ^{222}Rn to constrain planetary boundary layer mixing without the need for explicit transport models, although this method instead requires ^{222}Rn emissions to be well-known.

On the other hand, since the CO_{2ff} emission flux is arguably better known than that of any other species, $^{14}CO_2$ measurements have another potential use: to validate and improve atmospheric transport models. Combining $^{14}CO_2$ observations, particularly vertically resolved observations, with forward models that use bottom-up emission inventories as surface fluxes could identify model biases in the transport of emissions. Model-data evaluations using $^{14}CO_2$ are likely to be effective in regions with well-quantified bottom-up emissions, such as the USA (Gurney et al. 2009). Achieved improvements in model transport could then be employed in regions with less well-quantified emissions, such as Asia (Gregg et al. 2008).

Continued observations of $^{14}CO_2$ in background air are also needed. Such background observations provide information on fossil fuel emissions at global scales through the observation of the long-term trend or interhemispheric gradient of $^{14}CO_2$. Currently, uncertainties in the non-fossil contributions to the trend and gradient, primarily the oceanic contribution, limit the use of $^{14}CO_2$ to estimate global emissions (Levin et al. 2010). Improvements in quantifying non-fossil influences on background $^{14}CO_2$ could enable global-scale estimates of fossil fuel emissions to be achieved in the future.

4.10.3 Detection of Climate-Related Changes in Air–Land or Air–Ocean Carbon Exchanges

The feedback between anthropogenic climate change and the global C cycle represents one of the largest uncertainties in future projections of global warming (Jones et al. 2013). Several possible mechanisms have been proposed that would reduce the uptake of anthropogenic CO_2 or release currently stored C as the climate warms, thereby enhancing the accumulation of atmospheric CO_2 and the increase in surface temperatures in the future. However, the potential strength of these feedback mechanisms is not well understood. At least two of the proposed mechanisms would significantly perturb ^{14}C exchange with the atmosphere, so that observations of $^{14}CO_2$ may provide a method for detecting climate C feedbacks.

Thawing of arctic permafrost has the potential to release large amounts of long-sequestered C to the atmosphere in the form of CO_2 and CH_4. Since this C is likely to be old, $^{14}CO_2$ and $^{14}CH_4$ measurements could be used to quantify these releases through the observation of an anomalous negative isoflux in the northern high latitudes. The geographic separation should allow "old" permafrost emissions, concentrated in northern high latitudes, to be distinguished from fossil fuel emissions, concentrated in northern mid-latitudes. Coordinated observations of $^{14}CO_2$, $^{14}CH_4$, and other trace gases may further refine the separation of permafrost from fossil sources.

Climate change may also release old CO_2 from the deep ocean through enhancement in the wind-driven upwelling of the Southern Ocean, a process that has been proposed to have already begun, in part because of strengthening of Southern Hemisphere winds by stratospheric ozone depletion (Le Quere et al. 2007; Lovenduski et al. 2007). Enhanced upwelling of deep waters in the Southern Ocean causes an anomalous negative isoflux by exposing greater amounts of old water, containing older C, to the air. Observations of enhanced local gradients or trends in $^{14}CO_2$ over the Southern Ocean may therefore enable an anomalous CO_2 release to be quantified.

The centennial-scale variability in interhemispheric offset of atmospheric $^{14}CO_2$ (Hogg et al. 2002) also remains unexplained. Changes in ocean or atmospheric circulation have been suggested as causes (McCormac et al. 2002; Knox and McFadgen 2004), but whether the observed variations are consistent with what is known about ocean and atmosphere dynamics has not been evaluated.

The distribution of $^{14}CO_2$ in the atmosphere over the last few decades could provide more information on the ocean–atmosphere gas exchange velocity, particularly in the Southern Ocean (Krakauer et al. 2006). Most studies of ocean bomb ^{14}C uptake have used only seawater ^{14}C measurements to constrain ocean uptake, but measuring bomb ^{14}C in Southern Ocean water samples is particularly difficult because it is very dilute (Broecker et al. 1980; Peacock 2004; Levin et al. 2010). Several studies have shown that both the current rate of decline of atmospheric $^{14}CO_2$ and its gradient between the tropics and the Southern Ocean are sensitive to ocean–atmosphere gas exchange over the Southern Ocean, a process that is otherwise difficult to measure.

Radiocarbon provides an immensely powerful tool in understanding C-containing constituents of the atmosphere and their sources and variability. In the past atmosphere, ^{14}C measurements shed light on ocean C cycle processes as well as cosmic ray flux variability. Recent advances have made $^{14}CO_2$ measurements the tool of choice in constraining CO_{2ff} emissions in the modern atmosphere and potentially a key to understanding the fate of those emissions as they are taken up into thc ocean.

References

Boden, T.A., G. Marland, and R.J. Andres. 2012. Global, regional, and national fossil-fuel CO_2 emissions. Carbon Dioxide Information Analysis Center, Oak Ridge National Laboratory, U.S. Department of Energy, Oak Ridge, Tenn., U.S.A.

Bozhinova, D., M. Combe, S.W.L. Palstra, H.A.J. Meijer, M.C. Krol, and W. Peters. 2013. The importance of crop growth modeling to interpret the $^{14}CO_2$ signature of annual plants. *Global Biogeochemical Cycles* 27: 792–803.

Braziunas, T.F., I.Y. Fung, and M. Stuiver. 1995. The preindustrial atmospheric $^{14}CO_2$ latitudinal gradient as related to exchanges among atmospheric, oceanic, and terrestrial reservoirs. *Global Biogeochemical Cycles* 9: 565–584.

Broecker, W.S., T.-H. Peng, and R. Engh. 1980. Modeling the carbon system. *Radiocarbon* 22: 565–598.

Broecker, W.S., T.-H. Peng, H. Ostlund, and M. Stuiver. 1985. The distribution of bomb radiocarbon in the ocean. *Journal of Geophysical Research* C4: 6953–6970.

Bruns, M., I. Levin, K.O. Munnich, M.W. Hubberten, and S. Fillipakis. 1980. Regional sources of volcanic carbon dioxide and their influence on C14 content of present-day plant material. *Radiocarbon* 22: 532–536.

Caldeira, K., G.H. Rau, and P.B. Duffy. 1998. Predicted net efflux of radiocarbon from the ocean and increase in atmospheric radiocarbon content. *Geophysical Research Letters* 25: 3811–3814.

Ciais, P., C.L. Sabine, G. Bala, L. Bopp, V. Brovkin, J.G. Canadell, A. Chabbra, R. DeFries, J. Galloway, M. Heimann, C. Jones, C. Le Quere, R. Mymeni, S. Piao, and P. Thornton. 2013. Carbon and other biogeochemical cycles. In T.F. Stocker, D. Qin, G.-K. Plattner, M. Tignor, S. Allen, J. Boschung, A. Nauels, Y. Xia, V. Bex, and P. M. Midgley, eds. Climate Change 2013: The Physical Science Basis. Contribution of Working Group I to the Fifth Assessment Report of the Intergovernmental Panel on Climate Change. Cambridge University Press, Cambridge, United Kingdom and New York, NY, USA.

Currie, K.I., G. Brailsford, S. Nichol, A. J. Gomez, R.J. Sparks, K.R. Lassey, and K. Riedel. 2011. Tropospheric $^{14}CO_2$ at Wellington, New Zealand: the world's longest record. *Biogeochemistry* 104: 5–22. doi:10.1007/s10533-009-9352-6

De Jong, A.F., and W.G. Mook. 1982. An anomalous Suess effect above Europe. *Nature* 298: 641–644.

Denning, A.S., M. Holzer, K.R. Gurney, M. Heimann, R.M. Law, P.J. Rayner, I.Y. Fung, S. Fan, S. Taguchi, P. Friedlingstein, Y. Balkanski, J. Taylor, M. Maiss, and I. Levin. 1999. Three-dimensional transport and concentration of SF_6 a model intercomparison study (TransCom 2). *Tellus* 51B: 266–297.

Djuricin, S., D.E. Pataki, and X. Xu. 2010. A comparison of tracer methods for quantifying CO_2 sources in an urban region. *Journal of Geophysical Research* 115.

Friedrich, M., S. Remmele, B. Kromer, J. Hofmann, M. Spurk, K.F. Kaiser, C. Orcel, and M. Küppers. 2004. The 12,460-year hohenheim oak and pine tree-ring chronology from Central Europe: A unique annual record for radiocarbon calibration and paleoenvironment reconstructions. *Radiocarbon* 46: 1111–1122.

Frost, T., and R.C. Upstill-Goddard. 1999. Air-sea gas exchange into the millennium: Progress and uncertainties. *Oceanography and Marine Biology* 37(37): 1–45.

Galli, I., S. Bartalini, S. Borri, P. Cancio, D. Mazzotti, P. De Natale, and G. Giusfredi. 2011. Molecular gas sensing below parts per trillion: Radiocarbon-dioxide optical detection. *Physical Review Letters* 107: 270802.

Graven, H.D., and N. Gruber. 2011. Continental-scale enrichment of atmospheric $14CO_2$ from the nuclear power industry: Potential impact on the estimation of fossil fuel-derived CO_2. *Atmospheric Chemistry and Physics* 11: 12339–12349.

Graven, H.D., T.P. Guilderson, and R.F. Keeling. 2007. Methods for high-precision ^{14}C AMS measurement of atmospheric CO_2 at LLNL. *Radiocarbon* 49: 349–356.

Graven, H.D., B.B. Stephens, T.P. Guilderson, T.L. Campos, D.S. Schimel, J.E. Campbell, and R. F. Keeling. 2009. Vertical profiles of biospheric and fossil fuel-derived CO_2 and fossil fuel CO_2:CO ratios from airborne measurements of $\Delta^{14}C$, CO_2 and CO above Colorado, USA. *Tellus B* 61: 536–546.

Graven, H.D., N. Gruber, R. Key, S. Khatiwala, and X. Giraud. 2012a. Changing controls on oceanic radiocarbon: new insights on shallow-to-deep ocean exchange and anthropogenic CO_2 uptake. *Journal of Geophysical Research* 117.

Graven, H.D., T.P. Guilderson, and R.F. Keeling. 2012b. Observations of radiocarbon in CO_2 at La Jolla, California, USA 1992–2007: analysis of the long-term trend. *Journal of Geophysical Research* 117.

Graven, H.D., T.P. Guilderson, and R.F. Keeling. 2012c. Observations of radiocarbon in CO_2 at seven global sampling sites in the scripps flask network: Analysis of spatial gradients and seasonal cycles. *Journal of Geophysical Research* 117.

Graven, H.D., X. Xu, T. Guilderson, R.F. Keeling, S.E. Trumbore, and S. Tyler. 2013. Comparison of independent $\Delta 14CO_2$ records at point barrow, Alaska. *Radiocarbon* 55: 1541–1545.

Gregg, J.S., R.J. Andres, and G. Marland. 2008. China: Emissions Pattern of the world leader in CO_2 emissions from fossil fuel consumption and cement production. *Geophysical Research Letters* 35.

Gurney, K.R., R.M. Law, A.S. Denning, P.J. Rayner, D. Baker, P. Bousquet, L. Bruhwiler, Y.-H. Chen, P. Ciais, S. Fan, I.Y. Fung, M. Gloor, M. Heimann, K. Higuchi, J. John, T. Maki, S. Maksyutov, K.A. Masarie, P. Peylin, M. Prather, B.C. Pak, J. Randerson, J.L. Sarmiento, S. Taguchi, T. Takahashi, and C.-W. Yuen. 2002. Towards robust regional estimates of CO_2 sources and sinks using atmospheric transport models. *Nature* 415: 626–630.

Gurney, K.R., D.L. Mendoza, Y. Zhou, M.L. Fischer, C.C. Miller, S. Geethakumar, and S. de la Rue du Can. 2009. High resolution fossil fuel combustion CO_2 emission fluxes for the United States. *Environmental Science and Technology* 43: 5535–5541.

Heimann, M., and E. Maier-Reimer. 1996. On the relations between the oceanic uptake of CO_2 and its carbon isotopes. *Global Biogeochemical Cycles* 10: 89–110.

Hesshaimer, V., and I. Levin. 2000. Revision of the stratospheric bomb $^{14}CO_2$ inventory. *Journal of Geophysical Research* 105: 11641–11658.

Hesshaimer, V., M. Heimann, and I. Levin. 1994. Radiocarbon evidence for a smaller oceanic carbon dioxide sink than previously believed. *Nature* 201–203.

Hogg, A., F. McCormac, T. Higham, P. Reimer, M. Baillie, and J. Palmer. 2002. High-precision radiocarbon measurements of contemporaneous tree-ring dated wood from the British Isles and New Zealand: A.D. 1850–1950. *Radiocarbon* 44: 633–640.

Hogg, A.G., C. Bronk Ramsey, C. Turney, and J. Palmer. 2009a. Bayesian evaluation of the Southern Hemisphere radiocarbon offset during the holocene. *Radiocarbon* 51: 1165–1176.

Hogg, A.G., J. Palmer, G. Boswijk, P. Reimer, and D. Brown. 2009b. Investigating the interhemispheric ^{14}C offset in the 1st millennium A.D. and assessment of laboratory bias and calibration errors. *Radiocarbon* 51: 1177–1186.

Holton, J.R., P.H. Haynes, M.E. McIntyre, A.R. Douglass, R.B. Rood, and L. Pfister. 1995. Stratosphere-troposphere exchange. *Reviews of Geophysics* 33: 403–439.

Houghton, R.A. 2008. Carbon flux to the atmosphere from land-use changes: 1850–2005. TRENDS: A compendium of data on global change. Carbon dioxide information analysis center, Oak Ridge National Laboratory, US Department of Energy, Oak Ridge, TN, USA.

Hsueh, D.Y., N.Y. Krakauer, J.T. Randerson, X. Xu, S.E. Trumbore, and J.R. Southon. 2007. Regional patterns of radiocarbon and fossil fuel-derived CO_2 in surface air across North America. *Geophysical Research Letters* 34: L02816.

Hua, Q., and M. Barbetti. 2004. Review of tropospheric bomb C-14 data for carbon cycle modeling and age calibration purposes. *Radiocarbon* 46: 1273–1298.

Hua, Q., and M. Barbetti. 2007. Influence of atmospheric circulation on regional (CO_2)-^{14}C differences. *Journal of Geophysical Research-Atmospheres* 112.

Hua, Q., M. Barbetti, U. Zoppi, D. Fink, M. Watanasak, and G. Jacobsen. 2004. Radiocarbon in tropical tree rings during the little ice age. *Nuclear Instruments and Methods in Physics Research Section B-Beam Interactions With Materials and Atoms* 223: 489–494.

Hua, Q., M. Barbetti, D. Fink, K.F. Kaiser, M. Friedrich, B. Kromer, V.A. Levchenko, U. Zoppi, A.M. Smith, and F. Bertuch. 2009. Atmospheric C-14 variations derived from tree rings during the early younger dryas. *Quaternary Science Reviews* 28: 2982–2990.

Hua, Q., M. Barbetti, and A.Z. Rakowski. 2013. Atmospheric radiocarbon for the period 1950–2010. *Radiocarbon* 55: 2059–2072.

Hughen, K., S. Lehman, J. Southon, J. Overpeck, O. Marchal, C. Herring, and J. Turnbull. 2004. C-14 activity and global carbon cycle changes over the past 50,000 years. *Science* 303: 202–207.

Jacob, D.J. 1999. *Introduction to atmospheric chemistry*. Princeton, NJ, USA: Princeton University Press.

Johnston, C.A. 1994. Ecological engineering of wetlands by beavers. In *Global Wetlands: Old World and New*, ed. W.J. Mitsch, 379–384. Amsterdam: Elsevier.

Jones, C., E. Robertson, V. Arora, P. Friedlingstein, E. Shevliakova, L. Bopp, V. Brovkin, T. Hajima, E. Kato, M. Kawamiya, S. Liddicoat, K. Lindsay, C.H. Reick, C. Roelandt, J. Segschneider, and J. Tjiputra. 2013. Twenty-first-century compatible CO_2 emissions and airborne fraction simulated by CMIP5 earth system models under four representative concentration pathways. *Journal of Climate*, 26: 4398–4413.

Jöckel, P., M.G. Lawrence, and C.A.M. Brenninkmeijer. 1999. Simulations of cosmogenic ^{14}CO using the three-dimensional atmospheric transport model MATCH: Effects of ^{14}C production distribution and the solar cycle. *Journal of Geophysical Research*, 104(D9): 11733–11743.

Jöckel, P., and C.A.M. Brenninkmeijer. 2002. the seasonal cycle of cosmogenic ^{14}CO at the surface level: A solar cycle adjusted, zonal-average climatology based on observations. *Journal of Geophysical Research* 107: 4656.

Keeling, C.D., and T. Whorf. 2005. Atmospheric CO_2 records from sites in the SIO air sampling network. Trends: A compendium of data of global change. Carbon Dioxide Information Analysis Center, Oak Ridge National Laboratory, Oak RIdge, Tenn., USA.

Key, R.M., A. Kozyr, C.L. Sabine, K. Lee, R. Wanninkhof, J.L. Bullister, R.A. Feely, F.J. Millero, C. Mordy, and T.H. Peng. 2004. A global ocean carbon climatology: results from global data analysis project (GLODAP). *Global Biogeochemical Cycles* 18(4). doi:10.1029/2004gb002247 (4).

Kirschke, S., P. Bousquet1, P. Ciais, M. Saunois, J.G. Canadell, E.J. Dlugokencky, P. Bergamaschi, D. Bergmann, D.R. Blake, L. Bruhwiler, P. Cameron-Smith, S. Castaldi, F. D.R. Chevallier, L. Feng, A. Fraser, M. Heimann, E.L. Hodson, A. Houweling, B.A. Josse, P. J. Fraser, P.B. Krummel, J.-F.O. Lamarque, R.L. Langenfelds, C.L. Quéré, V. Naik, S. O'Doherty, P.I. Palmer, I. Pison, D. Plummer, B. Poulter, R.G. Prinn, M. Rigby, B. Ringeval, M. Santini, M. Schmidt, D.T. Shindell, I.J. Simpson, R. Spahni, L.P. Steele, S.A. Strode, K. Sudo, S. Szopa, G. R.V.D. Werf, A. Voulgarakis, M.V. Weele, R.F. Weiss, J.E. Williams, and G. Zeng. 2013. Three decades of global methane sources and sinks. *Nature Geoscience* 6: 813–823.

Kitigawa, H., H. Mukai, Y. Nojiri, Y. Shibata, T. Kobayashi, and T. Nojiri. 2004. Seasonal and secular variations of atmospheric $^{14}CO_2$ over the Western Pacific since 1994. *Radiocarbon* 46: 901–910.

Kjellstrom, E., J. Feichter, and G. Hoffman. 2000. Transport of SF_6 and (CO2)-C^{14} in the atmospheric general circulation model ECHAM4. *Tellus Series B-Chemical and Physical Meteorology* 52: 1–18.

Knox, F., and B. McFadgen. 2004. Radiocarbon/tree ring calibration, solar activity, and upwelling of ocean water. *Radiocarbon* 46: 987–995.

Kohler, P., R. Muscheler, and H. Fischer. 2006. A model-based interpretation of low-frequency changes in the carbon cycle during the last 120,000 years and its implications for the reconstruction of atmospheric delta C-14. *Geochemistry Geophysics Geosystems* 7.

Krakauer, N., J. Randerson, F. Primeau, N. Gruber, and D. Menemenlis. 2006. Carbon isotope evidence for the latitudinal distribution and wind speed dependence of the air-sea gas transfer velocity. *Tellus Series B-Chemical and Physical Meteorology* 58: 390–417.

Krol, M.C., and J. Lelieveld. 2003. Can the variability in tropospheric OH be deduced from measurements of 1,1,1-trichloroethane (methyl chloroform)? *Journal of Geophysical Research* D108.

Krol, M.C., J.F. Meirink, P. Bergamaschi, J.E. Mak, D. Lowe, P. Jockel, S. Houweling, and T. Rockmann. 2008. What can ^{14}CO measurements tell us about OH? *Atmospheric Chemistry and Physics* 8: 5033–5044.

Kuc, T., K. Rozanski, M. Zimnoch, J. Necki, L. Chmura, and D. Jelen. 2007. Two decades of regular observations of $^{14}CO_2$ and $^{13}CO_2$ content in atmospheric carbon dioxide in central europe: long-term changes of regional anthropogenic fossil CO_2 emissions. *Radiocarbon* 49: 807–816.

Lal, D. 1988. Theoretically expected variations in the terrestrial cosmic-ray production rates of isotopes. In: Proceedings of the International school of Physics, Solar-Terrestrial Relationships and the Earth Environment in the Last Millennia, pp. 215–233.

Lal, D., and Rama. 1966. Characteristics of global tropospheric mixing based on man-made C14 H3 and Sr90. *Journal of Geophysical Research* 71: 2865.

Land, C., J. Feichter, and R. Sausen. 2002. Impact of vertical resolution on the transport of passive tracers in the ECHAM4 model. *Tellus B* 54: 344–360.

Lassey, K.R., D.C. Lowe, and A.M. Smith. 2007. The atmospheric cycling of radiomethane and the "fossil fraction" of the methane source. *Atmospheric Chemistry and Physics* 7: 2141–2149.

Le Quere, C., C. Rodenbeck, E.T. Buitenhuis, T.J. Conway, R. Langenfelds, A. Gomez, C. Labuschagne, M. Ramonet, T. Nakazawa, N. Metzl, N. Gillett, and M. Heimann. 2007. Saturation of the southern ocean CO_2 sink due to recent climate change. *Science* 316: 1735–1738.

Lerman, J.C., W.G. Mook, and J.C. Vogel. 1970. C-14 in tree rings from different localities. Radiocarbon variations and absolute chronology. In: Proceedings of the Twelfth Nobel Symposium held at the Institute of Physics at Uppsala University. Wiley Interscience Division, New York.

Levin, I., R. Boesinger, G. Bonani, R.J. Francey, B. Kromer, K.O. Muennich, M. Suter, N.B. A. Trivett, and W. Wolfli. 1992. Radiocarbon in atmospheric carbon dioxide and methane global distribution and trends. In: Taylor, R.E., A. Long and R. S. Kra eds. Radiocarbon After Four Decades: An Interdisciplinary Perspective; Meeting, Lake Arrowhead, California, USA, 4–8 June 1990. Xviii + 596 p. New York, USA; Berlin, Germany: Springer. Illus: 503–518.

Levin, I., and V. Hesshaimer. 2000. Radiocarbon: A unique tracer of global carbon cycle dynamics. *Radiocarbon* 42: 69–80.

Levin, I., and U. Karstens. 2007. Inferring high-resolution fossil fuel CO_2 records at continental sites from combined $^{14}CO_2$ and CO observations. *Tellus* 59B: 245–250.

Levin, I., and B. Kromer. 2004. The tropospheric $^{14}CO_2$ level in mid-latitudes of the Northern Hemisphere (1959–2003). *Radiocarbon* 46: 1261–1272.

Levin, I., J. Schuchard, B. Kromer, and K.O. Munnich. 1989. The continental European Suess effect. *Radiocarbon* 31: 431–440.

Levin, I., B. Kromer, M. Schmidt, and H. Sartorius. 2003. A novel approach for independent budgeting of fossil fuel CO_2 over Europe by $^{14}CO_2$ observations. *Geophysical Research Letters* 30: 2194.

Levin, I., S. Hammer, B. Kromer, and F. Meinhardt. 2008. Radiocarbon observations in atmospheric CO_2: determining fossil fuel CO_2 over Europe using Jungfraujoch observations as background. *Science of the Total Environment* 391: 211–216.

Levin, I., T. Naegler, B. Kromer, M. Diehl, R.J. Francey, A.J. Gomez-Pelaez, L.P. Steele, D. Wagenbach, R. Weller, and D.E. Worthy. 2010. Observations and modelling of the global distribution and long-term trend of atmospheric $^{14}CO_2$. *Tellus B* 62: 26–46.

Lovenduski, N.S., N. Gruber, S.C. Doney, and I.D. Lima. 2007. Enhanced CO_2 outgassing in the southern ocean from a positive phase of the southern annular mode. *Global Biogeochemical Cycles* 21.

Manning, M.R., D.C. Lowe, W.H. Melhuish, R.J. Sparks, G. Wallace, C.A.M. Brenninkmeijer, and R.C. McGill. 1990. The use of radiocarbon measurements in atmospheric sciences. *Radiocarbon* 32: 37–58.

Manning, M.R., D.C. Lowe, R.C. Moss, G.E. Bodeker, and W. Allan. 2005. Short-term variations in the oxidizing power of the atmosphere. *Nature* 436: 1001–1004.

Marland, G. 2010. Accounting for carbon dioxide emissions from bioenergy systems. *Journal of Industrial Ecology* 14: 866–869.

Marland, G., T.A. Boden, and R.J. Andres. 2006. Global, regional and national CO_2 emissions. Trends: A compedium of data on global change. Carbon Dioxide Information Analysis Center, Oak Ridge National Laboratory, US Department of Energy, Oak Ridge, TN.

Masarik, J., and J. Beer. 1999. Simulation of particle fluxes and cosmogenic nuclide production in the earth's atmosphere. *Journal of Geophysical Research-Atmospheres* 104: 12099–12111.

Mazaud, A., C. Laj, E. Bard, M. Arnold, and E. Tric. 1991. Geomagnetic-field control of C-14 production over the Last 80 Ky—Implications for the radiocarbon timescale. *Geophysical Research Letters* 18: 1885–1888.

McCormac, F., A. Hogg, T. Higham, J. Lynch-Stieglitz, W. Broecker, M. Baillie, J. Palmer, L. Xiong, J. Pilcher, D. Brown, and S. Hoper. 1998. Temporal variation in the interhemispheric C-14 Offset. *Geophysical Research Letters* 25: 1321–1324.

McCormac, F., P. Reimer, A. Hogg, T. Higham, M. Baillie, J. Palmer, and M. Stuiver. 2002. Calibration of the radiocarbon time scale for the Southern Hemisphere: A.D. 1850–1950. *Radiocarbon* 44: 641–651.

Meijer, H.A.J., H.M. Smid, E. Perez, and M.G. Keizer. 1996. Isotopic characterization of anthropogenic CO_2 emissions using isotopic and radiocarbon analysis. *Physical Chemistry of the Earth* 21: 483–487.

Miller, J.B., C. Wolak, S.J. Lehman, C.E. Allison, H.D. Graven, T.P. Guilderson, R.F. Keeling, H. A.J. Meijer, T. Nakamura, T. Nakazawa, R.E. Neubert, A.M. Smith, J.R. Southon, and X. Xu. 2010. Preliminary results of from the first intercomparison of accelerator mass spectrometry atmospheric $^{14}CO_2$ measurements. In: W. Brand, ed. 15th WMO/IAEA Meeting of Experts on Carbon Dioxide Concentration and Related Measurement Techniques. World Meteorological Organization, Geneva.

Meijer, H.A.J., M.H. Pertuisot, and J. van der Plicht. 2006. High-Accuracy C-14 measurements for atmospheric CO_2 samples by AMS. *Radiocarbon* 48: 355–372.

Miller, J.B., S.J. Lehman, S.A. Montzka, C. Sweeney, B.R. Miller, C. Wolak, E.J. Dlugokencky, J. R. Southon, J.C. Turnbull, and P.P. Tans. 2012. Linking emissions of fossil fuel CO_2 and other anthropogenic trace gases using atmospheric $^{14}CO_2$. *Journal of Geophysical Research* 117.

Miller, J.B., S. Lehmann, C. Wolak, J. Turnbull, G. Dunn, H. Graven, R. Keeling, H.A.J. Meijer, A.T. Aerts-Bijma, S.W.L. Palstra, A.M. Smith, C. Allison, J. Southon, X. Xu, T. Nakazawa, S. Aoki, T. Nakamura, T. Guilderson, B. LaFranchi, H. Mukai, Y. Terao, M. Uchida, and M. Kondo. 2013. Initial results of an intercomparison of AMS-based atmospheric $^{14}CO_2$ measurements. *Radiocarbon* 55: 1475–1483.

Muscheler, R., J. Beer, G. Wagner, C. Laj, C. Kissel, G. Raisbeck, F. Yiou, and P. Kubik. 2004. Changes in the carbon cycle during the last deglaciation as indicated by the comparison of ^{10}Be and ^{14}C records. *Earth and Planetary Science Letters* 219: 325–340.

Naegler, T., and I. Levin. 2006. Closing the global radiocarbon budget 1945–2005. *Journal of Geophysical Research-Atmospheres* 111.

Naegler, T., and I. Levin. 2009a. Biosphere-atmosphere gross carbon exchange flux and the delta (CO_2)-^{13}C and Delta(CO_2)-^{14}C disequilibria constrained by the biospheric excess radiocarbon inventory. *Journal of Geophysical Research-Atmospheres* 114.

Naegler, T., and I. Levin. 2009b. Observation-based global biospheric excess radiocarbon inventory 1963–2005. *Journal of Geophysical Research* 114.

Nakamura, T., T. Nakazawa, H. Honda, H. Kitagawa, T. Machida, A. Ikeda, and E. Matsumoto. 1994. Seasonal variations in ^{14}C concentrations of stratospheric CO_2 measured with accelerator mass spectrometry. *Nuclear Instruments and Methods* B92: 413–416.

Nakamura, T., T. Nakazawa, N. Nakai, H. Kitigawa, H. Honda, T. Itoh, T. Machida, and E. Matsumoto. 1992. Measurement of ^{14}C concentrations of stratospheric CO_2 by accelerator mass spectrometry. *Radiocarbon* 34: 745–752.

Nydal, R. 1963. Increase in radiocarbon from most recent series of thermonuclear tests. *Nature* 200: 212.

Nydal, R. 1968. Further investigation on transfer of radiocarbon in nature. *Journal of Geophysical Research* 73: 3617.

Nydal, R., and K. Lovseth. 1965. Distribution of Radiocarbon from Nuclear Tests. *Nature* 206: 1029.

Nydal, R., and K. Lovseth. 1983. Tracing bomb ^{14}C in the atmosphere 1962–1980. *Journal of Geophysical Research-Oceans and Atmospheres* 88: 3621–3642.

O'Brien, K. 1979. Secular variations in the production of cosmogenic isotopes in the earth's atmosphere. *Journal of Paleolimnology* 84: 423.

Oeschger, H., U. Siegenthaler, U. Schotterer, and A. Gugelmann. 1975. A box diffusion model to study the carbon dioxide exchange in nature. *Tellus* XXVII:168–192.

Pacala, S.W., C. Breidenich, P.G. Brewer, I.Y. Fung, M.R. Gunson, G. Heddle, B.E. Law, G. Marland, K. Paustian, M. Prather, J.T. Randerson, P.P. Tans, and S.C. Wofsy. 2010. *Verifying greenhouse gas emissions: methods to support international climate agreements*. Committee on Methods for Estimating Greenhouse Gas Emissions: National Research Council.

Palstra, S.W., U. Karstens, H.-J. Streurman, and H.A.J. Meijer. 2008. Wine ethanol ^{14}C as a tracer for fossil fuel CO_2 emissions in Europe: Measurements and model comparison. *Journal of Geophysical Research* 113.

Pasquier-Cardin, A., P. Allard, T. Ferreira, E.C. Hatt, R. Coutinho, M. Fontugne, and M. Jaudon. 1999. Magma derived CO_2 emissions recorded in ^{14}C and ^{13}C content of plants growing in furnas caldera, Azores. *Journal of Volcanology and Geothermal Research* 92: 195–207.

Peacock, S. 2004. Debate over the ocean bomb radiocarbon sink: Closing the gap. *Global Biogeochemical Cycles* 18.

Petrenko, V.V., A.M. Smith, E.J. Brook, D.C. Lowe, K. Riedel, G. Brailsford, Q. Hua, H. Schaefer, N. Reeh, R.F. Weiss, D.M. Etheridge, and J.P. Severinghaus. 2009. $^{14}CH_4$ measurements in greenland ice: Investigating the last glacial termination CH_4 sources. *Science* 324: 506–508.

Peylin, P., S. Houweling, M. Krol, U. Karstens, C. Rodenbeck, C. Geels, A. Vermeulen, B. Badawy, C. Aulagnier, T. Pregger, F. Delage, G. Pieterse, P. Ciais, and M. Heimann. 2011. Importance of fossil fuel emission uncertainties over Europe for CO_2 modeling: Model intercomparison. *Atmospheric Chemistry and Physics* 11: 6607–6622.

Quay, P.D., J. Stutsman, D. Wilbur, A.K. Snover, E.J. Dlugokencky, and T.A. Brown. 1999. The isotopic composition of atmospheric methane. *Global Biogeochemical Cycles* 13: 445–461.

Rafter, T.A., and G.J. Fergusson. 1957. "Atom bomb effect"—Recent increase of carbon-14 content of the atmosphere and biosphere. *Science* 126: 557–558.

Randerson, J., I. Enting, E. Schuur, K. Caldeira, and I. Fung. 2002. Seasonal and Latitudinal variability of troposphere $\Delta(CO_2)$-^{14}C: Post bomb contributions from fossil fuels, oceans, the stratosphere, and the terrestrial biosphere. *Global Biogeochemical Cycles* 16.

Rayner, P.J., M.R. Raupach, M. Paget, P. Peylin, and E. Koffi. 2010. A new global gridded data set of CO_2 emissions from fossil fuel combustion: Methodology and evaluation. *Journal of Geophysical Research* 115: D19306.

Reimer, P.J. 2013. IntCal13 and marine13 radiocarbon age calibration curves 0–50,000 years CAL BP. *Radiocarbon* 55: 1869–1887.

Reimer, P.J., M.G.L. Baillie, E. Bard, A. Bayliss, J.W. Beck, P.G. Blackwell, C.B. Ramsey, C.E. Buck, G.S. Burr, R.L. Edwards, M. Friedrich, P.M. Grootes, T.P. Guilderson, I. Hajdas, T. J. Heaton, A.G. Hogg, K.A. Hughen, K.F. Kaiser, B. Kromer, F.G. McCormac, S.W. Manning, R.W. Reimer, D.A. Richards, J.R. Southon, S. Talamo, C.S.M. Turney, J. van der Plicht, and C.E. Weyhenmeye. 2009. IntCal09 and marine09 radiocarbon age calibration curves, 0–50,000 years CAL BP. *Radiocarbon* 51: 1111–1150.

Revelle, R., and H.E. Suess. 1957. Carbon dioxide exchange between atmosphere and ocean and the question of an increase of atmospheric CO_2 during the past decades. *Tellus* 9: 18–27.

Riley, W.G., D.Y. Hsueh, J.T. Randerson, M.L. Fischer, J. Hatch, D.E. Pataki, W. Wang, and M. L. Goulden. 2008. Where do fossil fuel carbon dioxide emissions from California go? An analysis based on radiocarbon observations and an atmospheric transport model. *Journal of Geophysical Research* 113.

Rodgers, K.B., S.E. Mikaloff-Fletcher, D. Bianchi, C. Beaulieu, E.D. Galbraith, A. Gnanadesikan, A.G. Hogg, D. Iudicone, B.R. Lintner, T. Naegler, P.J. Reimer, J.L. Sarmiento, and R.D. Slater. 2011. Interhemispheric gradient of atmospheric radiocarbon reveals natural variability of southern ocean winds. *Climate of the Past* 7: 1123–1138.

Saurer, M., P. Cherubini, G. Bonani, and R. Siegwolf. 2003. Tracing carbon uptake from a natural CO2 spring into tree rings: An isotope approach. *Tree Physiology* 23: 997–1004.

Schuur, E.A.G., et al. 2015. Climate change and the permafrost carbon feedback. *Nature*, 520 (7546): 171–179, doi:10.1038/nature14338

Scott, E.M., G.T. Cook, P. Naysmith, C. Bryant, and D. O'Donnell. 2007. A report on phase 1 of the 5th international radiocarbon intercomparison (VIRI). *Radiocarbon* 49: 409–426.

Shibata, S., E. Kawano, and T. Nakabayashi. 2005. Atmospheric $[^{14}C]CO_2$ variations in Japan during 1982–1999 based on ^{14}C measurements of rice grains. *Applied Radiation and Isotopes* 63: 285–290.

Slater, J., L.A. Currie, J. Dibb, and B.A.J. Benner. 2002. Distinguishing the relative contribution of fossil fuel and biomass combustion aerosols deposited at summit, greenland through isotopic and molecular characterization of insoluble carbon. *Atmospheric Environment* 36: 4463–4477.

Stephens, B., K. Gurney, P. Tans, C. Sweeney, W. Peters, L. Bruhwiler, P. Ciais, M. Ramonet, P. Bousquet, T. Nakazawa, S. Aoki, T. Machida, G. Inoue, N. Vinnichenko, J. Lloyd, A. Jordan, M. Heimann, O. Shibistova, R. Langenfelds, L. Steele, R. Francey, and A. Denning. 2007. Weak northern and strong tropical land carbon uptake from vertical profiles of atmospheric CO_2. *Science* 316: 1732–1735.

Stuiver, M., and T.F. Braziunas. 1993. Sun, ocean, climate and atmospheric $^{14}CO_2$: An evaluation of causal and spectral relationships. *Holocene* 3: 289–305.

Stuiver, M., and T.F. Braziunas. 1998. Anthropogenic and solar components of hemispheric ^{14}C. *Geophysical Research Letters* 25: 329–332.

Stuiver, M., and P.D. Quay. 1981. Atmospheric ^{14}C changes resulting from fossil-fuel CO_2 release and cosmic-ray flux variability. *Earth and Planetary Science Letters* 53: 349–362.

Suess, H.E. 1955. Radiocarbon concentration in modern wood. *Science* 122: 415–417.

Sweeney, C., E. Gloor, A. Jacobson, R. Key, G. McKinley, J. Sarmiento, and R. Wanninkhof. 2007. Constraining global air-sea gas exchange for CO_2 with recent bomb ^{14}C measurements. *Global Biogeochemical Cycles* 21.

Szidat, S., T.M. Jenk, H.-A. Synal, M. Kalberer, L. Wacker, I. Hajdas, A. Kasper-Giebl, and U. Baltensperger. 2006. Contributions of fossil fuel, biomass-burning, and biogenic emissions to carbonaceous aerosols in zurich as traced by ^{14}C. *Journal of Geophysical Research* 111.

Takahashi, T., S.C. Sutherland, R. Wanninkhof, C. Sweeney, R.A. Feely, D.W. Chipman, B. Hales, G. Friederich, F. Chavez, C. Sabine, A. Watson, D.C.E. Bakker, U. Schuster, N. Metzl, H. Yoshikawa-Inoue, M. Ishii, T. Midorikawa, Y. Nojiri, A. Kortzinger, T. Steinhoff, M. Hoppema, J. Olafsson, T.S. Arnarson, B. Tilbrook, T. Johannessen, A. Olsen, R. Bellerby, C.S. Wong, B. Delille, N.R. Bates, and H.J.W. de Baar. 2009. Climatological mean and decadal change in surface ocean pCO_2, and net sea-air CO_2 flux over the global oceans. *Deep-Sea Research Part I-Oceanographic Research Papers* 56: 554.

Tans, P.P., A.F. De Jong, and W.G. Mook. 1979. Natural atmospheric ^{14}C variation and the Suess effect. *Nature* 280: 826–828.

Telegadas, K. 1971. The seasonal atmospheric distribution and inventories of excess ^{14}C from March 1955 to July 1969. *Health and Safety Laboratory Environmental Quarterly* 243.

Townsend-Small, A., S.C. Tyler, D.E. Pataki, X. Xu, and L.E. Christensen. 2012. Isotopic measurements of atmospheric methane in Los Angeles, California, USA: Influence of "Fugitive" fossil fuel emissions. *Journal of Geophysical Research: Atmospheres* 117: n/a.

Trumbore, S.E., J.B. Gaudinski, P.J. Hanson, and J.R. Southon. 2002. quantifying ecosystem-atmosphere carbon exchange with a ^{14}C Label. *EOS transactions* 83: 265–268.

Turnbull, J.C., D. Guenther, A. Karion, C. Sweeney, E. Anderson, A.E. Andrews, J. Kofler, N.L. Miles, T. Newberger, S.J. Richardson, and P.P. Tans. 2012. An integrated flask sample collection system for greenhouse gas measurements. *Atmospheric Measurement Techniques* 5: 2321–2327.

Turnbull, J.C., H. Graven, J. Miller, S. Lehmann, and Workshop Participants. 2013. Atmospheric Radiocarbon Workshop Report. *Radiocarbon* 55: 1470–1474.

Turnbull, J.C., A. Karion, M.L. Fischer, I. Faloona, T. Guilderson, S.J. Lehman, B.R. Miller, J.B. Miller, S. Montzka, T. Sherwood, S. Saripalli, C. Sweeney, and P.P. Tans. 2011. Assessment of fossil fuel carbon dioxide and other anthropogenic trace gas emissions from airborne measurements over Sacramento, California in spring 2009. *Atmospheric Chemistry and Physics* 11: 705–721.

Turnbull, J.C., E.D. Keller, W.T. Baisden, G. Brailsford, T. Bromley, M. Norris, and A. Zondervan. 2014. Atmospheric measurement of point source fossil fuel CO_2 emissions. *Atmospheric Chemistry and Physics* (in press).

Turnbull, J.C., S.J. Lehman, J.B. Miller, R.J. Sparks, J.R. Southon, and P.P. Tans. 2007. A new high precision $^{14}CO_2$ time series for North American continental air. *Journal of Geophysical Research* 112: D11310.

Turnbull, J.C., J.B. Miller, S.J. Lehman, P.P. Tans, R.J. Sparks, and J.R. Southon. 2006. Comparison of $^{14}CO_2$, CO and SF_6 as tracers for determination of recently added fossil fuel CO_2 in the atmosphere and implications for biological CO_2 exchange. *Geophysical Research Letters* 33: L01817.

Turnbull, J.C., J.B. Miller, S.J. Lehman, D.F. Hurst, W. Peters, P.P. Tans, J.R. Southon, S.A. Montzka, J.W. Elkins, D.J. Mondeel, P.A. Romashkin, N.F. Elansky, and A. Shkorokhod. 2009a. Spatial distribution of $\Delta^{14}CO_2$ across Eurasia: Measurements from the TROICA-8 expedition. *Atmospheric Chemistry and Physics* 9: 175–187.

Turnbull, J.C., P.J. Rayner, J.B. Miller, T. Naegler, P. Ciais, and A. Cozic. 2009b. On the use of $^{14}CO_2$ as a tracer for fossil fuel CO_2: Quantifying uncertainties using an atmospheric transport model. *Journal of Geophysical Research* 114: D22302.

UNSCEAR, 2000. Sources and Effects of Ionizing Radiation. UNSCEAR 2000 Report to the General Assembly, with Scientific Annexes, 1, Annex C. United Nations, New York.

Usoskin, I.G., and B. Kromer. 2005. Reconstruction of the ^{14}C production rate from measured relative abundance. *Radiocarbon* 47: 31–37.

Van Der Laan, S., U. Karstens, R.E.M. Neubert, I.T. Van Der Laan-Luijkx, and H.A.J. Meijer. 2010. Observation-based estimates of fossil fuel-derived CO_2 emissions in the Netherlands using $\Delta^{14}C$, CO and ^{222}radon. *Tellus B* 62: 389–402.

Vogel, F.R., S. Hammer, A. Steinhof, B. Kromer, and I. Levin. 2010. Implication of weekly and diurnal ^{14}C calibration on hourly estimates of CO-based fossil fuel CO_2 at a moderately polluted site in south-western Germany. *Tellus* 62: 512–520.

Vogel, J.C., A. Fuls, E. Visser, and B. Becker. 1993. Pretoria calibration curve for short-lived samples, 1930–3350 B.C. *Radiocarbon* 35: 73–85.

Weinstock, B., and H. Niki. 1972. Carbon Monoxide Balance in Nature. *Science* 176: 290–292.

Zondervan, A., and H.A.J. Meijer. 1996. Isotopic Characterisation of CO_2 sources during regional pollution events using isotopic and radiocarbon analysis. *Tellus* 48B: 601–612.

Chapter 5
Radiocarbon in the Oceans

E.R.M. Druffel, S.R. Beaupré and L.A. Ziolkowski

5.1 Introduction

Oceanic systems are a relatively heterogeneous component of the Earth System because of the low rate of mixing compared to that of the atmosphere. Most of the mixing in the ocean occurs horizontally, though some occurs vertically. Mixing in the surface ocean (0–100 m depth) is controlled by the winds that produce currents (speed 5–50 cm s^{-1}) that are faster that those in the deep ocean (>1000 m depth, speed <5 cm s^{-1}), which are controlled by small gradients in density. Surface ocean water in the mid-ocean basins of the Atlantic, Pacific, and Indian oceans circulates laterally in anti-cyclonic, mid-ocean *gyres* that are fueled by the trade and westerly winds, the Coriolis effect, and the placement of the continents. For example, lateral currents that make up the North Pacific gyre centered at about 20°N are the North Equatorial Current (south), the Kuroshio Current (west), the North Pacific Drift (north), and the California Current (east). The movement of water in these gyres is generally lateral allowing the water to stay near the surface for a considerable time (weeks to months). Surface ocean water in the equatorial regions and on the eastern coasts of continents also moves laterally in currents, but they

E.R.M. Druffel (✉)
Department of Earth System Science, University of California, Irvine, Irvine, CA, USA

S.R. Beaupré
Department of Geology and Geophysics, Woods Hole Oceanographic Institution, Woods Hole, MA, USA

S.R. Beaupré
School of Marine and Atmospheric Sciences, Stony Brook University, Stony Brook, NY, USA

L.A. Ziolkowski
Marine Science Program and Department of Earth and Ocean Sciences, University of South Carolina, Columbia, SC, USA

E.A.G. Schuur et al. (eds.), *Radiocarbon and Climate Change*,
DOI 10.1007/978-3-319-25643-6_5

contain a component of water that originates from deeper than 100 m, because of the divergence of water at the surface. This vertical movement of water is called upwelling, and brings with it cold, high-nutrient water, such as that found off the Pacific coasts of North and South America.

Most deep water is formed in the surface region near Iceland in a mixture of cool waters from the north and saline waters from the south and is called North Atlantic Deep Water. Dry winter winds from the Canadian Arctic cause massive evaporation leaving the water very dense, causing it to sink and then move southward between the depths of 1500 and 4000 m throughout the western Atlantic (Broecker et al. 1985). This deep water mixes with Antarctic Ocean water, becomes entrained in the deep eastward current that surrounds Antarctica, and flows north into the deep Indian and Pacific oceans. This pattern of deep circulation is termed the *deep conveyor belt* (Broecker 1991).

Radiocarbon (^{14}C) has helped to understand the magnitude of transport by the conveyor and is discussed in this chapter. Most of the ^{14}C contained in seawater enters via the atmosphere through gas exchange of CO_2. After quick hydration (minutes) to carbonic acid and dissociation to bicarbonate and carbonate ions, the C is taken up by plants during photosynthesis and converted to organic matter.

This chapter begins by examining dissolved inorganic carbon (DIC), the largest reservoir of C in ocean water (36,000 Pg C). Its associated ^{14}C content is lower than that of the atmosphere and ^{14}C age measurements have been used to infer the transport rate of water in the subsurface and deep oceanic reservoirs. Particulate and dissolved organic carbon (POC, DOC) in seawater are much smaller C reservoirs (<700 Pg C). Their origin is primarily from biological processes within the ocean, including photosynthesis, respiration, and microbial utilization. A minor component originates on land and is transported to the ocean via rivers and the atmosphere. Coastal zones have higher primary production rates, and thus, their surrounding sediments are higher in organic matter content, compared to sediments beneath the gyres whose overlying surface waters are nutrient-poor and have low productivity. The turnover and cycling of organic matter in seawater and sediment is the focus of the second section of this chapter.

For much of the first part of the chapter, our interpretation of ^{14}C measurements uses either natural or bomb-produced ^{14}C for assessing ocean mixing processes on the order of annual-to-millennial timescales. In addition, accelerator mass spectrometry (AMS) now allows the use of low-level ^{14}C labeling to understand C cycling processes that occur on much shorter timescales. These new labeling methods are described for the terrestrial biota and soils in Chap. 6 and for the cycling of CH_4 in ocean water in Chap. 9. We introduce the use of ^{14}C to observe the cycling of individual organic compounds in oceanic systems, such as in DOC and sediments. In the last section of this chapter, we introduce new research that reveals the concentration and $\Delta^{14}C$ values of black carbon (BC) in DOC, which is the refractory material produced during biomass burning and fossil fuel combustion.

In summary, this chapter focuses on examples of how ^{14}C has been used to determine the cycling time of water within the world ocean and to study the sources and turnover times of organic matter within the coastal and open ocean water column and sediments.

5.2 Dissolved Inorganic Carbon Cycling in the World Ocean

5.2.1 Radiocarbon Age of the Surface Ocean

Dissolved inorganic carbon is defined as the collective abundance of CO_2 and the carbonate species that it forms upon hydration in seawater: carbonic acid (H_2CO_3), bicarbonate ion (HCO_3^-), and carbonate ion (CO_3^{2-}). The majority of this C enters the surface ocean (0–100 m depth) via dissolution of atmospheric CO_2 and is subsequently redistributed to deeper waters. This results in $\Delta^{14}C$ values that decrease with depth as a consequence of deep water (>500 m depth) isolation from the atmosphere and net radioactive decay. Eventual migration of ^{14}C-depleted deep water back to the surface and mixing with recently incorporated atmospheric CO_2 produces surface DIC that appears older than expected (i.e., lower than atmospheric $\Delta^{14}C$). This phenomenon is known as the "reservoir effect," while the *differences* between conventional ^{14}C ages of DIC and that of materials grown contemporaneously in the atmosphere are known as "reservoir ages" (Stuiver et al. 1986). Thus, the ^{14}C ages of DIC in surface waters (0–100 m depth) typically range from 300 ^{14}C years in the mid-ocean gyres to 700 ^{14}C years in the eastern tropical Pacific (Broecker et al. 1960; Bien et al. 1963; Druffel 1981; Stuiver et al. 1986). For example, the ^{14}C age of a fish living in the surface ocean is several hundred years older than that of a co-living, terrestrial organism. The reservoir age of polar surface waters is older, as old as 1400 ^{14}C years in the Southern Ocean (Rafter 1968) because of intense mixing between surface and deep (low ^{14}C) waters. These ranges are dependent on the amount of upwelling and downwelling present in the surface waters. For example, mid-gyres are dominated by downwelling, thus the water stays at the surface and accumulates $^{14}CO_2$ from the atmosphere for a longer period of time than that at equatorial regions that are dominated by divergence and upwelling of low ^{14}C subsurface waters.

In general, the abundance of ^{14}C in surface DIC is an indicator of upwelling strength and water mass movement, and only minimally controlled by CO_2 exchange with the atmosphere. At a rate of CO_2 gas exchange of 10–20 mol m^{-2} $year^{-1}$, it would take 10 years for the DIC in the surface ocean to exchange with CO_2 in the atmosphere (both reservoirs contain about the same amount of DIC, ~750 Pg C). This is much longer than the time it takes to exchange surface water with subsurface water by mixing (<1 year). This is the reason that surface DIC is hundreds of ^{14}C years old, because it has mixed with deeper, older waters that have been depleted in ^{14}C due to isolation from the atmosphere and radioactive decay. In this way, the surface ocean represents a mixture of two source pools with distinct ^{14}C: the atmosphere and the deeper ocean.

5.2.2 Bomb Radiocarbon in the Surface Ocean

As we have learned, the production of bomb ^{14}C in the late 1950s and early 1960s doubled the amount of ^{14}C in the atmosphere, which slowly decreased thereafter due to exchange with the C on land and in the surface ocean (Fig. 5.1). The maximum in the surface ocean occurred ~10 years after that in the atmosphere, demonstrating that the turnover time of $^{14}CO_2$ in the atmosphere with respect to transfer to the surface ocean was about 10 years (Broecker et al. 1982). The $\Delta^{14}C$ values rose in the surface ocean by about 200 ‰, compared to nearly 1000 ‰ in the atmosphere. The reason for this difference is the high impedance for the exchange of CO_2 across the air–sea interface; in other words, the surface ocean is so concentrated in DIC because of the formation of H_2CO_3, that it takes a long time to replace all of the DIC in the surface ocean.

5.2.3 Radiocarbon in the Deep Ocean

Early work on the distribution of $\Delta^{14}C$ measurements in deep waters of the world ocean (2000–4000 m) showed that values are highest in the northern North Atlantic (−70 ‰) (Broecker et al. 1960), lowest in the North Pacific (−240 ‰) (Bien et al. 1963), and intermediate in the deep Antarctic (−160 ‰) and Indian (−190 ‰) oceans. Radiocarbon was the definitive tool used to place a time frame on the circulation of the deep ocean and led to an early estimate for the transit time of deep water from the North Atlantic (600 ^{14}C years) to the North Pacific (2200 ^{14}C years) of approximately 1600 ^{14}C years.

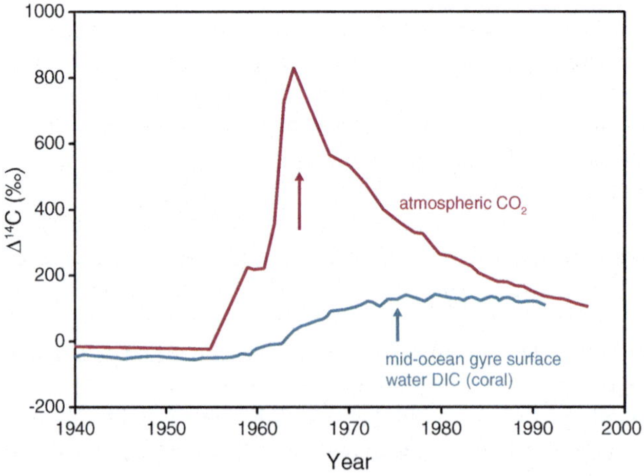

Fig. 5.1 $\Delta^{14}C$ in atmospheric CO_2 (*red*) (Nydal and Lovseth 1983; Levin and Kromer 2004) and surface corals from the southwest Pacific (*blue*) (Druffel et al. 1995). *Arrows* indicate maximum values

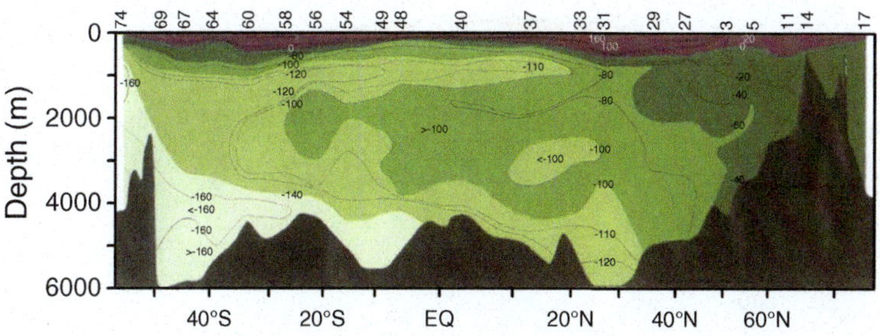

Fig. 5.2 $\Delta^{14}C$ values (‰) in the western Atlantic Ocean as measured during GEOSECS 1971–1973. Numbers at the *top* indicate station numbers and *black* area represents the seafloor. Figure modified from Stuiver and Ostlund (1980)

In the 1970s, the Geochemical Ocean Section Study (GEOSECS) was conducted to make an oceanic inventory of chemicals and radioisotopes, including ^{14}C, in the world oceans. As a result, the first detailed distribution of ^{14}C in the major oceans was produced (Ostlund and Stuiver 1980; Stuiver and Ostlund 1980). The presence of bomb ^{14}C ($\Delta^{14}C > -40$ ‰) was restricted to the upper 1000 m of the water column (except for high-latitude locations) (Fig. 5.2), illustrating the slow mixing that occurs between the surface and the deep ocean. By the 1980s, bomb ^{14}C was detected at ~30°N in the deep Atlantic, because of the fast transport of North Atlantic Deep Water southward (Ostlund and Rooth 1990; Druffel et al. 1992). Stuiver et al. (1983) used GEOSECS $\Delta^{14}C$ results and a box model to estimate that the replacement times for Atlantic, Indian, and Pacific oceans' deep waters (>1500 m depth) are about 275, 250, and 510 ^{14}C years, respectively (Stuiver et al. 1983). Attempts to quantify the distribution of bomb ^{14}C in the Atlantic and Pacific oceans had large errors because of the difficulty in separating the natural and bomb-produced portions of this isotope. Two methods were used to separate the two: the first relied on the linear correlation between dissolved silica and $\Delta^{14}C$ values (Broecker et al. 1995) and the second, more accurate, method was based on the strong linear correlation between potential alkalinity and $\Delta^{14}C$ values (Rubin and Key 2002).

More recently, Schlitzer (2007) used $\Delta^{14}C$ values (Fig. 5.3) along with chlorofluorocarbon (CFC-11 and CFC-12) measurements and a coarse-resolution global model to constrain deep and bottom water transport rates and spreading pathways. The recent Global Ocean Data Analysis Project (GLODAP) dataset of Key et al. (2004) was used for this, which contains $\Delta^{14}C$ values for nearly 1200 stations from World Ocean Circulation Experiment (WOCE) and pre-WOCE cruises. The model correctly reproduces the deep ocean ^{14}C field and the concentration gradients between different basins. Schlitzer (2007) concluded that the rates of equator-ward deep and bottom water transports from the North Atlantic and

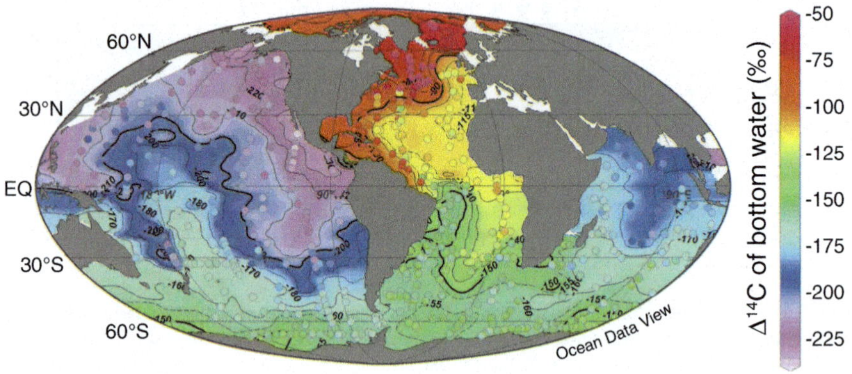

Fig. 5.3 Bottom water $\Delta^{14}C$ values simulated by a coarse-resolution global model (*color-shaded field*) and from data (*colored dots*). Figure modified from Schlitzer (2007)

Southern Ocean are of similar magnitude (16–18 × 10^6 m^3 s^{-1} or 16–18 Sverdrups), and that deep and bottom water formation in the Southern Ocean is not confined to the Weddell Sea. To put this flow into perspective, 1 Sverdrup is the combined flow of all the world's rivers to the ocean.

5.2.4 Climate Change and Radiocarbon in Corals

The ocean is a global reservoir of heat, freshwater, and CO_2, all of which are important parts of the Earth's climate system. These constituents exchange freely with the atmosphere and can cause the ocean to mix at variable (non-steady state) rates both vertically and horizontally. For example, decadal variability of salinity (salt content) was noticed in the western basins of the Atlantic Ocean (Curry et al. 2003). Between the 1950s and the 1990s, salinity decreased at all depths north of 45°N, and salinity increased in the upper few hundred meters of the water column between 40°N and 20°S. High-latitude freshening was attributed to increased river runoff to the Arctic, increased precipitation rates, and higher wind-driven exports of freshwater or ice from the Arctic to the North Atlantic (Dickson et al. 2002). Higher salinity in the tropical Atlantic was attributed to altered circulation, intensified trade winds that altered precipitation patterns, and higher evaporation due to a warmer surface ocean. These changes in the hydrologic cycle have been indicated to be the result of global warming (Curry et al. 2003).

Another example of non-steady-state mixing in the ocean involves inter-annual changes in sea surface temperature that accompany changes in climate, such as those associated with El Niño/Southern Oscillation (ENSO) events. The difference in sea surface temperature between the warm eastern equatorial Pacific and the cool western North Pacific oscillates on a multi-decadal timescale and is known as the Pacific Decadal Oscillation. When this temperature difference is large, ENSO events occur more frequently, such as in the time period from 1976 to 2002. When this temperature difference is small, ENSO events occur less often, such as in the time period from 1945 to 1976.

Radiocarbon varies in the surface ocean as a function of climate. One way to measure ^{14}C in the past ocean waters is to measure it in annually banded corals. Corals accrete aragonite (a form of calcium carbonate) using CO_3^{2-} ions in seawater DIC. Thus, the $\Delta^{14}C$ value of a coral band represents the $\Delta^{14}C$ value in DIC at the time the coral lived. Note that geochronological samples such as coral bands and tree rings are usually reported as Δ values as defined by Stuiver and Polach (1977). As long as the aragonite has not been altered or recrystallized, the $\Delta^{14}C$ value is retained and is a reliable recorder of DIC $\Delta^{14}C$. This is similar to the way a tree ring reflects the $\Delta^{14}C$ of the CO_2 in the air it used to photosynthesize organic matter. The record of bomb ^{14}C in the surface ocean is clearly evident in corals and starts to be detectable around 1957 (Druffel 1987) (Fig. 5.1). Maxima in $\Delta^{14}C$ values were achieved at different times, depending on the mixing rate of surface and deeper waters at the location of interest. Maximum $\Delta^{14}C$ values of +120–190 ‰ were found in the early 1970s for corals from the mid-ocean gyres (Fig. 5.1), whereas they were lower (+65–100 ‰) and were achieved >10 years later in the equatorial and coastal upwelling zones (Fig. 5.4). An example of how variation in climate causes changes in surface ocean ^{14}C is found in the seasonal $\Delta^{14}C$ record in a Galapagos coral (Guilderson and Schrag 1998) (Fig. 5.4). Guilderson and Schrag found that $\Delta^{14}C$ was higher than normal during El Niño events, and that winter-time $\Delta^{14}C$ values rose markedly (by 20–60 ‰) after the Pacific Decadal Oscillation climate shift of 1976; they attributed this to shallower, higher $\Delta^{14}C$ waters that mixed into the tropical Pacific after this time. Another example is the periodic appearance of low $\Delta^{14}C$ values (−55 to −60 ‰) during the past century in Bermuda corals (Druffel 1997). This study used a multibox, isopycnal

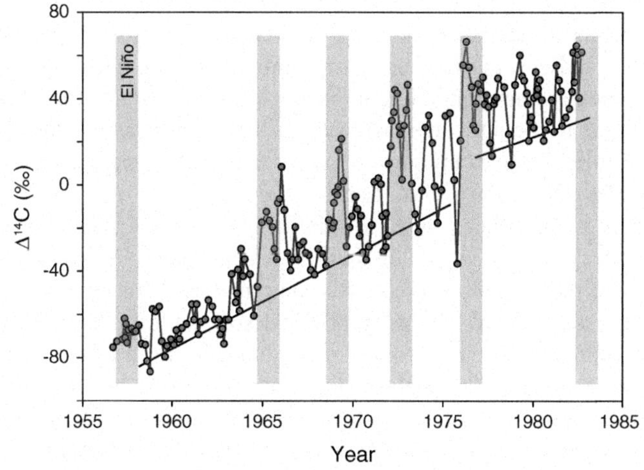

Fig. 5.4 $\Delta^{14}C$ of monthly coral samples from a Galapagos coral during the post-bomb period. *Shaded areas* indicate El Niño events. Figure modified from Guilderson and Schrag (1998)

(along lines of constant density) mixing model to demonstrate that the ventilation rate (water mass renewal rate) between the surface and the water from as deep as 700 m was unusually high during seven time periods over the past 100 years.

5.2.5 *Daily Variability of Surface Radiocarbon*

Not only does $\Delta^{14}C$ change annually in the surface ocean, it also changes on a daily basis. Surface samples collected daily during cruises to the North Pacific, the Sargasso Sea, and the Southern Ocean displayed ranges of $\Delta^{14}C$ values from 11 to 30 ‰, much larger than the total uncertainty (7.8 ‰, 2-sigma) of the measurements (McDuffee and Druffel 2007; Druffel and Griffin 2008; Druffel et al. 2010). The variability is attributed to the changes in the upper water masses; even though the ship from which samples are collected remains at the same geographic location, water moves laterally past the ship. Thus, the patchiness that exists in the DIC $\Delta^{14}C$ values of the surface ocean needs to be taken into account when interpreting a DIC $\Delta^{14}C$ measurement of a single water sample.

5.3 Organic Carbon Cycling in the Coastal and Open Ocean

Organic matter is loosely defined as any compound whose primary molecular structure is built upon a skeleton of C atoms. Despite this elemental limitation, the bonding properties of C combined with associated heteroatoms (e.g., hydrogen, oxygen, nitrogen, phosphorus, and sulfur) enable the natural formation of countless unique molecules. They are principally derived from photosynthetic reduction of CO_2 (or DIC) and ensuing metabolic transformations through progressive trophic levels. Accordingly, they range in complexity from individual structures as simple as formaldehyde to the intricate machinery of living organisms. Collectively, these organic molecules reside as mixtures in both seawater and marine sediments as two of the principle reservoirs of C on Earth.

The variety of molecules in these mixtures is both the potential key and the principle challenge to understanding their biogeochemistry. Therefore, studies have examined more manageable subsets of organic matter characterized by local environment (e.g., seawater or sediment), physical size, and/or chemical composition. While these approaches are the bases of more than 100 fruitful years of research, the addition of bulk organic ^{14}C analyses (e.g., average value of a mixture) began in earnest in the 1960s with advances toward compound-class (total carbohydrates, hydrolyzable amino acids, lipids) and compound-specific ^{14}C analyses in the 1990s. Since 2000, analytical techniques have sufficiently matured to also examine the ^{14}C content of BC, a unique molecular assemblage ubiquitous (2–4 %) among all marine organic reservoirs that may be responsible for the excessive bulk age of DOC. Here, we provide a brief introduction to these ^{14}C analyses and their associated interpretations in the canonical organic reservoirs of the sea.

5.3.1 Dissolved Organic Carbon

Marine DOC is operationally defined as the organic C in seawater that is small enough to pass through a filter with pore size diameters of ~0.2–1 μm (Fig. 5.5). The consequences of this simple sized-based definition are far reaching and can be grouped into three perspectives. First, constituents of this pool are sufficiently small to resist sinking and, in the absence of particle/solute interactions, are transported through the ocean primarily as solutes in seawater (Duursma 1961). Second, defining this material based on physical dimensions rather than molecular structure necessarily creates a chemically heterogeneous reservoir of molecules that vary in size, structure, and reactivity (Benner 2002). Third, the mass of DOC is approximately 662 Pg globally, making it the largest reservoir of reduced C in seawater (Hansell et al. 2009). Despite its biogeochemical prominence, the published body of bulk marine DOC $\Delta^{14}C$ values is presently limited to only a few regions globally (Fig. 5.6). However, these measurements have illuminated our understanding of DOC in the oceans since the first reported measurements over 40 years ago (Williams et al. 1969).

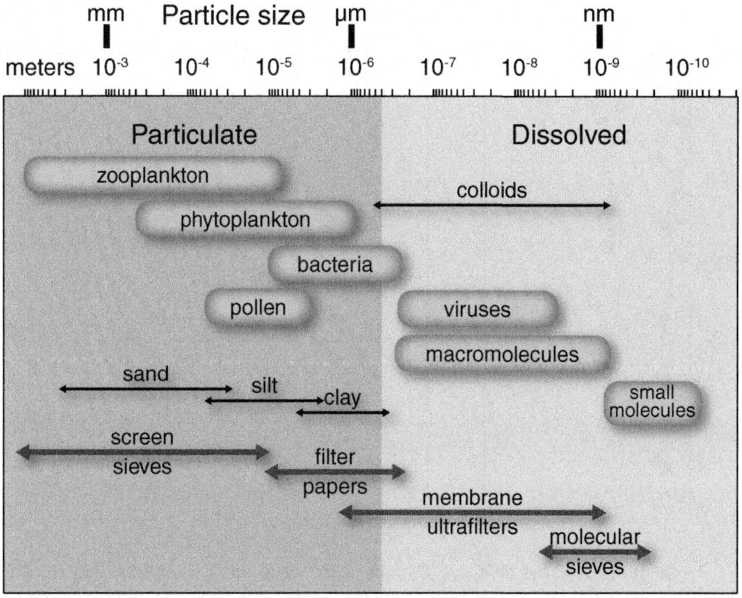

Fig. 5.5 Size spectrum of seawater constituents and the operationally defined separation between dissolved and particulate organic matter. Figure modified from Hedges (2002)

Our present view of marine DOC may be attributed to the following four characteristic patterns common to open ocean (e.g., oligotrophic) bulk DOC depth profiles (Williams and Druffel 1987; Bauer et al. 1992; Druffel and Bauer 2000). First, DOC concentrations and $\Delta^{14}C$ values are high in surface waters, decrease monotonically with depth until becoming practically invariant below 1000 m (Fig. 5.6). For example, DOC collected from the Central North Pacific in the mid-1980s decreased from 72 μM and −179 ‰ in the surface ocean to ~35 μM and −525 ‰ at depth, which translates to bulk ages spanning 1620–6020 ^{14}C years, respectively. Second, DOC $\Delta^{14}C$ profiles are almost uniformly depleted by ~300 ‰ compared to contemporaneous DIC $\Delta^{14}C$ profiles. Third, both concentrations and $\Delta^{14}C$ values are higher in the deep North Atlantic Ocean (43 μM and −390 ‰, i.e., ~4000 ^{14}C years) than the deep North Pacific Ocean (35 μM and −525 ‰, i.e., ~6000 ^{14}C years). Fourth, the ^{14}C age difference between the deep waters of these two basins (~2000 ^{14}C years) is on the order of thermohaline circulation (~1600 ^{14}C years), while the actual bulk ^{14}C ages (4000–6000 ^{14}C years) of deep DOC are more than double this timescale (Stuiver et al. 1983). Combined, these observations suggest that the majority of DOC is derived from the surface ocean, primarily by photosynthesis of ambient DIC, and then redistributed via processes similar to those that control DIC. However, the old ages of deep DOC (~4000–6000 ^{14}C years) compared to DIC (~900–2300 ^{14}C years) and their

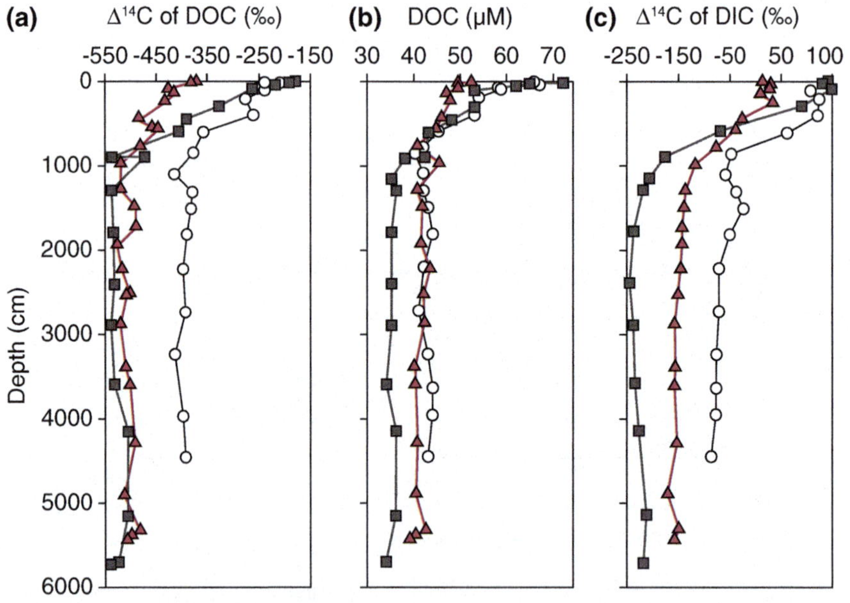

Fig. 5.6 Depth profiles of (**a**) DOC $\Delta^{14}C$ values, (**b**) DOC concentrations, and (**c**) DIC $\Delta^{14}C$ values in the North Atlantic (*white circles*), North central Pacific (*gray squares*), and Southern (*pink triangles*) Oceans. Figure modified from Bauer (2002), after Bauer et al. (1992), Druffel and Bauer (2000), Williams and Druffel (1987)

vertical uniformity over thousands of meters (Fig. 5.6) imply that a proportion of these molecules is highly persistent and has become well-mixed throughout the ocean. Consequently, the blend of relict and recently synthesized DOC in the upper ocean is consistent with the appearance of bulk ^{14}C values that are depleted relative to ambient DIC. This broad interpretation of the global dataset is consistent with oceanographic processes, demonstrates the importance of this reservoir for long-term storage of C, and highlights future research directions.

Given the variety of organic molecules and their associated reactivities, perhaps the most unexpected feature of DOC depth profiles is consistency with a simple two-component mixing model (Williams and Druffel 1987; Mortazavi and Chanton 2004; Beaupre and Druffel 2009; Beaupre and Aluwihare 2010). Each of these profiles can be explained by a highly persistent background (bg) component that is uniformly distributed with depth, to which a comparatively reactive component of recent origin was added in excess (xs). If $\Delta^{14}C$ variability due to radioactive decay is negligible compared to either the timescale of mixing or the magnitude of measurement uncertainty, then DOC concentrations ([DOC]) and $\Delta^{14}C$ measurements at any depth can be described by the following approximate equations for conservation of mass:

$$[\mathrm{DOC}] = [\mathrm{DOC_{bg}}] + [\mathrm{DOC_{xs}}] \tag{5.1}$$

$$\Delta^{14}\mathrm{C} = \left(\Delta^{14}\mathrm{C_{bg}}[\mathrm{DOC_{bg}}] + \Delta^{14}\mathrm{C_{xs}}[\mathrm{DOC_{xs}}]\right)/[\mathrm{DOC}] \tag{5.2}$$

Substituting Eq. 5.1 into Eq. 5.2 and rearranging produces a standard "Keeling plot" equation that describes the observed linear relationship between depth profiles of bulk $\Delta^{14}C$ values and corresponding inverse DOC concentrations (Keeling 1958; Mortazavi and Chanton 2004); see review in Beaupré and Aluwihare (2010) for applicability to ^{14}C, DOC, and marine systems.

$$\Delta^{14}\mathrm{C} = \left(\Delta^{14}\mathrm{C_{bg}} - \Delta^{14}\mathrm{C_{xs}}\right)[\mathrm{DOC_{bg}}](1/[\mathrm{DOC}]) + \Delta^{14}\mathrm{C_{xs}} \tag{5.3}$$

Although high coefficients of determination ($r^2 > 0.82$) on plots of $\Delta^{14}C$ versus 1/[DOC] suggest consistency with binary mixtures (Fig. 5.7), this model does not assume $\Delta^{14}C$ values or concentrations for either component. Rather, the intercept ($\Delta^{14}C_{xs}$) on a highly correlated plot of $\Delta^{14}C$ versus 1/[DOC] reveals the $\Delta^{14}C$ value of the excess component and can be used to constrain dominant sources of DOC. For example, $\Delta^{14}C_{xs}$ values from Keeling plots of open ocean profiles were similar to $\Delta^{14}C$ values of surface marine DIC, while $\Delta^{14}C_{xs}$ values from the Mid-Atlantic Bight indicated that nearshore DOC sources varied seasonally between riverine DOC ($\Delta^{14}C$ = *ca.* +200 ‰) and marine DIC ($\Delta^{14}C$ = +50–80 ‰) (Bauer et al. 2001; Mortazavi and Chanton 2004). While mathematically robust, the traditional Keeling plot is a correlation-based model that does not explicitly consider mechanisms by which DOC can be redistributed in the ocean. Alternatively, the Keeling plot may be derived based upon the mechanism of mixing various volumes (v) of water from two different end-member solutions (subscripts 1 and 2) via the corresponding approximate equations for the conservation of mass (Eqs. 5.4–5.6).

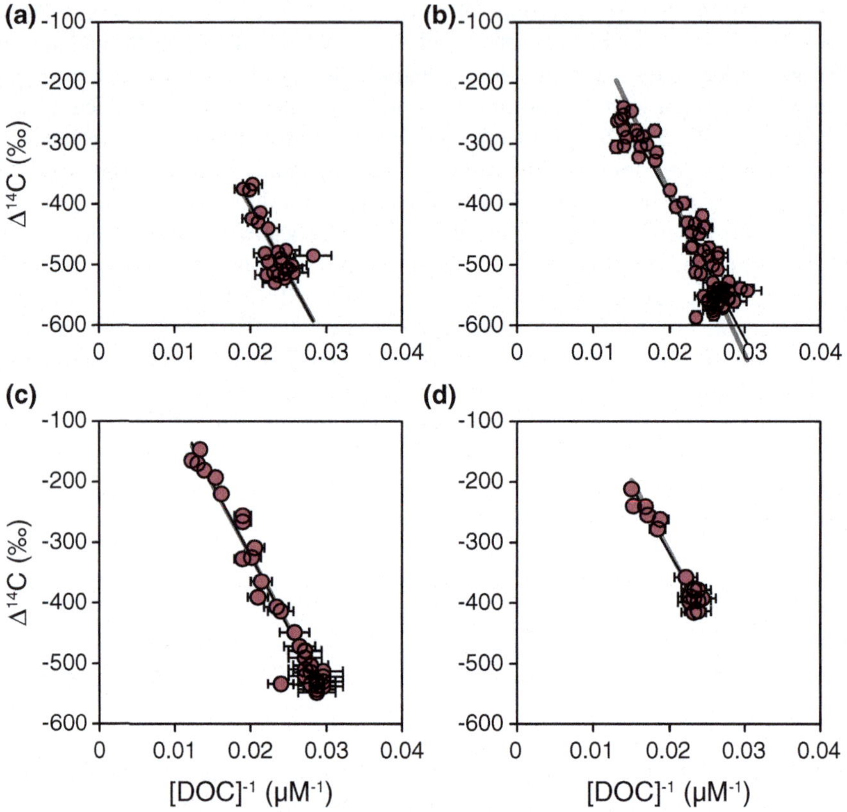

Fig. 5.7 DOC Keeling plots, geometric mean best-fit lines (*black lines*), and solution mixing models (*gray lines*) from (**a**) the Southern Ocean, (**b**) eastern North Pacific, (**c**) central North Pacific, and (**d**) Sargasso Sea. *Gray lines* predicted by solution mixing are partially obscured by the *black lines* in all panels due to their close agreement with geometric mean linear regressions. Figure modified from Beaupré and Aluwihare (2010)

$$v = v_1 + v_2 \tag{5.4}$$

$$[DOC] = ([DOC_1]v_1 + [DOC_2]v_2)/v \tag{5.5}$$

$$\Delta^{14}C = \left(\Delta^{14}C_1[DOC_1]v_1 + \Delta^{14}C_2[DOC_2]v_2\right)/([DOC]v) \tag{5.6}$$

Combining Eqs. 5.4–5.6 and rearranging reveals the linear relationship between bulk $\Delta^{14}C$ values and 1/[DOC] (Eq. 5.7) expected from mixing two solutions.

$$\Delta^{14}C = \text{slope}(1/[DOC]) + \text{intercept} \tag{5.7}$$

Although this relationship is analogous to traditional Keeling plots (Eq. 5.3), its derivation provides concordance between traditional binary mixing and a likely mechanism for DOC redistribution throughout the ocean (Fig. 5.6), i.e., circulation and mixing of unique water masses (Beaupre and Aluwihare 2010). Despite their utility in constraining DOC biogeochemistry, Keeling plots are in essence two-component models and therefore unable to reconcile fully representative age distributions of organic molecules in seawater.

More detailed age distributions and biogeochemical insights have been obtained by measuring the ^{14}C content of groups of DOC molecules defined by their physical dimensions. These DOC "size fractions" are isolated by flowing seawater through a series of ever-smaller filters (e.g., ultra filtration) and examining the organic material retained on their surfaces. This simple, yet powerful approach has two merits. First, it reveals a general pattern of decreasing $\Delta^{14}C$ values with successively smaller DOC size fractions (Fig. 5.8) which may be interpreted as evidence for the production of smaller, more persistent molecules from the degradation of larger, more reactive molecules (Guo et al. 1996). Second, filtration sufficiently concentrates DOC to permit compound-class and compound-specific ^{14}C analyses (Benner et al. 1992). For example, Repeta and Aluwihare (2006) observed $\Delta^{14}C$ values in neutral sugars of ultra-filtered DOC from 600 m depth that were slightly enriched (ranging from −133 to −108 ‰ among four different compounds, with a reported mean value of −123 ± 10 ‰) relative to ambient DIC (−155 ± 7 ‰). If these sugars were both advected to this depth (i.e., $\Delta^{14}C$ equal to ambient DIC, −155 ± 7 ‰) and delivered by sinking particles as newly synthesized molecules

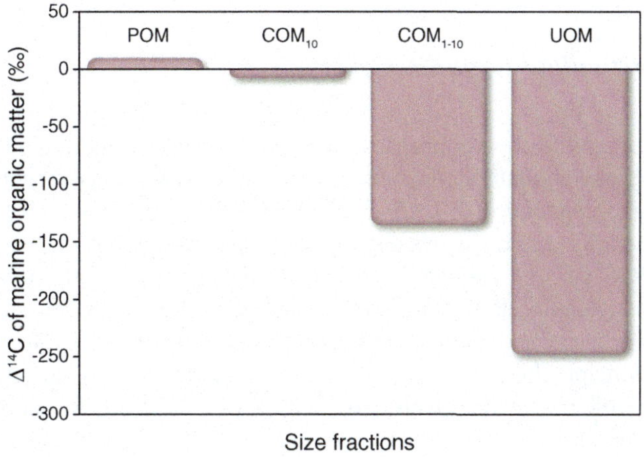

Fig. 5.8 Radiocarbon distribution among size fractions of marine organic matter collected from the Mid-Atlantic Bight, defined as follows: POM (sinking particulate organic matter, collected in sediment traps), COM_{10} (colloidal organic matter, where 10 kDa ≤ COM_{10} ≤ 0.2 μm), COM_{1-10} (colloidal organic matter, where 1 kDa ≤ COM_{1-10} ≤ 10 kDa), and ultra-filtered organic matter (UOM, <1 kDa). Note, UOM in this context refers to organic C in the unretained filtrate, with $\Delta^{14}C$ values determined by mass balance. Figure modified from Guo et al. (1996)

from the surface ocean (i.e., $\Delta^{14}C$ values equal to surface DIC, 89 ± 7 ‰), then by isotopic mass balance (e.g., Equations 5.1 and 5.2) the majority (~85 %) must have arrived by advection (Repeta and Aluwihare 2006).

In addition to illuminating the biogeochemistry of marine DOC, ^{14}C analyses have been used to constrain fluxes of DOC originating from non-marine sources (i.e., allochthonous DOC). For example, the estimated global riverine input of ~0.25 Pg DOC per year is sufficient to support the ^{14}C-based oceanic residence time of the marine DOC pool, yet molecular composition and stable C (^{13}C) isotopic values suggest the majority of marine DOC ultimately originates from primary production in the euphotic zone (Williams and Gordon 1970; Meybeck 1982; Meyers-Schulte and Hedges 1986; Opsahl and Benner 1997; Cauwet 2002). If this discrepancy cannot be dismissed by uncertainty in the calculated fluxes or by measurements that are insufficiently representative of marine DOC throughout the world ocean, then it must be attributed to either a rapid loss of terrestrially derived DOC or transformations in which the imprint of a terrestrial origin is diminished. Another interesting source of allochthonous material can be found at the bottom of the ocean. Marine DOC derived from off-axis hydrothermal vents, CH_4 hydrate bearing seeps, and hydrocarbon seeps has been observed with ages (e.g., 11,800 to 14,400 ^{14}C years) far exceeding those typically found in deep seawater (Wang et al. 2001; McCarthy et al. 2011; Pohlman et al. 2011). Estimating the influence of these allochthonous sources on the marine DOC reservoir will require additional surveys of their global distributions, lifetimes, fluid delivery rates, associated DOC concentrations and isotopic compositions, and an understanding of the degradation kinetics of the molecules released into the ocean.

5.3.2 Particulate Organic Carbon

Particulate organic carbon is primarily derived from photosynthesis in the surface ocean and is operationally defined as the organic C in seawater retained by filters with pore size diameter of ~0.45–1 μm. Like DOC, this simple size-based definition necessarily creates a chemically heterogeneous reservoir. Unlike DOC, however, POC is large enough to sink through the ocean on timescales ranging from months ("sinking POC") to years ("suspended POC") (Bacon and Anderson 1982). This property is of paramount importance to the global C cycle, because it is one of the principle fluxes of C out of the surface ocean. First, it fuels heterotrophic respiration while descending, with only ~10 % of exported POC surviving transit below the euphotic zone (~100 m) and only ~1 % surviving 1000 s of meters into the deep ocean. On long timescales, the CO_2 produced at depth by mineralization of POC contributes to the large-scale gradients of DIC concentrations that increase with depth, as well as with lateral distance from the downwelling origins of deepwater masses. Second, long-term storage of the 0.1 % of POC flux that is eventually buried in marine sediments modulates variations in atmospheric O_2 and CO_2 on geological timescales (Schlesinger 1997).

The global POC pool of 30 Pg C is only 5 % that of DOC. However, studies of temporal variability indicate that it is more dynamic. For example, most of the sinking POC collected in sediment traps moored at 3650 m depth at Station M in the North Pacific between 1993 and 1998 had post-bomb marine $\Delta^{14}C$ values (≥−50 ‰) (Hwang et al. 2004). This isotopic enrichment, even at depth, suggests that POC cycles on timescales shorter than the number of years elapsed since the marine DIC bomb spike maximum (≤30 years). Furthermore, variability in $\Delta^{14}C$ values of sinking POC (~80 ‰) collected at this depth exceeded that of surface DIC (~40 ‰) in overlying waters (Hwang et al. 2004; Masiello and Druffel 1998). This observation precludes the simplest model that assumes conservative transport of POC from the surface ocean to depth. Rather, it implies that the isotopic value of sinking POC is influenced by processes in addition to variations in surface water masses and their associated DIC $\Delta^{14}C$ values.

For example, vertical transit times based on decreasing $\Delta^{14}C$ values from surface to depth (~500–1000 ^{14}C years; Fig. 5.9) significantly exceed estimates from observed particle sinking velocities (Druffel and Williams 1990; Hwang et al. 2004). This apparent paradox reveals much about POC biogeochemistry when considering neither radioactive decay nor isotopic fractionation (see Chap. 3) during particle descent can account for the observed $\Delta^{14}C$ depletion. It can, however, be explained by an increasing proportion of ^{14}C-depleted constituents in bulk POC with increasing depth (Druffel and Williams 1990). Hypothesized mechanisms for assimilation of older C at depth include biological and abiotic exchange with constituents of the DOC pool, incorporation of ambient DIC during metabolic processes (i.e., anaplerotic reactions), and the addition of terrestrially-derived particles or resuspended sedimentary organic C that has been transported laterally from

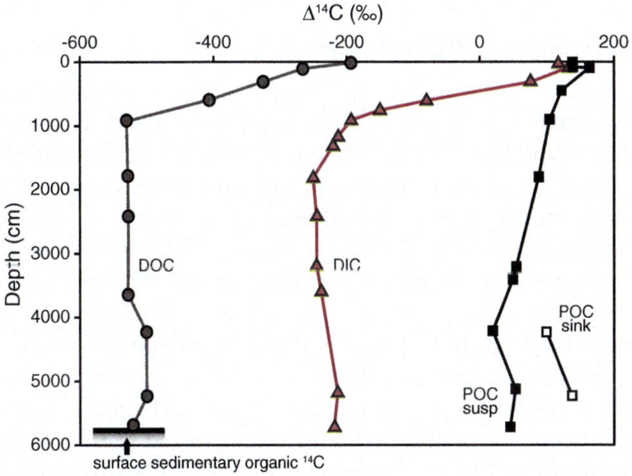

Fig. 5.9 Depth profiles of suspended POC, sinking POC, DIC, and DOC $\Delta^{14}C$ values from the North Central Pacific (31°00′N, 159°00′W), sampled in 1985 and 1987. Figure modified from Druffel and Williams (1990)

the coastal ocean (Rau et al. 1986; Druffel et al. 1992, 1996, 1998; Hwang et al. 2005, 2010). Consequently, POC $\Delta^{14}C$ depth profiles imply both compositional variability and a spectrum of chemical resilience throughout the water column.

Hwang and Druffel (2003) verified the complexity of POC cycling by examining the distribution of $\Delta^{14}C$ values among four compound classes: extractable lipids, total hydrolyzable amino acids (THAA), total hydrolyzable neutral sugars (TCHO), and an acid-insoluble fraction. The acid-insoluble fraction consisted of molecules that did not partition into the other fractions by conventional separation techniques and therefore may be considered "molecularly uncharacterized." This material had $\Delta^{14}C$ and $\delta^{13}C$ values that were simultaneously indistinguishable from those of the lipid fraction ($\delta^{13}C$ *ca.* −25 to −23 ‰, $\Delta^{14}C$ *ca.* −60 to +10 ‰) and significantly lower than the THAA and TCHO fractions ($\delta^{13}C$ *ca.* −23 to −17 ‰, $\Delta^{14}C$ *ca.* 0 to +90 ‰). Unlike all other fractions, however, the acid-insoluble material constituted an increasing proportion (>40 %) of bulk POC with depth. Combined, these results suggest that the decrease in bulk POC $\Delta^{14}C$ values with depth is likely due to the accumulation of the low $\Delta^{14}C$, lipid-like acid-insoluble fraction (Hwang and Druffel 2003). Most recently, Hwang et al. (2010) found a strong correlation between POC $\Delta^{14}C$ values and aluminum content in sinking particles, consistent with the incorporation of aged organic C associated with resuspended lithogenic sediments. Together, selective preservation of organic matter combined with the incorporation of resuspended ^{14}C-depleted sediments has been hypothesized to produce the low $\Delta^{14}C$ values of the acid-insoluble fraction of bulk POC. Additional molecular-level POC $\Delta^{14}C$ distributions and observations of organic matter provenance will help to fully explain the development of these patterns and hence the biogeochemical cycling of POC.

5.3.3 Sedimentary Organic Carbon

Sedimentary organic C is an important long-term storage pool in terms of long-term C cycling and the climate. Sedimentary organic C is the accumulation of many constituents, including POC deposited on the seafloor and porewater that contains DOC. Organic C burial is closely linked to atmospheric oxygen concentrations and the cycling of major and minor elements (Berner 1982; Hedges and Keil 1995). While marine sedimentary organic C originates from both terrigenous and marine organic matter, most of this material is ultimately removed from the water column and incorporated into the DOC pool before it can be deposited on the seafloor. To understand this removal process, the sources, sinks, and controls of organic C preservation in sediments have been widely studied. The preservation of specific compounds in sedimentary organic C, the source of old C from river systems to marine sedimentary organic C, and recycling of sedimentary organic C are specific studies that will be discussed. BC is a component of marine sedimentary organic C that, for the most part, is resistant to degradation and can thus represent a significant portion of the marine sedimentary organic C pool.

5.3.3.1 Compound-Specific Radiocarbon Studies

Processes that control the relative inputs of terrestrial- and marine-produced organic C to the sediments remain a long-standing area of study in oceanography. The pioneering study using ^{14}C to address this issue employed compound-specific ^{14}C analyses to determine the relative contributions of *n*-alkanes from ^{14}C-free petroleum ($\Delta^{14}C = -1000$ ‰), ^{14}C-free shale ($\Delta^{14}C = -1000$ ‰), and modern plant waxes ($\Delta^{14}C = 0$ ‰ and +250 ‰ in pre- and post-bomb) to sediments in the Santa Monica Basin, California (Pearson and Eglinton 2000). Using the known chemical preservation index values and $\Delta^{14}C$ values of each of the end members, the origins of *n*-alkanes in surface sediments were 12 % petroleum, 80 % plant waxes, and 8 % shale (Pearson and Eglinton 2000). The origins of *n*-alkanes in deeper sediments (2.5–7.5 cm) were 5 % petroleum, 87 % plant waxes, and 8 % shale. They modeled the $\Delta^{14}C$ value of the terrestrial end member and found that terrestrial alkanes had a residence time of decades prior to delivery to the sediments. Since then, other studies tracing the fate of terrestrial organic C in sediments using alkanes and end-member mixing models have been published. More recent work has focused on if the terrestrial plant waxes (i.e., *n*-alkanes) were pre-aged on land before being deposited to marine sediments (e.g., Kusch et al. 2010).

5.3.3.2 Terrestrial Input by Rivers

Understanding the fate of terrestrial C to marine sediments is becoming increasingly important due to global change. The fate of terrestrial C in the Arctic is important because large quantities of organic C from thawing permafrost may be remobilized and delivered to the Arctic Ocean (Frey and Smith 2005). Radiocarbon is a useful tool for tracing the fate of thawing C in the Arctic because C recently liberated from permafrost is more depleted in ^{14}C than in living plants (see Sect. 6.4.1). Siberian rivers have provided insight into C export from permafrost landscapes. Organic C in the Kolyma River mouth is more depleted in ^{14}C than sedimentary organic C collected offshore in the East Siberian Sea (−553 and −424 ‰ respectively, Vonk et al. 2010). When sedimentary organic C $\Delta^{14}C$ results were coupled with geochemical proxies that indicate C sources (i.e., *n*-alkanes and *n*-alkanoic acids), the predominant source of C was coastal erosion rather than terrestrial organic C (Vonk et al. 2010). When a two-end-member mixing model was applied to the Kolyma River system, it confirmed the geochemical proxy data in that the soil organic C from erosion ($\Delta^{14}C = -1000$ ‰) is selectively preserved over the more modern soil-derived organic C (i.e., $\Delta^{14}C = -60$ ‰). In another study on a suite of five major Siberian rivers, the $\Delta^{14}C$ of the bulk soil organic C widely varied (−74 to −609 ‰) and the $\Delta^{14}C$ values of terrigenous soil were relatively low (−500 to −818 ‰, Gustafsson et al. 2011). The rivers in western Siberia contain older terrigenous C, suggesting that western Siberia may be warming more quickly than the east.

5.3.3.3 Resuspension, Lateral Transport, and Redeposition

After deposition of organic C to the sediment, other processes such as resuspension, lateral transport, and redeposition may occur. These processes enhance remineralization of more easily degradable components because of extended exposure to oxygen. This concept was illustrated by using ^{14}C to characterize the composition of river-exported organic C and resuspended sediments along a narrow continental shelf and slope off central California (Hwang et al. 2005; Komada et al. 2005). Sedimentary organic C was separated into extractable lipids, acid–hydrolyzable, and acid-insoluble fractions. In the river, the $\Delta^{14}C$ values of the lipids (−757 to −863 ‰) and acid-insoluble fraction (−470 to −649 ‰) were lower than the coastal marine sediment lipids (−506 to −664 ‰) and acid-insoluble fraction (−236 to −299 ‰), suggesting that sedimentary shale was the primary source of C to these fractions (Fig. 5.10, Komada et al. 2005). Surface sediments further off-shore also contained material with a wide range of $\Delta^{14}C$ values. Sediment samples in the shallowest water (100 m) contained acid-hydrolyzable material that contained post-bomb C (+12 and +45 ‰), while lipids and acid-insoluble fractions at this same depth were distinctly ^{14}C-poor (−535 and −236 ‰, respectively, Hwang et al.

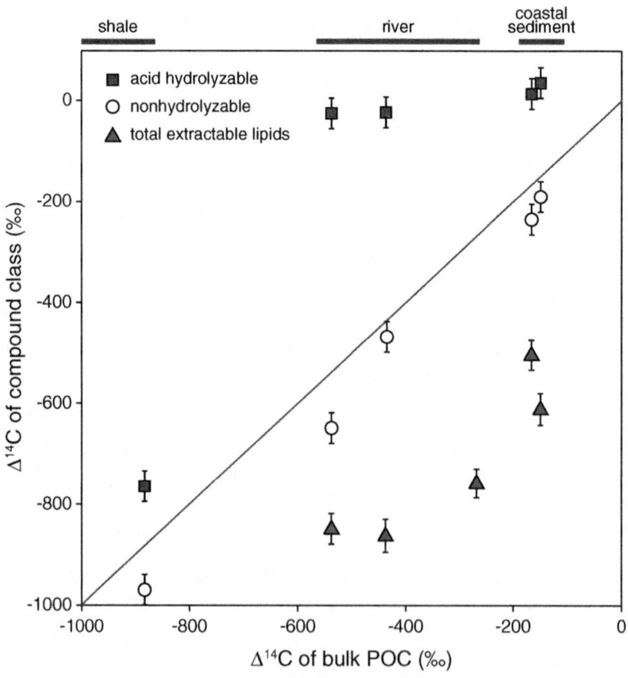

Fig. 5.10 $\Delta^{14}C$ values of organic C compound classes in shale, river POC, and coastal sedimentary organic C as a function of corresponding bulk values. *Diagonal line* indicates a 1:1 relationship. Figure modified from Komada et al. (2005)

2005). Further offshore, in deeper water, the lipids became less depleted in ^{14}C (−319 to −404 ‰) indicating a shift in source of C to the lipids (Hwang et al. 2005). Combined, these studies concluded that relict organic C exported by rivers was an important source of ^{14}C-poor lipids and acid-insoluble material to sedimentary organic C on the continental margin.

5.3.3.4 Porewater Dissolved Organic Carbon

Sedimentary porewater DOC appears to be disconnected from overlying DOC in seawater, both in its concentration and $\Delta^{14}C$ value. The concentration of DOC in sedimentary porewater is typically higher than that in the overlying water, while the difference in the $\Delta^{14}C$ of the porewater DOC and the overlying water is greatly influenced by the oxygen content of the sediment. When oxygen is abundant, the $\Delta^{14}C$ value of porewater DOC appears to be related to that of the DOC in the overlying water column. In oxic basins, sedimentary organic C is remineralized to DOC and thus, the $\Delta^{14}C$ values of sedimentary organic C and porewater DOC are similar (−232 and −220 ‰, respectively), which are higher than $\Delta^{14}C$ of the DOC in the overlying waters (Bauer et al. 1995). However, Bauer et al. (1995) showed that in an anoxic basin, the $\Delta^{14}C$ values of sedimentary organic C were higher than the porewater DOC $\Delta^{14}C$ values (−4 and −260 ‰, respectively). This difference in $\Delta^{14}C$ values was likely due to a higher proportion of SOC preserved in the anoxic sediments (Burdige 2002).

5.3.4 Black Carbon

BC is a chemically heterogeneous substance that is predominantly produced through the combustion of organic matter. Chemically, BC comprises a continuum of combustion products ranging from slightly charred, degradable biomass, to highly condensed, refractory soot. All components of this continuum are enriched in C and dominated by aromatic structures. Charred BC is formed at low temperatures (<300 °C) while soot BC is typically smaller and formed as a secondary product that condenses from hot gases produced during high-temperature (>600 °C) combustion of solid and liquid fuels. Organic matter that has undergone geologic thermogenesis will also be highly aromatic and analytically may appear as BC. The properties that affect marine BC distribution are the particle size, transport mechanism, and reactivity.

Global annual production of aerosol BC (0.014 Pg C $year^{-1}$) is composed of about equal proportions of fossil fuel and biomass burning (Kuhlbusch 1998). The atmospheric lifetime of BC aerosols ranges from hours to a month, allowing time for BC to be delivered to the most remote oceanic locations. Currently, the global BC cycle has many unknowns because the sizes of all BC pools are not thoroughly quantified and the sinks of BC are not well understood (Masiello 2004). Stored in soils and lacustrine sediments where it may slowly degrade, BC enters the ocean via aerosol and river deposition and sorption to POC (Masiello and Druffel 2001; Dickens et al. 2004; Flores-Cervantes et al. 2009). A long-term storage reservoir of BC is in oceanic sedimentary organic C, and BC is believed to be an important component in DOC (Ziolkowski and Druffel 2010; Coppola et al. 2014). Radiocarbon has been used to quantify the sources and storage terms within the global BC budget.

5.3.4.1 Sources of Black Carbon to the Ocean: Aerosols and Rivers

In the atmosphere, the relative contribution of BC from fossil fuels ($\Delta^{14}C = -1000$ ‰) or biomass burning (prebomb $\Delta^{14}C = 0$ ‰) can be determined using the BC $\Delta^{14}C$ value. In southern Asia, emission inventories suggest biofuel/biomass burning accounts for most of the aerosol C, whereas elemental composition of ambient aerosols points to fossil fuel combustion as the primary source of aerosol C. Reported $\Delta^{14}C$ values of BC determined the biomass contribution to atmospheric BC in southern Asia (Gustafsson et al. 2009). Two BC techniques were used, one that isolates more recalcitrant BC (soot BC via chemothermal oxidation) than the other (elemental BC via thermo-optical transmission). Even though the variation in the $\Delta^{14}C$ values was low (−145 to −239 ‰) for both sites, the $\Delta^{14}C$ values of the elemental BC ranged up to 200 ‰. The soot BC was consistently higher in $\Delta^{14}C$ (−92 to −267 ‰), while the elemental C values were lower in $\Delta^{14}C$ (−319 to −594 ‰). The authors found that the elemental BC and the soot BC contained 46 ± 8 % and 68 ± 6 % biomass C, respectively. The authors concluded that the source of biomass/biofuels may be larger than previously believed.

Rivers are believed to be a major source of BC to the ocean. Arctic rivers may be particularly important because of large and frequent vegetation fires and large drainage basins that can capture atmospheric BC. While this BC may be initially stored in soils, through various processes it may eventually be transported to the ocean. A recent study used BC $\Delta^{14}C$ values in estuarine sediments from five Siberian and two North American rivers to apportion soot BC between biomass burning and fossil fuel combustion (Fig. 5.11). BC in pan-arctic sediments had a wide range of $\Delta^{14}C$ values, from +51 ‰ to −905 ‰ (Elmquist et al. 2008). Assuming that BC formed from biomass burning has a $\Delta^{14}C$ value of +230 ‰ and BC formed during fossil fuel combustion has a $\Delta^{14}C$ of −1000 ‰, these pan-arctic sediments contained a wide range of BC sources. BC exported to the Arctic Ocean originated mostly (80 %) from ^{14}C-free sources such as fossil fuel and relict BC in uplifted source rocks, and a small part (20 %) from biomass burning. Sediments at the outflow of these Arctic rivers contained BC that are likely a combination of combusted fossil fuels (^{14}C-free), recently photosynthesized biomass/vegetation burned during fires (modern ^{14}C), and semi-relict biomass such as peat with some erosion of relict BC recycled in uplifted rocks.

Coastal marine sediments with significant riverine input could contain a combination of BC from combustion-derived sources and the aromatic C in uplifted

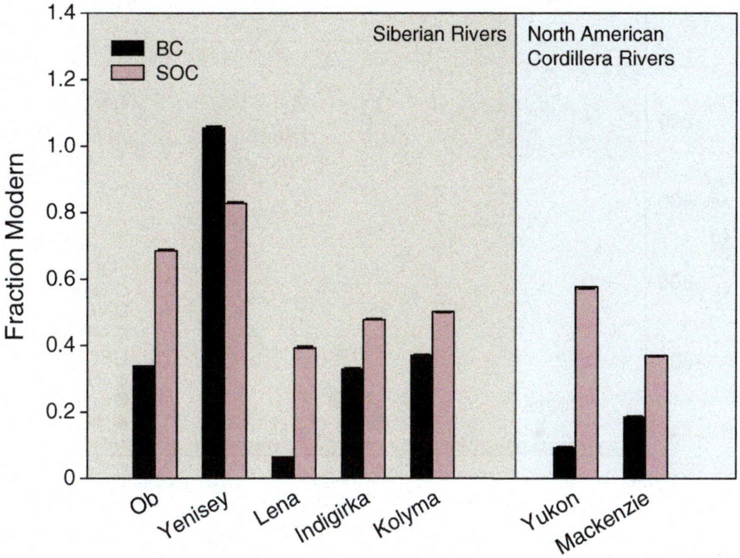

Fig. 5.11 The source classification of soot BC in or near pan-arctic rivers. To the *left* of the line are Siberian rivers while to the *right* are North American Cordillera Rivers. The Fraction Modern of soot BC is shown in *black bars* while the sedimentary organic C is in *pink bars*. Figure modified from Elmquist et al. (2008)

rocks. Unraveling the relative contribution of these two types of aromatic C is not straight forward. In one study, using a dual-isotope (^{13}C and ^{14}C) approach, it was found that the source of aromatic C in pre-industrial coastal sediment was not combustion but graphite from uplifted rocks (Dickens et al. 2004). The rock-derived material, termed graphitic BC, was terrestrially-derived and depleted in ^{14}C (Fig. 5.12). Because these sediment samples were taken from pre-industrial horizons, fossil fuel-derived soot was not a significant graphitic BC source. These findings also indicated that sedimentary BC may contain a significant portion of graphitic BC rather than combustion-derived BC.

5.3.4.2 Black Carbon in Sedimentary Organic Carbon

Because we cannot balance the global BC cycle, there is much interest in identifying the long-term storage pools of BC. One potentially important pool of BC is sedimentary organic C. Since BC is decomposed more slowly than the majority of other components in sedimentary organic C, BC can become a large proportion of the sedimentary organic C pool. On average, sedimentary BC is 6 % of sedimentary organic C but it can be up to 50 % of sedimentary organic C in some continentally influenced sites (Suman 1983; Goldberg 1985; Verardo and Ruddiman 1996). Radiocarbon has been a useful tool for studying the source of BC to open ocean

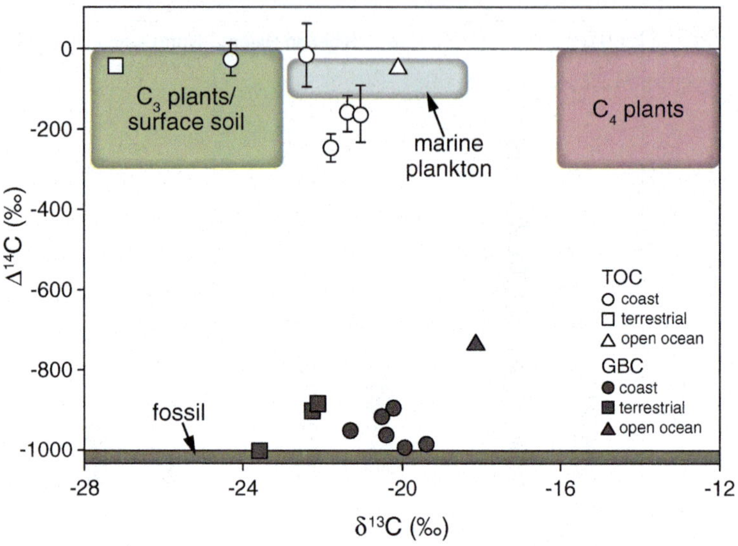

Fig. 5.12 Plot of $\Delta^{14}C$ versus $\delta^{13}C$ for all total organic C (*open symbols*) and graphitic BC (*closed symbols*) samples. Approximate ranges of the isotopic values for possible end members are indicated as *rectangles* and are based on a review of recent literature. Figure modified from Dickens et al. (2004)

sediments. At two open ocean sites that have little riverine input, the ratio of BC to non-BC sedimentary organic C increased with depth in the sediment indicating preferential remineralization of non-BC sedimentary organic C (Masiello and Druffel 1998). The fact that these sites were far from riverine input is important because historically it is generally assumed that the major source of BC to the ocean is via rivers. A significant ^{14}C age offset between the BC and the non-BC sedimentary organic C ranging from 2400 ^{14}C years (Northeast Pacific Ocean) to 13,000 ^{14}C years (Southern Ocean) (Fig. 5.13) was found in these samples (Masiello and Druffel 1998). This ^{14}C age offset suggests that the BC is pre-aged in a separate C pool before sedimentary deposition. Based on mass balances of ^{14}C and BC, Masiello and Druffel (1998) suggested that the BC is aged in the oceanic DOC pool before deposition to the sedimentary organic C.

5.3.4.3 Black Carbon in Dissolved and Particulate Organic Carbon

Based on the hypothesis that BC in marine sediment is pre-aged before deposition and the fact that bulk ^{14}C ages of deep DOC are more than double that of thermohaline circulation, there has been much interest in studying the source of BC in marine DOC. Recently, the ^{14}C content of BC was isolated from high molecular weight (>1000 Da) DOC (Ziolkowski and Druffel 2010). Using the benzene polycarboxylic acid (BPCA) method, they found that BC exported from rivers is

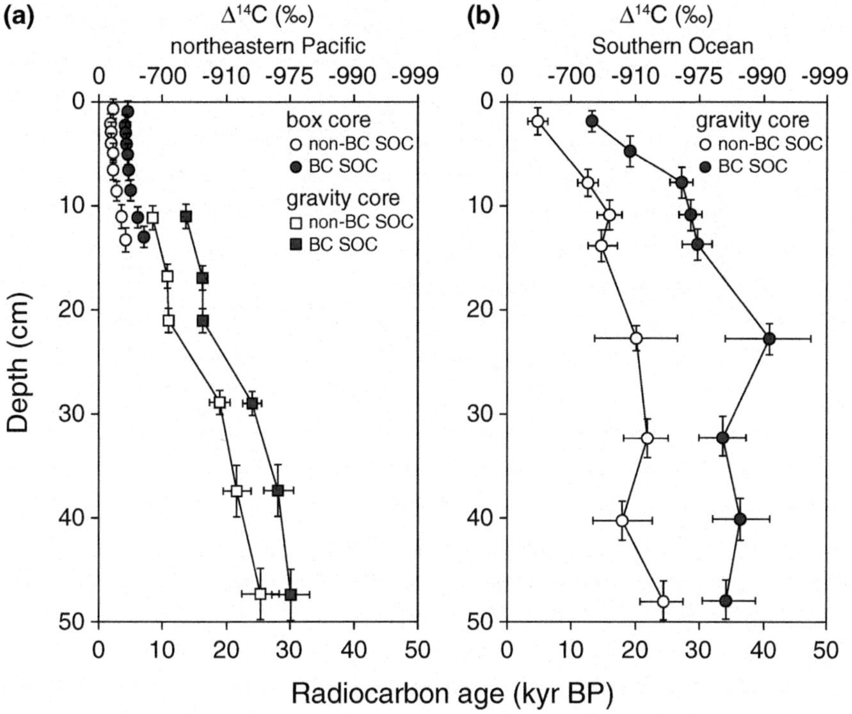

Fig. 5.13 $\Delta^{14}C$ and ^{14}C ages of BC and non-BC sedimentary organic C in deep-sea sediments from (**a**) a northeastern Pacific box core and gravity core, and (**b**) a Southern Ocean gravity core. *Error bars* are based on the reproducibility of the chromic and sulfuric acid extraction on replicate samples, and when not visible are smaller than the data point. Figure modified from Masiello and Druffel (1998)

highly aromatic and was <500 ^{14}C years old, while open ocean samples contained less aromatic BC and had an average age of 18,000 ± 3000 ^{14}C years (Fig. 5.14). The low abundance of BC in the high molecular weight DOC (0.5–3.5 %) suggests that BC is more easily degraded in the ocean than presently believed and/or the low molecular weight DOC contains a larger proportion of aged BC. The old ^{14}C age of the BC in marine DOC also suggests that there is a fossil source of aromatic C to the marine DOC pool.

Another mode of transporting BC to the sediment is via absorption to sinking POC. This process would be higher in coastal regions where the POC fluxes are high. In the Gulf of Maine, the BC flux was 1–20 % of the POC flux and all samples indicated a fossil fuel combustion component (Flores-Cervantes et al. 2009). Using ^{14}C to partition the biomass and fossil fuel sources, they found that the highest fossil fuel contribution was closest to Boston Harbor, whereas the samples with less fossil fuel contribution were further from shore (i.e., Georges Bank). Energy emissions data support their findings indicating that the contribution

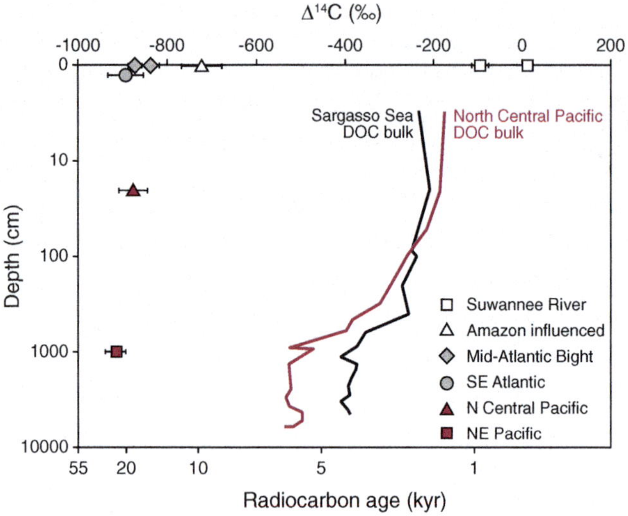

Fig. 5.14 $\Delta^{14}C$ and ^{14}C ages of BC and bulk DOC as a function of depth. The depth profiles of DOC are from the Sargasso Sea and North Central Pacific as reported in Druffel et al. (1992). Figure modified from Ziolkowski and Druffel (2010)

from biomass burning to the BC in sedimentary organic C is roughly 30 % in New England. A recent example is also found in Coppola et al. (2014) who show that POC may be the main transport mechanism of BC to sediments in the northeastern Pacific.

5.4 Conclusions and Future Directions

The use of natural and bomb ^{14}C in oceanic systems is a powerful tool for revealing the mixing time of subsurface, intermediate, deep, and bottom water masses. The use of ^{14}C in corals and shells reveals changes in the rates of water mass mixing and climate change during the past. Advances in AMS and analytical techniques are allowing for new applications of ^{14}C in marine biogeochemistry.

The observed distribution of bulk $\Delta^{14}C$ values has been integral to the development of our modern understanding of marine DOC. However, the published body of measurements is of insufficient spatiotemporal resolution to further constrain the relative importance of processes hypothesized to act upon it. Higher resolution surveys of bulk DOC, concerted isotopic and molecular characterization across all size fractions of organic matter, contemporaneous observations of solutes that trace both biogeochemical pathways and hydrography (e.g., salinity, nutrients, DIC, minerals, and radiogenic isotopes), and observations of process-induced modifications are the most promising approaches to further constrain marine DOC biogeochemistry.

Compound-class and compound-specific ^{14}C analyses of POC and sedimentary organic C have demonstrated that organic matter is often composed of isotopically heterogeneous materials that cycle on widely varying timescales and originate from different sources. While the use of compound-specific ^{14}C on marine organic matter has resulted in fewer unresolved terms, many unknowns remain. Notably, the sources and sinks of marine BC remain somewhat elusive and ^{14}C is a powerful tool for exploring the global BC cycle.

References

Bacon, M.P., and R.F. Anderson. 1982. Distribution of thorium isotopes between dissolved and particulate forms in the deep-sea. *Journal of Geophysical Research-Oceans and Atmospheres* 87: 2045–2056.

Bauer, J.E. 2002. Carbon isotopic composition of DOM. In *Biogeochemistry of Marine Dissolved Organic Matter*, ed. D.A. Hansell, and C.A. Carlson, 59–90. San Diego: Academic Press.

Bauer, J.E., P.M. Williams, and E.R.M. Druffel. 1992. C-14 activity of dissolved organic-carbon fractions in the north-central Pacific and Sargasso Sea. *Nature* 357: 667–670.

Bauer, J.E., C.E. Reimers, E.R.M. Druffel, and P.M. Williams. 1995. Isotopic constraints on carbon exchange between deep-ocean sediments and seawater. *Nature* 373: 686–689.

Bauer, J.E., E.R.M. Druffel, D.M. Wolgast, and S. Griffin. 2001. Sources and cycling of dissolved and particulate organic radiocarbon in the northwest Atlantic continental margin. *Global Biogeochemical Cycles* 15: 615–636.

Beaupre, S.R., and L. Aluwihare. 2010. Constraining the 2-component model of marine dissolved organic radiocarbon. *Deep-Sea Research Part II-Topical Studies in Oceanography* 57: 1494–1503.

Beaupre, S.R., and E.R.M. Druffel. 2009. Constraining the propagation of bomb-radiocarbon through the dissolved organic carbon (DOC) pool in the northeast Pacific Ocean. *Deep-Sea Research Part I-Oceanographic Research Papers* 56: 1717–1726.

Benner, R. 2002. Chemical composition and reactivity. In *Biogeochemistry of Marine Dissolved Organic Matter*, ed. D.A. Hansell, and C.A. Carlson, 59–90. San Diego: Academic Press.

Benner, R., J.D. Pakulski, M. McCarthy, J.I. Hedges, and P.G. Hatcher. 1992. Bulk chemical characteristics of dissolved organic matter in the ocean. *Science* 255: 1561–1564.

Berner, R.A. 1982. Burial of organic carbon and pyrite sulfur in the modern ocean—its geochemical and environmental significance. *American Journal of Science* 282: 451–473.

Bien, G.S., N.W. Rakestraw., and H.E. Suess. 1963. Radiocarbon dating of deep water of the Pacific and Indian Oceans. International Atomic Energy Agency.

Broecker, W.S. 1991. The great ocean conveyor. *Oceanography* 4: 79–89.

Broecker, W., R. Gerard, M. Ewing, and B.C. Heezen. 1960. Natural radiocarbon in the Atlantic Ocean. *Journal of Geophysical Research* 65: 2903–2931.

Broecker, W.S., and T.-H. Peng. 1982. Tracers in the Sea. Palisades, New York: Eldigio Press.

Broecker, W.S., D.M. Peteet, and D. Rind. 1985. Does the ocean-atmosphere system have more than one stable mode of operation? *Nature* 315: 21–26.

Broecker, W.S., S. Sutherland, W. Smethie, T.H. Peng, and G. Ostlund. 1995. Oceanic radiocarbon—separation of the natural and bomb components. *Global Biogeochemical Cycles* 9: 263–288.

Burdige, D.J. 2002. Sediment pore waters. In *Biogeochemistry of marine dissolved organic matter*, ed. D.A. Hansell, and C.A. Carlson, 612–652. Amsterdam/Boston: Academic Press.

Cauwet, G. 2002. DOM in the coastal zone. In *Biogeochemistry of marine dissolved organic matter*, ed. C.A. Carlson, and D.A. Hansell, 579–609. Amsterdam/Boston: Academic Press.

Coppola, A.I., L.A. Ziolkowski, C.A. Masiello, and E.A. Druffel. 2014. Aged black carbon in marine sediments and sinking particles. *Geophysical Research Letters* 41: 2427–2433.

Curry, R., B. Dickson, and I. Yashayaev. 2003. A change in the freshwater balance of the Atlantic Ocean over the past four decades. *Nature* 426: 826–829.

Dickens, A.F., Y. Gelinas, C.A. Masiello, S. Wakeham, and J.I. Hedges. 2004. Reburial of fossil organic carbon in marine sediments. *Nature* 427: 336–339.

Dickson, B., I. Yashayaev, J. Meincke, B. Turrell, S. Dye, and J. Holfort. 2002. Rapid freshening of the deep north Atlantic Ocean over the past four decades. *Nature* 416: 832–837.

Druffel, E.M. 1981. Radiocarbon in annual coral rings from the eastern tropical Pacific Ocean. *Geophysical Research Letters* 8: 59–62.

Druffel, E.R. 1987. Bomb radiocarbon in the Pacific: Annual and seasonal timescale variations. *Journal of Marine Research* 45: 667–698.

Druffel, E.R.M. 1997. Pulses of rapid ventilation in the north Atlantic surface ocean during the past century. *Science* 275: 1454–1457.

Druffel, E.R.M., and J.E. Bauer. 2000. Radiocarbon distributions in Southern Ocean dissolved and particulate organic matter. *Geophysical Research Letters* 27: 1495–1498.

Druffel, E.R.M., and S. Griffin 1995. Regional variability of surface ocean radiocarbon from Southern Great Barrier Reef corals. *Radiocarbon* 37: 517–524.

Druffel, E.R.M., and S. Griffin. 2008. Daily variability of dissolved inorganic radiocarbon at three sites in the surface ocean. *Marine Chemistry* 110: 185–189.

Druffel, E.R.M., and P.M. Williams. 1990. Identification of a deep marine source of particulate organic carbon using bomb ^{14}C. *Nature* 347: 172–174.

Druffel, E.R.M., P.M. Williams, J.E. Bauer, and J.R. Ertel. 1992. Cycling of dissolved and particulate organic matter in the open ocean. *Journal of Geophysical Research* 97: 15639–15659.

Druffel, E.R.M., J.E. Bauer, P.M. Williams, S. Griffin, and D. Wolgast. 1996. Seasonal variability of particulate organic radiocarbon in the northeast Pacific Ocean. *Journal of Geophysical Research* 101: 20543–20552.

Druffel, E.R.M., S. Griffin, J.E. Bauer, D.M. Wolgast, and X.-C. Wang. 1998. Distribution of particulate organic carbon and radiocarbon in the water column from the upper slope to the abyssal NE Pacific Ocean. *Deep-Sea Research Part II—Topical Studies in Oceanography* 45: 667–687.

Druffel, E.R.M., S. Beaupre, S. Griffin, and J. Hwang. 2010. Variability of dissolved inorganic radiocarbon at a surface site in the northeast Pacific Ocean. *Radiocarbon* 52: 1150–1157.

Duursma, E.K. 1961. Dissolved organic carbon, nitrogen and phosphorous in the sea. *Netherlands Journal of Sea Research* 1: 1–141.

Elmquist, M., I. Semiletov, L.D. Guo, and O. Gustafsson. 2008. Pan-Arctic patterns in black carbon sources and fluvial discharges deduced from radiocarbon and PAH source apportionment markers in estuarine surface sediments. *Global Biogeochemical Cycles* 22, GB2018, doi:10.1029/2007GB002994.

Flores-Cervantes, D.X., D.L. Plata, J.K. MacFarlane, C.M. Reddy, and P.M. Gschwend. 2009. Black carbon in marine particulate organic carbon: Inputs and cycling of highly recalcitrant organic carbon in the Gulf of Maine. *Marine Chemistry* 113: 172–181.

Frey, K.E., and L.C. Smith. 2005. Amplified carbon release from vast west Siberian peatlands by 2100. *Geophysical Research Letters* 32: 1–4.

Goldberg, E.D. 1985. *Black carbon in the environment: Properties and distribution*. New York: Wiley.

Guilderson, T.P., and D.P. Schrag. 1998. Abrupt shift in subsurface temperatures in the tropical Pacific associated with changes in El Nino. *Science* 281: 240–243.

Guo, L., P.H. Santschi, L.A. Cifuentes, S.E. Trumbore, and J. Southon. 1996. Cycling of high-molecular-weight dissolved organic matter in the Middle Atlantic Bight as revealed by carbon isotopic (^{13}C and ^{14}C) signatures. *Limnology and Oceanography* 41: 1242–1252.

Gustafsson, O., M. Krusa, Z. Zencak, R.J. Sheesley, L. Granat, E. Engstrom, P.S. Praveen, P.S. P. Rao, C. Leck, and H. Rodhe. 2009. Brown clouds over South Asia: Biomass or fossil fuel combustion? *Science* 323: 495–498.

Gustafsson, O., B. van Dongen, J. Vonk, O. Dudarev, and I. Semiletov. 2011. Widespread release of old carbon across the Siberian arctic echoed by its large rivers. *Biogeosciences* 8: 1445–1461.

Hansell, D.A., C.A. Carlson, D.J. Repeta, and R. Schlitzer. 2009. Dissolved organic matter in the ocean: A controversy stimulates new insights. *Oceanography* 22: 202–211.

Hedges, J. 2002. Why dissolved organics matter. In *Biogeochemistry of marine dissolved organic matter*, ed. D.A. Hansell, and C.A. Carlson, 1–33. San Diego: Academic Press.

Hedges, J.I., and R.G. Keil. 1995. Sedimentary organic-matter preservation—An assessment and speculative synthesis. *Marine Chemistry* 49: 81–115.

Hwang, J.S., and E.R.M. Druffel. 2003. Lipid-like material as the source of the uncharacterized organic carbon in the ocean? *Science* 299: 881–884.

Hwang, J., E.R.M. Druffel, S. Griffin, K.L. Smith, R.J. Baldwin, and J.E. Bauer. 2004. Temporal variability of delta C-14, delta C-13, and C/N in sinking particulate organic matter at a deep time series station in the northeast Pacific Ocean. *Global Biogeochemical Cycles* 18, doi:10.1029/2004GB002221.

Hwang, J., E.R.M. Druffel, and T. Komada. 2005. Transport of organic carbon from the california coast to the slope region: A study of delta C-14 and delta C-13 signatures of organic compound classes. *Global Biogeochemical Cycles* 19, doi:10.1029/2004GB002422.

Hwang, J., E.R. Druffel, and T.I. Eglinton. 2010. Widespread influence of resuspended sediments on oceanic particulate organic carbon: Insights from radiocarbon and aluminum contents in sinking particles. *Global Biogeochemical Cycles* 24, doi:10.1029/2010GB003802.

Keeling, C.D. 1958. Widespread influence of resuspended sediments on oceanic particulate organic carbon: Insights from radiocarbon and aluminum contents in sinking particles. *Geochimica Et Cosmochimica Acta* 13: 322–334.

Key, R.M., A. Kozyr, C.L. Sabine, K. Lee, R. Wanninkhof, J.L. Bullister, R.A. Feely, F.J. Millero, C. Mordy, and T.H. Peng. 2004. A global ocean carbon climatology: Results from Global Data Analysis Project (GLODAP). *Global Biogeochemical Cycles* 18, doi:10.1029/2004GB002247.

Komada, T., E.R.M. Druffel, and J. Hwang. 2005. Sedimentary rocks as sources of ancient organic carbon to the ocean: An investigation through delta C-14 and delta C-13 signatures of organic compound classes. *Global Biogeochemical Cycles* 19, doi:10.1029/2004GB002347.

Kuhlbusch, T. 1998. Black carbon and the carbon cycle. *Science* 280: 1903–1904.

Kusch, S., J. Rethemeyer, E. Schefuss, and G. Mollenhauer. 2010. Controls on the age of vascular plant biomarkers in black sea sediments. *Geochimica Et Cosmochimica Acta* 74: 7031–7047.

Levin, I., and B. Kromer. 2004. The tropospheric (CO_2)-C-14 level in mid-latitudes of the northern hemisphere (1959–2003). *Radiocarbon* 46: 1261–1272.

Masiello, C.A. 2004. New directions in black carbon organic geochemistry. *Marine Chemistry* 92: 201–213.

Masiello, C.A., and E.R.M. Druffel. 1998. Black carbon in deep-sea sediments. *Science* 280: 1911–1913.

Masiello, C.A., and E.R.M. Druffel. 2001. Carbon isotope geochemistry of the Santa Clara River. *Global Biogeochemical Cycles* 15: 407–416.

McCarthy, M., S. Beaupre, B. Walker, I. Voparil, T. Guilderson, and E. Druffel. 2011. Chemosynthetic origin of C-14-depleted dissolved organic matter in a ridge-flank hydrothermal system. *Nature Geoscience* 4: 32–36.

McDuffee, K., and E. Druffel. 2007. Short-term variability of dissolved inorganic radiocarbon in Sargasso Sea surface waters. *Marine Chemistry* 106: 513–518.

Meybeck, M. 1982. Carbon, nitrogen, and phosphorus transport by world rivers. *American Journal of Science* 282: 401–450.

Meyers-Schulte, K.J., and J.I. Hedges. 1986. Molecular evidence for a terrestrial component of organic-matter dissolved in ocean water. *Nature* 321: 61–63.

Mortazavi, B., and J.P. Chanton. 2004. Use of keeling plots to determine sources of dissolved organic carbon in nearshore and open ocean systems. *Limnology and Oceanography* 49: 102–108.

Nydal, R., and K. Lovseth. 1983. Tracing bomb C-14 in the atmosphere 1962–1980. *Journal of Geophysical Research* 88, No. C6: 3621–3642.

Opsahl, S., and R. Benner. 1997. Distribution and cycling of terrigenous dissolved organic matter in the ocean. *Nature* 386: 480–482.

Ostlund, H.G., and C.G.H. Rooth. 1990. The north-Atlantic tritium and radiocarbon transients 1972–1983. *Journal of Geophysical Research* 95, No. C11: 20147–20165.

Ostlund, H.G., and M. Stuiver. 1980. GEOSECS Pacific radiocarbon. *Radiocarbon* 22: 25–53.

Pearson, A., and T.I. Eglinton. 2000. The origin of N-Alkanes in Santa Monica Basin surface sediment: A model based on compound-specific delta C-14 and delta C-13 data. *Organic Geochemistry* 31: 1103–1116.

Pohlman, J., J. Bauer, W. Waite, C. Osburn, and N. Chapman. 2011. Methane hydrate-bearing seeps as a source of aged dissolved organic carbon to the oceans. *Nature Geoscience* 4: 37–41.

Rafter, T.A. 1968. Carbon-14 variations in nature. 3. 14C measurements in South Pacific and Antarctic Oceans. *New Zealand Journal of Science* 11: 551–588.

Rau, G.H., D.M. Karl, and R.S. Carney. 1986. Does inorganic carbon assimilation cause ^{14}C depletion in deep-sea organisms? *Deep-Sea Research* 33: 349–357.

Repeta, D.J., and L.I. Aluwihare. 2006. Radiocarbon analysis of neutral sugars in high-molecular-weight dissolved organic carbon: Implications for organic carbon cycling. *Limnology and Oceanography* 51: 1045–1053.

Rubin, S.I., and R.M. Key. 2002. Separating natural and bomb-produced radiocarbon in the ocean: The potential alkalinity method. Global Biogeochemical Cycles 16, doi:10.1029/2001GB001432.

Schlesinger, W.H. 1997. *Biogeochemistry: An analysis of global change.* San Diego: Adademic Press.

Schlitzer, R. 2007. Assimilation of radiocarbon and chlorofluorocarbon data to constrain deep and bottom water transports in the world ocean. *Journal of Physical Oceanography* 37: 259–276.

Stuiver, M., and H.G. Ostlund. 1980. GEOSECS Atlantic radiocarbon. *Radiocarbon* 22: 1–24.

Stuiver, M., P.D. Quay, and H.G. Ostlund. 1983. Abyssal water C-14 distribution and the age of the world oceans. *Science* 219: 849–851.

Stuiver, M., G.W. Pearson, and T. Braziunas. 1986. Radiocarbon age calibration of marine samples back to 9000 cal yr BP. *Radiocarbon* 28: 980–1021.

Suman, D.O. 1983. Agricultural burning in Panama and Central America: Burning parameters and the coastal sedimentary record, doctoral thesis, University of California at San Diego.

Verardo, D., and W. Ruddiman. 1996. Late pleistocene charcoal in tropical Atlantic deep-sea sediments: Climatic and geochemical significance. *Geology* 24: 855–857.

Vonk, J.E., L. Sanchez-Garcia, I. Semiletov, O. Dudarev, T. Eglinton, A. Andersson, and O. Gustafsson. 2010. Molecular and radiocarbon constraints on sources and degradation of terrestrial organic carbon along the Kolyma Paleoriver transect, east Siberian Sea. *Biogeosciences* 7: 3153–3166.

Wang, X., R. Chen, J. Whelan, and L. Eglinton. 2001. Contribution of "old" carbon from natural marine hydrocarbon seeps to sedimentary and dissolved organic carbon pools in the Gulf of Mexico. *Geophysical Research Letters* 28: 3313–3316.

Williams, P.M., and E.R.M. Druffel. 1987. Radiocarbon in dissolved organic matter in the central north Pacific Ocean. *Nature* 330: 246–248.

Williams, P.M., and L.I. Gordon. 1970. C-13: C-12 Ratios in dissolved and particulate organic matter in sea. *Deep-Sea Research* 17: 19–27.

Williams, P.M., H. Oeschger, and P. Kinney. 1969. Natural radiocarbon activity of dissolved organic carbon in north-east Pacific Ocean. *Nature* 224: 256–258.

Ziolkowski, L.A., and E.R.M. Druffel. 2010. Aged black carbon identified in marine dissolved organic carbon. Geophysical Research Letters 37. doi:10.1029/2010GL043963.

Chapter 6
Radiocarbon in Terrestrial Systems

E.A.G. Schuur, M.S. Carbone, C.E. Hicks Pries, F.M. Hopkins and S.M. Natali

6.1 Introduction

Terrestrial systems are arguably the most heterogeneous component of the Earth system because C is transferred from one ecosystem to another very slowly as compared to the rate of C mixing by winds in the atmosphere and by currents in the ocean. In this way, most of the ^{14}C contained in a particular terrestrial ecosystem enters via the atmosphere as plants capture CO_2 in the process of photosynthesis (primary productivity). Carbon that enters a plant is allocated toward a number of physiological processes to support respiration, growth, reproduction, and mainte-

E.A.G. Schuur (✉)
Center for Ecosystem Sciences and Society, and Department of Biological Sciences, Northern Arizona University, Flagstaff, AZ, USA

M.S. Carbone
Department of Geography, University of California, Santa Barbara, CA, USA

M.S. Carbone
National Center for Ecological Synthesis and Analysis, University of California, Santa Barbara, CA, USA

F.M. Hopkins
Department of Earth System Science, University of California, Irvine, Irvine, CA, USA

S.M. Natali
Woods Hole Research Center, Falmouth, MA, USA

E.A.G. Schuur · C.E. Hicks Pries
Department of Biology, University of Florida, Gainesville, FL, USA

M.S. Carbone
Earth Systems Research Center, University of New Hampshire, Durham, NH, USA

C.E. Hicks Pries
Climate and Ecological Sciences Division, Lawrence Berkeley National Laboratory, Berkeley, CA, USA

E.A.G. Schuur et al. (eds.), *Radiocarbon and Climate Change*,
DOI 10.1007/978-3-319-25643-6_6

nance of fungal symbionts. Each of these functions has a demand and a fate for C that in turn controls how fast those C atoms are retained within the organism. This chapter begins by examining how C, with its associated ^{14}C ratio reflective of the atmosphere at the time of photosynthesis, is allocated to wood, roots, non-structural carbohydrate storage pools, and mycorrhizae and can provide information about the life span of whole organisms and individual plant parts. Age measurements have then been used to infer the dynamics of ecosystems and of C cycling within those ecosystem as plants and their individual tissues live and die.

Once plants die that C is still contained within the ecosystem for some time period as dead organic matter. This is the fuel for other organisms, in particular microorganisms that break down organic C as an energy source, while liberating nutrients contained in dead biomass for further recycling within the ecosystem. This cycle of organic matter inputs from plants and organic matter release by microorganisms controls the amount of dead organic C stored within ecosystems. The rate of cycling can be determined by tracking the movement of ^{14}C in and out of the soil organic matter (SOM) pool. Because dead organic matter comprises a range of different plant inputs incorporated into the mineral soil matrix and modified by the action of soil biota, this turnover process has been intensively studied using ^{14}C, which is an ideal tool to understand the minutes to millennia cycling rate of different components of organic matter. The turnover and cycling of SOM is the focus of the second section of this chapter, while the third section examines the return pathway of C back out of ecosystems, focusing primarily on the largest return pathway—that of CO_2. While C inputs occur via one process, the return process—respiration—has multiple sources as nearly every living organism converts organic reduced C into oxidized C as a matter of maintaining metabolism. As a result, separating the individual sources of respiration within an ecosystem is an important topic where ^{14}C offers unique insight. This insight has also been applied to other pathways of ecosystem C loss such as fire, CH_4, and hydrologic export. These are typically much smaller pathways for ecosystem C loss, but can be important for determining the net C balance for ecosystems.

Lastly, this chapter introduces several new ^{14}C approaches. For much of the first part of the chapter, our interpretation of ^{14}C measurements and ecosystem dynamics relies on either the natural production and decay of ^{14}C for assessing long timescales, or the human-induced production of ^{14}C during the atmospheric weapons testing period for assessing years to decade timescales (Fig. 6.1). The advent of accelerator mass spectrometry (AMS) for measuring ^{14}C opened up the possibility of low-level ^{14}C labeling to understand C cycling processes occurring on shorter timescales of minutes to months. While these experiments are based on a long history of using ^{14}C labeling in biological studies, AMS allows the use of very low-level labeling, thus introducing the possibility of doing these experiments outside in the natural environment. We present several such experiments where ^{14}C was enriched within ecosystems at low levels, either accidentally through the incineration of medical waste or purposefully in ecosystem experiments either as part of CO_2 enrichment experiments, where added fossil fuel ^{14}C-depleted CO_2 became a useful isotope tracer, or in systems at ambient CO_2. In summary, this chapter focuses on how ^{14}C is used both as a tracer of source pools and for

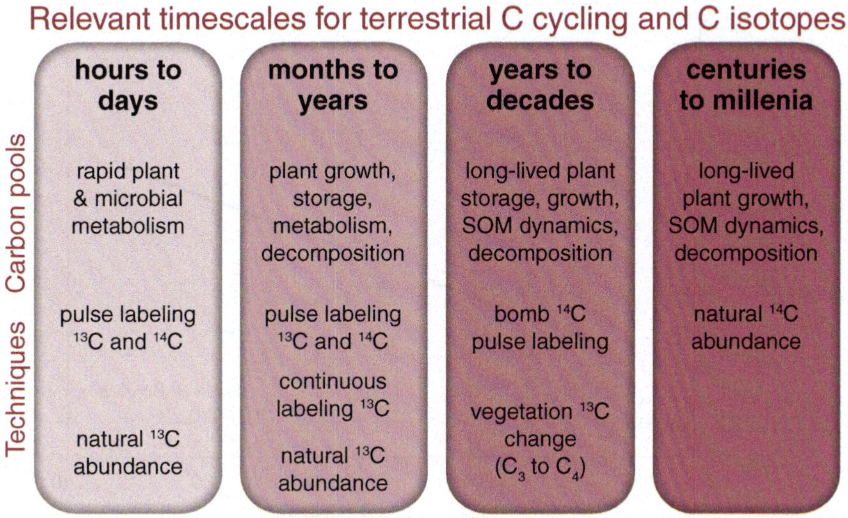

Fig. 6.1 Isotopes are used to track location and movement of C within organisms and ecosystems. Relevant timescales and approaches used with stable ^{13}C and ^{14}C depend on the rate of C movement through organism and ecosystem pools

determining age on multiple timescales, together providing a powerful approach for understanding the dynamics of terrestrial ecosystems within the Earth system.

6.2 Ages of Living Things in an Ecological Context

6.2.1 Age of Trees

Radiocarbon in tree rings is a fundamental measurement for determining atmospheric variation in ^{14}C in both the modern period (Chap. 4), and over the last ~13 kyr for calibrating calendar ages from ^{14}C ages (Chap. 7). As a result, tree ring measurements have played a basal role in shaping the field of ^{14}C science. In this section, we examine a new application for ^{14}C in tree rings—specifically to understand ages of trees or other long-lived organisms where annual growth rings are not apparent. Previous research counted annual rings to determine the age of the particular tree ring sample, and thus, the ^{14}C content of the atmosphere in that year could be determined by measurement. In this new application, the ^{14}C content of trees could be used to determine the age of the organism by comparing tree ^{14}C to historical records of atmospheric ^{14}C, and by doing so opened up new understanding into the ecology of long-lived organisms.

Some of the early work applying ^{14}C measurements to forest ecology arose from the fact that many tropical trees either lack growth rings entirely or that growth rings occur with random (non-annual) periodicity. One of the early direct measurements of a

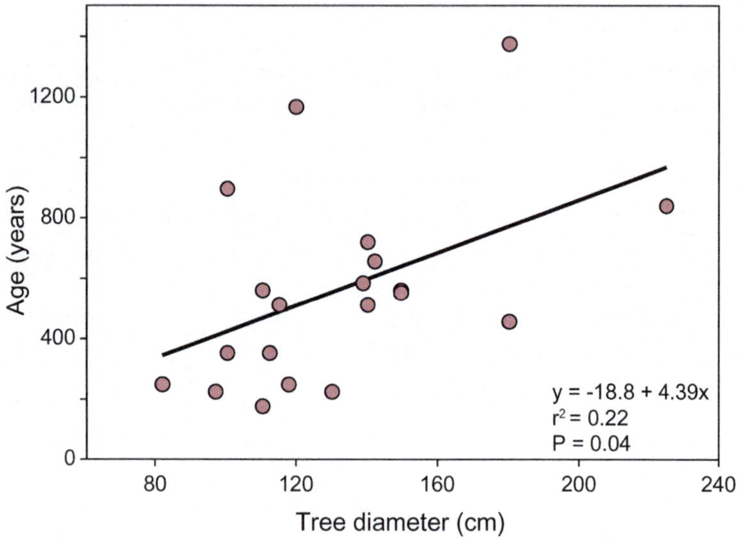

Fig. 6.2 Tree age inferred from ^{14}C measurement of the trunk center as a function of tree size for a variety of tropical tree species. Figure modified from Chambers et al. (1998)

tropical tree age using ^{14}C was that of a Brazil nut tree (*Bertholletia excelsa*) from the Amazon estimated to be 440 ± 60 years by measuring the inner most part of the tree bole (Camargo et al. 1994). By measuring the tree circumference and assuming linear growth the authors estimated the circumference growth rate to be 0.6 cm $year^{-1}$, which was about half that of other contemporary measurements based on more standard forest biometric approaches. The authors further speculated based on the size of the tree they had measured that even larger Brazil nut trees whose diameters had been reported could be as old as 800–1000 years, if the assumption of linear growth rate still applied. This, however, was significantly older than extrapolated tropical tree ages based on an earlier ^{14}C study (Mozeto et al. 1988).

The discovery of ancient tropical trees (>1000 years) was further bolstered by another study of tropical tree ages from the Amazon that reported emergent tree ages ranging from 200 years to as old as 1400 years (Chambers et al. 1998). Here, it was shown that tree size (diameter) was a weak but significant predictor of tree age (Fig. 6.2). Reported tree ages estimated with ^{14}C measurements of the center wood from tree boles significantly altered the perception of tropical tree longevity (by a factor of 0.5–2) based on contemporary measurements of growth rates or mortality rates of these giant trees. Determining ages of long-lived trees and other plants using this methodology has subsequently been applied, for example, to trees in Asian tropical forests where Dipterocarps in Borneo were estimated to be over 1000 years old (Kurokawa et al. 2003), *Acacia* spp. trees in Africa estimated to be 200–650 years old (Andersen and Krzywinski 2007), emergent trees from Central America estimated to be 200–600 years old (Fichtler et al. 2003), as well as to confirm the annual periodicity of growth rings in sagebrush (*Artemesia* spp.) in the American Southwest (Biondi et al. 2007).

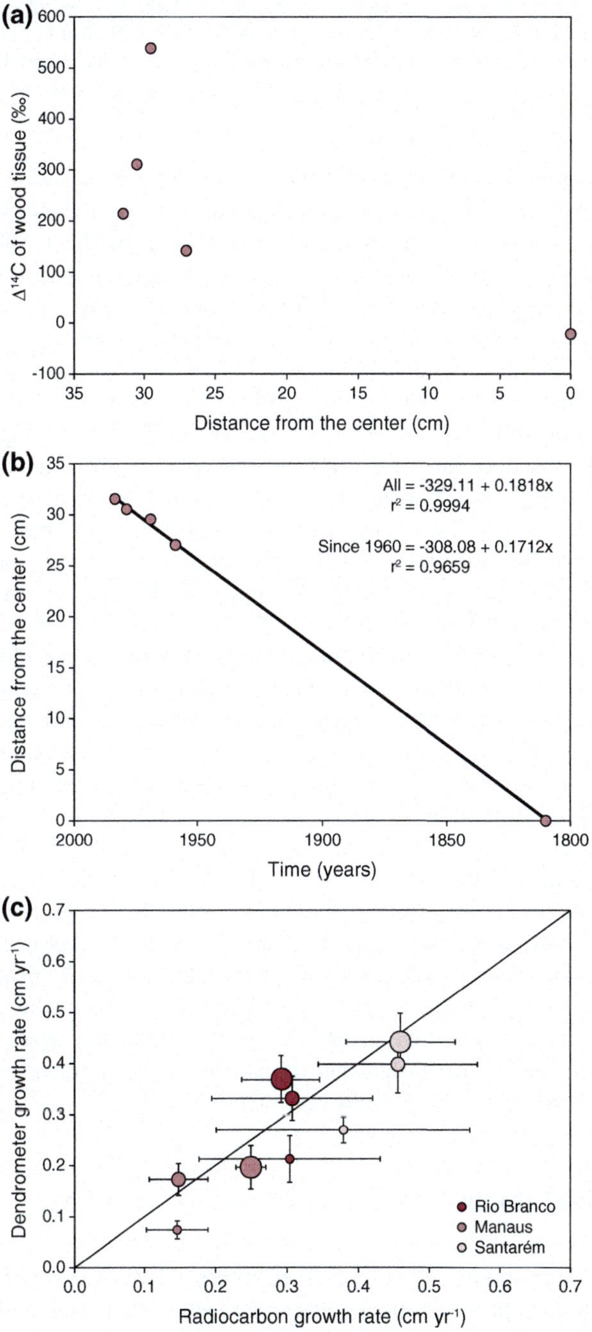

(a)
Δ14C of wood tissue (‰)
Distance from the center (cm)
(b)
All = -329.11 + 0.1818x
r² = 0.9994
Since 1960 = -308.08 + 0.1712x
r² = 0.9659
Distance from the center (cm)
Time (years)
(c)
Dendrometer growth rate (cm yr-1)
Radiocarbon growth rate (cm yr-1)
Rio Branco
Manaus
Santarém

◀ **Fig. 6.3** Tree ring measurements reflect atmospheric ^{14}C during the year of growth. (**a**) $\Delta^{14}C$ values measured at different distances along a tree core and (**b**) calibrated to calendar age allow determination of (**c**) mean tree growth rate as measured by ^{14}C, which is correlated with direct repeated measurements of tree growth. Figure modified from Vieira et al. (2005). Copyright (2005) National Academy of Sciences, USA

Determining the ages of individual ancient tropical trees was revealing but also raised questions about whole forest stand dynamics. Specifically, how common were these ancient trees (Martinez-Ramos and Alvarez-Buylla 1998)? If rare, the older ages of tropical trees would have a relatively smaller impact toward revising our understanding of overall tropical forest stand dynamics (Worbes and Junk 1999). Further studies have applied ^{14}C measurements to determine stand ages of tropical trees, focusing on the distribution of ages within a forest rather than solely on the giant emergent trees that were the target of previous studies (Vieira et al. 2005). Determining the full range of tree ages within a stand highlighted an issue that was acknowledged in the earlier tree age studies. Due to non-directional variation in atmospheric ^{14}C content during the period between 50 and 350 years BP (Chap. 4), single ^{14}C measurements from the centers of trees could not be resolved to calendar age if the actual tree age fell within this range of dates. However, calendar age for this ambiguous time period was at least partly resolved by the measurement of multiple ^{14}C values from different sections of the tree bole for trees. Because tree boles are continuously growing outward, multiple ^{14}C dates from a single tree combined with 'wiggle' matching to observed patterns of atmospheric variation provides an approach for constraining the age and growth rate of intermediate-aged trees (Fig. 6.3) (Vieira et al. 2005). Comprehensive ^{14}C measurements of tropical tree ages from whole forest stands continue to revise the understanding of tropical forest stand dynamics; 17–50 % of the trees >10 cm diameter in a Central Amazonian rain forest had ages greater than 300 years and overall averaged very slow 0.1 cm year^{-1} growth rates (Vieira et al. 2005). This showed that the few emergent trees that comprise a large proportion of the forest biomass grew relatively faster, whereas a large number of small trees grew extremely slow, attaining ages of hundreds of years old that was surprising given their small stature. Since the issue of determining tree ages of intermediate-aged trees is resolvable, it is apparent that using ^{14}C to determine the ages of long-lived organisms provides unique insight into organismal and ecosystem dynamics into almost every ecosystem where it has been applied.

6.2.2 *Age of Roots*

Beyond considering the age of individual organisms, it is also possible to consider the replacement time of particular tissues or components within single organisms using ^{14}C. For example, a tree replaces leaves and fine roots rapidly, on the order of 1 year or more depending on the species, while maintaining some components such as large roots and branches for years or decades. The turnover of these tissues (especially leaves) can be measured directly by tagging and following cohorts of tissues within a forest stand. Tagging is most effective for short-lived tissues as compared to those that are replaced more slowly, and for aboveground tissues as compared to those difficult to access belowground. Recently, there have been studies that focused on the longevity of the fine-root pool, the residence time of which impacts C inputs to the soil system. Initial work showed that fine-root ^{14}C content was elevated relative to the atmosphere (Gaudinski et al. 2001), an initially puzzling finding for a plant component that was often thought to be replaced on annual or shorter timescale based on belowground camera observations. The live fine-root pool (<2 mm diameter) was estimated to have a mean age of 3–18 years in deciduous and evergreen forest stands in the northeastern USA, with the dead fine-root pool estimated to be slightly older. Apparent root ages measured using ^{14}C could be a function of long-lived roots, or alternatively, of roots grown using stored C pools with a measurable residence time within the tree. In the Gaudinski et al. (2001) study, newly grown roots were shown to be comparable in ^{14}C content to the contemporary atmosphere, suggesting that stored C had <2-year residence time and that it was the longevity of the fine-root pool itself that led to the older calculated ages. These ^{14}C-based estimates of >1-year residence times for the fine-root pool were subsequently confirmed by a study that used ^{13}C values from a CO_2 enrichment study in an eastern US pine plantation to estimate a 4–6-year residence time for the fine-root pool (Matamala et al. 2003).

Reconciling isotope observations with results based on direct viewing of roots using belowground cameras requires separating the fine-root pool into separate root pools with differing residence times (Trumbore and Gaudinski 2003). Indeed both methods have biases in their observation of these different components of the fine-root pool. The isotope observations may underestimate the influence of the finest roots that are difficult to remove from soil cores and thus to include in the measurement, whereas the root cameras may underestimate the influence of the longer-lived roots that are slower to recover from the disturbance associated with installing belowground cameras (Strand et al. 2008). The influence of multiple pools with differing residence times is a common theme for interpreting ^{14}C measurements (Chap. 3). Essentially, if the life span of the >2-mm fine-root pool does not cluster around a mean age with normal distribution, but instead comprises a large proportion of roots living and dying rapidly along with some persistent fine roots living for many years, this then reconciles multiple observation techniques that at first seem at odds (Fig. 6.4). The hypothesis of roots of multiple ages within the fine-root pool has subsequently been supported by measurements of ^{14}C in a temperate deciduous forest that was exposed to an inadvertent ^{14}C label. The inadvertent label study is discussed in more detail in a further section of this chapter, but the results supported an average age of the fine-root pool to be several years and demonstrated that some root cohorts lived and died quickly as the label entered the dead-root pool in 18 months (Joslin et al. 2006). And while new roots were constructed from stored C pools, much of this was C with very rapid turnover (<1 year) (Gaudinski et al. 2009). This reconciliation of methods supported by a skewed distribution of root life spans has also been corroborated in a deciduous forest that used bomb ^{14}C (Chap. 4) and root cameras in the same site (Tierney and Fahey 2002), and also in a tropical forest in Brazil (Trumbore 2006).

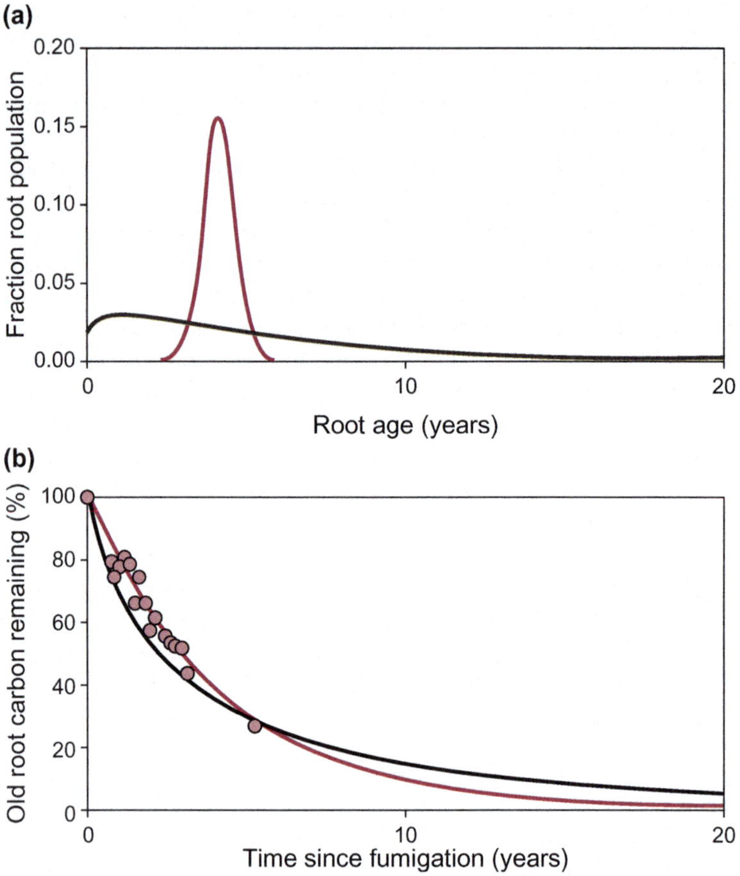

Fig. 6.4 The distribution of root age in a population affects estimates of root C dynamics. (**a**) The *red curve* assumes that root ages are normally distributed so that mean age equals mean residence time. The *black curve* has the same mean age as the *red curve*, but represents a right-skewed age distribution with many young roots and fewer old roots. This pattern is in better agreement with both minirhizotron and isotopic observations. (**b**) The two root populations would respond differently to an isotopic tracer introduced by fumigating a forest with CO_2. Figure modified from Trumbore and Gaudinski (2003)

Indeed, it makes biological sense that root longevity may correlate with other functional properties such as branching order, length, depth within the soil, or tissue nitrogen content, or that the <2-mm fine-root pool would need to be further subdivided in more diameter classes in order to understand and quantify tissue residence time. Further work along these lines have shown that smaller-diameter roots (<0.5 mm) had faster average residence times as compared to the 0.5–2 mm pool of roots (Joslin et al. 2006) and that roots deeper in the soil had older mean ages (Gaudinski et al. 2001). These multiple observations have been brought together in a model that required both two separate fine-root pools and two separate dead-root pools, with the prediction that one of the root pools had a residence time of <1 year whereas the second root pool had a residence time of $\sim$10 years to reconcile model output with observations (Riley et al. 2009). The existence of this longer-lived pool of roots suggests that a lower amount of belowground primary productivity is likely necessary to support the fine-root pool as compared to what is predicted by current models using a single pool of fine roots. Lastly, ^{14}C measurements spearheaded the advancement of our understanding of belowground tissue turnover in plants and this approach is likely to have applications to aboveground tissues as well. Residence times of leaves and fine branches are generally easier to measure with simple tagging techniques, but as in all scientific studies of relatively short duration ^{14}C might offer a fruitful approach for plant species or tissues with residence times on the order of a decade or more.

6.2.3 Consumer Organisms: Heterotrophs

Radiocarbon can be measured in living organisms other than plants, but when we consider consumers it is no longer an accurate metric of age as described in the discussion of primary producers (plants) above. Consumers have a wide diet range and thus the ^{14}C content reflects not only how long C was contained within its body, but also how long it was within the body of its prey. Radiocarbon of consumers has largely been used to interpret diet, much as how stable isotopes have been traditionally employed when prey have widely differing isotope values. A full description of ^{14}C in consumers is beyond the scope of this chapter, but some notable examples include the measurement of ^{14}C in fungi to determine which species were obtaining C from symbioses with plants and which were using soil C for growth (Hobbie et al. 2002). The former fungi had ^{14}C matching recent photosynthate, whereas the latter fungi were more enriched in bomb ^{14}C that was present in the senesced leaf litter and recently formed SOM. Another notable study detected old C from the erosion of peat C in northern Alaska entering the terrestrial food chain (Schell 1983). This was recorded in a seasonal depression of ^{14}C in fish and ducks with ^{14}C well below the atmospheric value in the year of sampling, and in some cases well below the pre-bomb level. In this case, peat C entering the food chain through aquatic freshwater invertebrates was the only source of old C that could have caused this ^{14}C value in higher consumers. While this example is likely more of an exception than a rule, it demonstrates how the age of the food source allows ^{14}C to be used as a source tracer for animal consumers.

6.2.4 Plant Mortality, Moss, and Peat Accumulation

Along with understanding the lifetime of plants and individual plant parts as described in the previous sections, ^{14}C has also been used to understand the other end of the age spectrum—as an estimate of mortality. Quantifying plant mortality can be used to understand the timing of ecological disturbance events that lead to the resetting of successional sequences that are present across all ecological systems. One example of this is from a tundra ecosystem in Alaska where permafrost has been observed to warm and thaw over the past several decades (Osterkamp et al. 2009; Schuur et al. 2009). One of the consequences of degrading permafrost is ground subsidence as ice within the soil melts and water drains away. Changes in surface hydrology have a direct impact on tundra plant community, in particular the non-vascular moss understory that is dependent on accessing water perched on the permafrost surface. Radiocarbon measurements made on dead *Sphagnum* spp. that are widespread at this site were used to determine when this hydrologic change took place, thus offering a chance to understand the timing of the disturbance several decades later. This study found a range of ^{14}C values for the dead moss at the surface. When compared with atmospheric records over the past decades, there was a clustering of ages around 1989 suggesting that permafrost degradation, if linked to moss mortality, had occurred one and a half decades prior to sampling (Fig. 6.5). This was validated by direct measurements of permafrost temperature measured at the same site since 1985.

Quantifying the age of moss was then used in a somewhat different way to assess the severity of an unprecedented tundra fire in Alaska that burned 1039 km^2 in the summer and fall of 2007 (Mack et al. 2011). This single fire tripled the cumulative area burned on the north slope of Alaska since 1950 when record collecting began

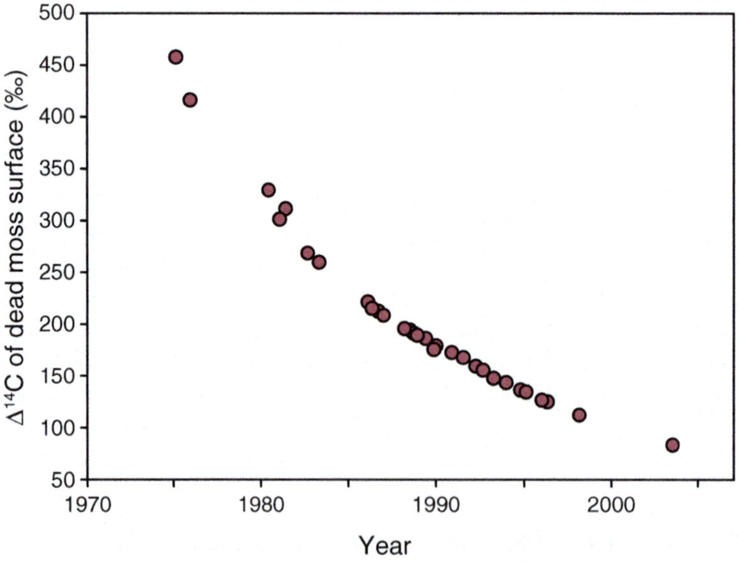

Fig. 6.5 $\Delta^{14}C$ values of surface *Sphagnum* moss that had died in response to permafrost thaw. Measurements are represented by the year where Northern Hemisphere atmosphere had matching ^{14}C values. Figure modified from Osterkamp et al. (2009)

and from remote sensing appeared to be of unprecedented severity for a tundra fire. Here, ^{14}C measurements of the remaining moss/peat layer exposed following combustion were used to estimate the time required to accumulate the soil C in those combusted organic soil layers. These measurements provided unique insight into the nature of this disturbance event. In contrast to the idea that this unusual fire combusted millennial-aged soil C that exists in deep peat layers in tundra ecosystems, isotope measurements revealed bomb C in all exposed moss/peat samples, with an average year of formation to be 1970. While this fire was unprecedented in area burned, ^{14}C measurements clearly showed that it did not burn down into ancient peat.

Lastly, this same approach of estimating year of death was also applied in a vastly different context—a tropical rainforest in Panama where ^{14}C in the outermost wood was measured in trees where the year of death was known, ranging from 1970 to 1989 (Horvitz and Sternberg 1999). This study found that ^{14}C content was negatively correlated with year of death due to declining ^{14}C in the atmosphere (Chap. 4). Significant local differences in this decline were noted when compared to tree ring records from other locations due to spatial differences in the atmospheric ^{14}C content that are especially pronounced in the post-bomb period. Tree death measurements are useful for quantifying the rate of patch dynamics that are the dominant disturbance agent in these forests.

Moss mortality can also be followed through time in peat- or moss-dominated ecosystems due to vertical accumulation of organic soil layers that preserves a record of mortality events over time. This application is a bridge between understanding ecological dynamics of long-lived organisms to understanding SOM turnover presented in more detail in the next section. More recent isotope studies estimating soil C accumulation and turnover in moss-dominated ecosystems are based on a relatively long history of use in peatland studies. Here, the issue of the heterogeneous nature of SOM is usually avoided by measuring ^{14}C in plant macrofossils, often the recognizable remains of poorly decomposable *Sphagnum* spp. mosses (Trumbore and Harden 1997; Blaauw 2010). Rather than only measuring surface moss samples, as in the tundra examples earlier, vertical accumulation rate of moss/peat is quantified by measuring the ^{14}C content of individual $\sim$1-cm soil layers extending down to the base of the peat deposit. This approach can be roughly separated into those studies of the upper moss/peat layer that may extend several decades to several hundred years, often used as a metric of C accumulation, from those studies focused on the long-term (centuries to millennia) variation in the accumulation of peat tied to climatic fluctuations over the Holocene. Both of these timescales benefit from the 'wiggle' matching approach that helps correlate individual moss ^{14}C measurements to past atmospheric ^{14}C since the vertical accumulation of moss provides a known sequence through time. While this work has a longer history in peatland studies, the main use of ^{14}C in those systems often has been primarily as a dating tool over centuries to millennia to understand the timing of peatland expansion (e.g., Gorham 1991; Smith et al. 2004), vegetation change (e.g., Kuhry 2008), climatic fluctuations (e.g., Barber et al. 2003; Blaauw and Christen 2005; Ellis and Rochefort 2006; Barber and Langdon 2007), and disturbances (e.g., Kuhry 1994; Tolonen and Turunen 1996). Determining the age of samples is the focus of Chap. 7; thus, these topics will not be covered in detail in this chapter.

The aspect of peatland research most tied to understanding ecosystem dynamics focuses on using ^{14}C to determine the net accumulation rate of C in peat soil. Initial

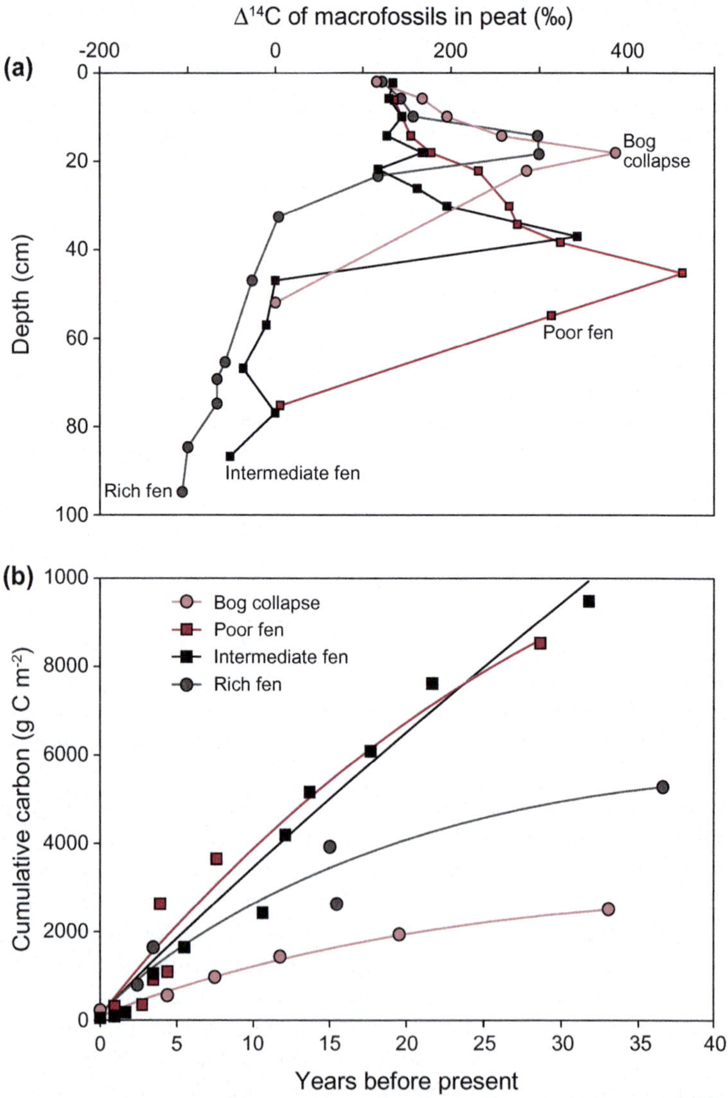

Fig. 6.6 (**a**) Depth profiles of $\Delta^{14}C$ values of macrofossils picked from peat at four wetland sites. (**b**) Total amount of C accumulated versus age estimated from ^{14}C of the depth above which the C was inventoried. Figure modified from Trumbore et al. (1999)

accumulation estimates were based on the basal age of the peat deposit divided by the total vertical height of organic matter and ranged from 0.2 to 0.8 mm $year^{-1}$ across a broad range of peatlands in Eurasia and North America (Gorham 1991). However, it was recognized that a single average accretion rate over millennia was not likely to describe the decrease in accumulation rate through time, and a more sophisticated model that accounted for inputs and losses (Chap. 3, Eq. 3.20) was used to describe decreasing peat accumulation over time (Clymo 1984). Calculations based on this model show that modern accumulation rates are lower than the past because of loss of C from deeper in the peat profile offsetting, in part, C gains at the surface (Gorham 1991; Tolonen and Turunen 1996). More recent work has extended this approach to assess both C inputs (plant growth) to and outputs (decomposition) from moss-dominated boreal forests and wetlands (Trumbore and Harden 1997; Trumbore et al. 1999). Those studies used multiple measurements of ^{14}C throughout the organic soil profile to determine the inputs and outputs for the organic soil and of net C accumulation when combined with estimates of deep C losses (Fig. 6.6). Net C accumulation values ranging from 25 to 180 g C m^{-2} $year^{-1}$ from boreal bogs and fens calculated with ^{14}C were similar to estimates based on biometrics, but higher than gas exchange measurements from those same systems. These accumulation rates are similar to ^{14}C measurements from a tundra ecosystem in Alaska where estimates of net C accumulation from replicate cores averaged 14.4 g C m^{-2} $year^{-1}$ (Hicks Pries et al. 2013). Here, net C accumulation rates were generally similar to gas exchange measurements. Radiocarbon estimates C accumulation rate averaged over multiple decades rather than for any given year. This can provide an important perspective of ecosystem C exchange for comparison with direct C exchange measurements that typically occur on a shorter timescale of years.

6.3 Soil Organic Matter Dynamics

Vertical accumulation as described for peatlands, however, is more the exception than the rule for SOM dynamics in terrestrial ecosystems. In most ecosystems, SOM is best viewed as an open system (Chap. 3) where new C is constantly being added via photosynthesis and organism death, while older C is constantly removed by microbial decomposition. These processes operate concurrently on the bulk SOM pool over a range of timescales from minutes to millennia. The exchange rate of C from organic soil pools is difficult to quantify due to this wide range of timescales, and ^{14}C provides one of the few tools for determining in situ soil C dynamics in both disturbed and undisturbed ecosystems. Measurements of natural abundance ^{14}C can be used to estimate dynamics of stable, long-term SOM pools that are old enough to have measurable radioactive decay, while the incorporation of more recent bomb ^{14}C into SOM pools provides a sensitive indicator of C exchange rates with the atmosphere that has occurred on annual to decadal timescales (Fig. 6.1). This section discusses the application of core mathematical models (Chap. 3) to estimate C residence time in soils and addresses the challenges posed by the dynamic and highly heterogeneous nature of C contained in SOM pools.

6.3.1 Heterogeneity of Soil Organic Matter Pools

Meaningful estimates of C residence time based on ^{14}C measurements and mathematical models require the separation of heterogeneous SOM into different pools that are relatively homogeneous in terms of decomposition rates. Soil heterogeneity affects flux and residence time estimates because the *average age* of a C atom in a heterogeneous pool is not the same as the *residence time* spent in that pool (Chap. 3 , Fig. 6.7). The heterogeneity of the total SOM pool can be observed by comparing ^{14}C content of bulk soil C with that of physically or chemically separated SOM fractions or with the ^{14}C content of CO_2 respired from laboratory incubations of soil (Fig. 6.8). While residence times of SOM lie along a time continuum from minutes

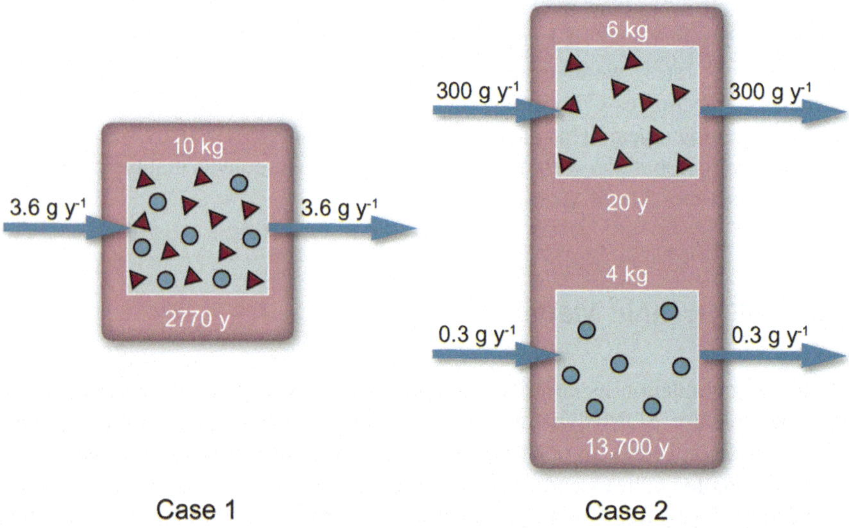

Fig. 6.7 Soil heterogeneity affects flux and residence time estimates. If a soil C pool containing 10 kg C m^{-2} with residence time of 2770 years (estimated from ^{14}C) is modeled as a homogeneous reservoir (*Case 1*), the annual flux in/out of this reservoir at steady state would be 3.6 g m^{-2} $year^{-1}$. However, if modeled as a two-component mixture with 60 % of the C in one pool and 40 % in the other (*Case 2*), the larger pool would have a residence time of 20 years and the smaller pool a residence time of 13,700 years. While the overall measured ^{14}C content does not change, annual flux in *Case 2* would be 300.3 g m^{-2} $year^{-1}$ and age of respired CO_2 ~20 years. Figure modified from Torn et al. (2009)

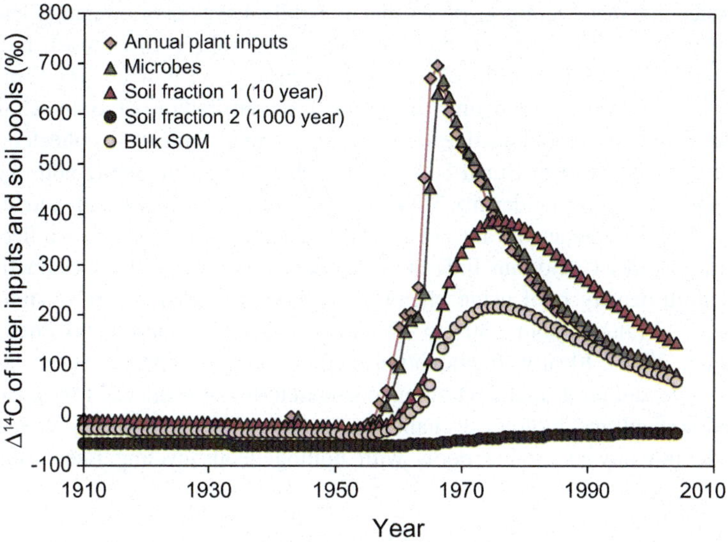

Fig. 6.8 Incorporation of $\Delta^{14}C$ in litter (*pink diamond*), bulk SOM (*pink circle*), SOM fractions with 10 (*pink triangle*) and 1000 year (*gray circle*) residence times, and $\Delta^{14}CO_2$ of microbial respiration (*gray triangle*). Figure modified from Trumbore (2009)

to thousands of years, ecosystem C models often group SOM into two or three broad pools (e.g., Parton et al. 1987; Jenkinson et al. 1991). These models typically include an *active* pool with a 1–2-year residence time, conceptualized as including root exudates, fresh plant litter, and microbial cell contents. Next, models include a *slow* pool with decadal-scale residence time and comprising plant structural components and organic compounds associated with soil aggregates, followed by a *passive* pool with a residence time on the order of hundreds to thousands of years and comprising stabilized organic matter strongly associated with mineral surfaces (Trumbore 2009).

One challenge in estimating SOM turnover is determining the best method to effectively separate SOM into these conceptual pools with relatively homogeneous residence times. Most studies accomplish this through a combination of physical and chemical fractionation methods, as well as incubations, which can be used to isolate active and slow-cycling C pools (Torn et al. 2005; Trumbore et al. 1989; Trumbore 1993, 2009; Trumbore and Zheng 1996; Gaudinski et al. 2000; Olk and Gregorich 2006; Paul et al. 2006; von Lutzowa et al. 2007; Castanha et al. 2008). Physical fractionation by means of density separation isolates high-density fractions that generally contain slower-cycling C from low-density organic matter that, with the exception of charcoal, turns over on annual to decadal timescales. The high-density C fractions associated with soil minerals can have a wide range of residence times and are often further separated using chemical fractionation methods. Low-density fractions and respired CO_2 from incubations both are dominated by fast-cycling C pools that typically have incorporated a large proportion of bomb ^{14}C and thus have positive $\Delta^{14}C$ values. In contrast, bulk soils, particularly deep in the profile, or separated fractions with slow-cycling C may have negative $\Delta^{14}C$ values reflecting the dominant contribution of older C that has undergone extensive radioactive decay. For example, bulk soil collected from a mature tropical forest in Brazil had a $\Delta^{14}C$ value of −88 ‰, while both incubated CO_2 (+160 ‰) and low-density SOM (+170 ‰) from the same soil were enriched in ^{14}C. The ^{14}C-depleted value of the high-density pool (−102 ‰) reflects slow turnover of older C and thus little incorporation of bomb ^{14}C. Further fractionation of the high-density pool using chemical methods revealed several distinct pools whose $\Delta^{14}C$ ranged from −305 to +72 ‰ (Fig. 6.9) (Trumbore and Zheng 1996; Trumbore 2009). Ideally, fractionation methods aim to separate SOM into functionally relevant pools that are homogeneous in terms of residence times. However, methodological and biological challenges persist, and operationally defined fractions do not always isolate pools with homogeneous cycling rates. But, taken

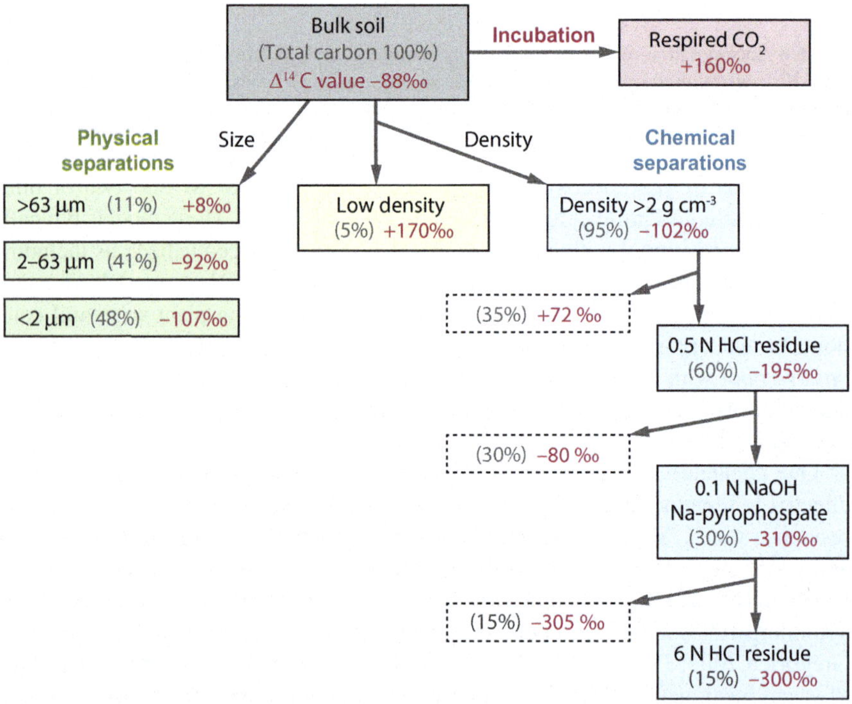

Fig. 6.9 Examples of different $\Delta^{14}C$ values for fractions of SOM separated by size, density, and chemistry. *Pink numbers* are $\Delta^{14}C$ values (‰), and *gray numbers* are proportion (%) of the total soil organic matter pool for each fraction. Data are from Trumbore and Zheng (1996) for mature forest Oxisol soils collected in 1992 from Paragominas, Brazil, at 30–35 cm depth. The $\Delta^{14}C$ for atmospheric CO_2 in the year of sampling was +143 ‰. Figure modified from Trumbore (2009)

together, SOM fractionation techniques in combination with ^{14}C modeling can be used to track C residence times in bulk soil pools (e.g., Torn et al. 2002, 2005) and in individual soil fractions (Trumbore 1993; Trumbore and Zheng 1996; Baisden et al. 2002; Fissore et al. 2009).

6.3.2 Modeling Soil Carbon Residence Times with Natural Abundance ^{14}C

Models incorporating natural abundance ^{14}C (Chap. 3) can be used to determine SOM turnover on centennial to millennial timescales since the recent input of bomb ^{14}C to the atmosphere has had little impact on these C pools with very slow exchange. Under steady-state conditions (C inputs equal outputs; Chap. 3), ^{14}C content of a homogeneous SOM pool reflects the rate of decomposition relative to radioactive decay (Torn et al. 2009). Turnover time, defined as the reciprocal of the first-order decomposition constant, k, can thus be directly calculated from measured SOM ^{14}C values (Chap. 3, Eq. 3.26). Mathematical models of natural ^{14}C abundance can be applied to soil samples collected prior to the bomb testing period (e.g., Tamm and Ostlund 1960; Trumbore et al. 1989) or to contemporary soil samples with long turnover times, which are assumed to contain minimal bomb C (Torn et al. 1997; Trumbore and Harden 1997). For example, a pre-bomb ^{14}C model was used to examine interactions between soil mineralogy and organic C storage along a 0.3–4100 kyr age chronosequence in Hawaii (Torn et al. 1997). The sites along the gradient were similar in terms of dominant tree species (*Metrosideros polymorpha*), elevation, climate, and parent material but differed in the length of time for soil formation processes to occur. Because these ecosystems were known to be accumulating C over long timescales, a non-steady-state pre-bomb model was used to determine SOM turnover. While soils from this study were collected decades after the bomb ^{14}C period, turnover times were estimated from soil samples deep in the profile, which presumably had minimal input of bomb ^{14}C. Along this chronosequence, soil C inventory increased with age up to the 150 kyr site (60 kg C m^{-2}) and then decreased by 50 % at the oldest site. Turnover time of the deepest soil horizon ($\sim$0.5–1.5 m) ranged from 1 to 170 kyr and also peaked at the intermediate-aged site (Fig. 6.10). Furthermore, turnover time was correlated with the percentage of non-crystalline minerals, which accounted for more than 40 % of the variation in organic C content across sites in these volcanic soils. Non-crystalline minerals, which have large and variable surface area and are particularly effective in sorbing organic C molecules, increased during the early stages of soil development but then decreased over longer timescales as they dehydrated to crystalline clays with a lower affinity for soil C. While this study was one of the first to demonstrate that dynamics and storage of deep soil C was driven by changes in the millennial-scale cycling of mineral-stabilized soil C pools rather than by

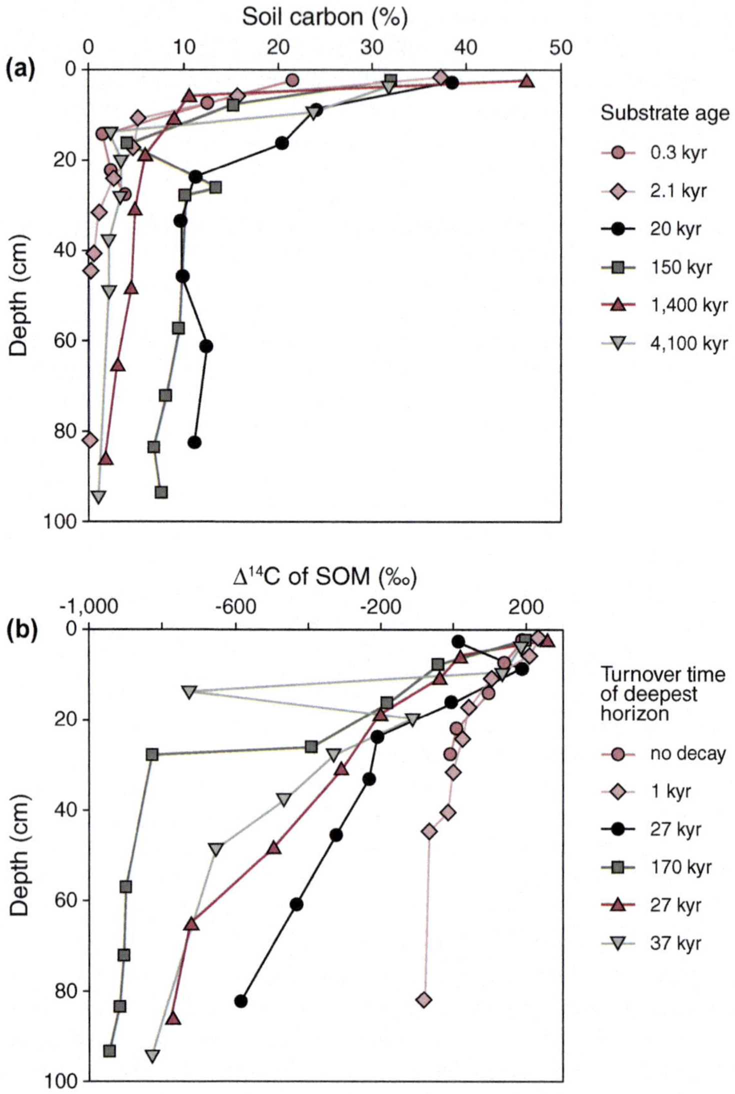

Fig. 6.10 Across a chronosequence in Hawaii (**a**) soil C was highest and (**b**) soil organic matter $\Delta^{14}C$ was lowest at mid-aged sites, which had slow turnover because organic matter was protected by high content of non-crystalline minerals in moderate-aged sites. Figure modified from Torn et al. (1997)

changes in the fast-cycling C pool, other natural abundance ^{14}C studies have subsequently demonstrated the importance of soil mineralogy as a driver of C turnover times in tropical savannas (Wattel-Koekkoek et al. 2003), forests, and agricultural fields in the USA (Paul et al. 2001).

6.3.3 Modeling Soil Carbon Residence Times with Bomb ^{14}C: Steady-State Systems

While natural abundance ^{14}C can estimate dynamics on the order of hundreds to thousands of years, the recent input of bomb ^{14}C into the atmosphere and its subsequent incorporation into SOM pools allows estimates of C turnover in soil pools that range from years to decades (Fig. 6.1). Because of the large annual variation in atmospheric ^{14}C since the mid-twentieth century, turnover rates of decadal-cycling SOM are estimated using time-dependent models that explicitly account for interannual variation in atmospheric ^{14}C content (Trumbore 1993). For each annual time step, modeled C inputs to soil are assigned ^{14}C values equal to atmospheric values in the year C was fixed via photosynthesis, whereas the ^{14}C content of SOM outputs is based on both decomposition rates and radioactive decay (Chap. 3). Because the change in atmospheric ^{14}C in the time since bomb enrichment has been nonlinear with a short period of increasing ^{14}C content followed by a longer period of decreasing ^{14}C content, a single measured SOM ^{14}C value will predict two possible turnover times with these mathematical models. In these instances, additional information such as ^{14}C measured on archived soils from the same site or independent estimates of C inventory and flux rates can be used to distinguish between the two possible outcomes (Torn et al. 2005). An additional uncertainty in the application of time-dependent ^{14}C models involves estimating the time interval between plant C fixation and its eventual assimilation into the SOM pool after senescence. Measurable time lags can be expected when the residence time of C in plant tissue is greater than one year (Gaudinski et al. 2000; Torn et al. 2005; Nowinski et al. 2008) or when modeling dissolved organic C (DOC) inputs into deeper soil horizons (Torn et al. 2005). The failure to account for time lags within plants themselves will result in overestimated SOM turnover times.

Models tracking bomb ^{14}C can be applied both to ecosystems at steady state and to those that are accumulating or losing SOM stocks. Steady-state models, which assume relatively constant C inventories, may be reasonably used for mature ecosystems that have not experienced recent disturbance. For example, a steady-state model was used to examine SOM turnover along an elevation-based temperature gradient (climosequence) in relatively undisturbed sites in the Sierra Nevada range in California (Trumbore et al. 1996). Radiocarbon was measured in three soil fractions: the low-density fraction comprised primarily of actively cycling SOM, high-density hydrolysable fraction, and the non-hydrolysable dense residue comprised primarily of passive SOM. These fractions were measured in soils collected at the time of the study, and also in soils that had been collected at the same sites decades earlier in order to address the multiple solution problem of interpreting bomb ^{14}C. More than half (50–90 %) of the SOM in the upper 20 cm of soil along this gradient was comprised of low-density C, whose residence time increased from 6 to 8 years at 17.8 °C (470 m elevation) to 53–71 years at 4 °C (2900 m elevation); however, there was no clear relationship between temperature

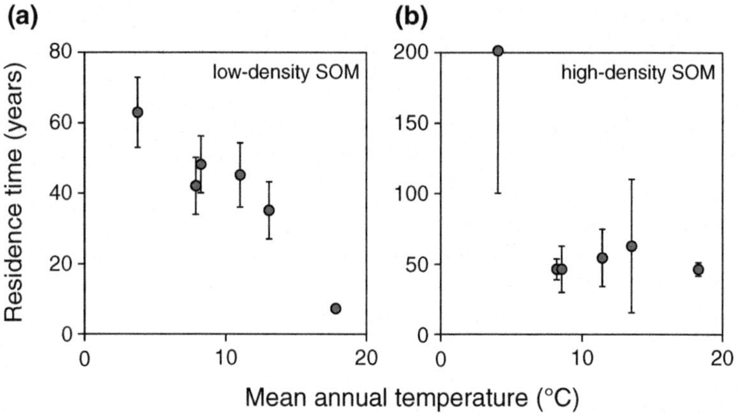

Fig. 6.11 Relationship between residence time and temperature along an elevation gradient in the Sierra Nevada, CA. (**a**) Residence time of actively cycling SOM fractions (low-density) decreased with increasing mean annual temperature, but there was no temperature effect on (**b**) slower cycling high-density SOM fractions. Figure modified from Trumbore et al. (1996)

and residence time of the high-density hydrolysable fraction, whose residence times ranged from 20 to 300 years (Fig. 6.11) (Trumbore et al. 1996).

Similar trends between temperature and C turnover time were found by a study that looked at active, slow, and passive SOM pools along a 22 °C mean annual temperature gradient in 22 mature pine and hardwood forest stands located in six states across the USA (Fissore et al. 2008). In this study, long-term incubations (525 days) provided estimates of the size of the active pool and residence times for both the active and slow pool. The size of the passive pool was determined from acid hydrolysis and size of the slow pool by difference. The residence time of the active pool ranged from 10 to 85 days and decreased with increasing temperature similar to the findings of the California climosequence. But, there was no effect of temperature on residence times of the slow (4–27 years) and the passive (380–5699 years) fractions, which together comprised 98 % of total SOM in these forests. Furthermore, passive C pools were almost twofold greater in hardwood than pine forest stands, suggesting that climate-induced shifts in vegetation from hardwood to pine may have a greater impact on total SOM pools than the direct effect of temperature. In sum, the role of temperature in bulk soil C dynamics will clearly be dependent upon the size of the actively cycling pool, as has been observed in several other ^{14}C modeling studies including an elevation gradient in Hawaii (Townsend et al. 1995), tropical forests in Brazil (Trumbore et al. 1995), and grasslands in the Swiss Alps (Leifeld et al. 2009).

In addition to mineralogy and temperature, other factors that vary across the landscape can profoundly influence soil C turnover. For example, ^{14}C modeling has been applied to demonstrate the importance of topographic and biological controls over soil C turnover. A two-pool ^{14}C model was used to study the effects of topography and large herbivores on turnover of stable (slow + passive pools) soil C from grassland in Yellowstone National Park (Frank et al. 2011). This study compared the soil collected inside and outside fenced plots (exclosures), which were erected in 1958–1962 and were located across toposequences at sites that ranged from hilltop, slope, to slope-bottom positions. The exclosures prevented grazing from large ungulates, such as elk, bison, and pronghorn, which may alter soil C through changes in both the plant community and soil environment. The size of the active pool was estimated as the sum of the microbial C pool, determined from chloroform fumigation, and the cumulative amount of C respired above background levels during a 50-day incubation period. The stable pool was then calculated as the difference in C content between total and active pools. Radiocarbon was measured on bulk SOM and, for the active fraction, from respired CO_2 collected from incubated soils. Assuming a one-year residence time for the active pool, a steady-state model was applied to solve for turnover of the stable SOM pool, which ranged from 37 to 653 years in the top 10 cm of soil and from 89 to 869 years in 10–20 cm soil. A surprising result from this study was the large effect of grazers on the residence time of stable SOM and the interaction between grazers and slope effects. In drier grasslands located in upslope sites, there was a 300-year increase in stable C residence time when grazers were present, while in more mesic slope-bottom sites, there was a 350-year decrease in residence time when grazers were present. The changes in SOM dynamics were attributed to changes in plant species and chemical composition. These studies illustrate the importance of isolating SOM fractions, not only for accurate residence time estimates, but also because biological and environmental controls over soil C dynamics vary across SOM fractions with different residence times.

6.3.4 Modeling Soil C Residence Times with Bomb ^{14}C: Non-Steady-State Systems

In many ecosystems, steady-state models are not applicable because SOM stocks are in disequilibrium due to ecosystem succession following a disturbance such as fire or land-use transition. When information about the magnitude of the change in the SOM pool is known, or can be inferred, a non-steady-state ^{14}C model (Chap. 3) can be applied to estimate rates of C accumulation/loss and SOM residence times. When a discrete disturbance event such as a fire or land-use transition occurs, C accumulation can be estimated based on contemporary C pools located above a visible disturbance marker in the soil profile, such as a plow or char line. This approach was used to model SOM dynamics in a deciduous forest in the northeastern USA, an area that has been accumulating SOM since abandonment from agriculture in the late 1800s (Gaudinski et al. 2000). Here, a non-steady-state ^{14}C model was applied to Oe + Oa and A horizons above the plow layer where SOM had accumulated since agricultural abandonment, while a steady-state model was applied to soil horizons below the plow layer based on the assumption that the land-use shift did not significantly affect soil C dynamics in this part of the profile. Modeling results of measured ^{14}C supported significant SOM accumulation above the plow layer, where 4.4 kg C m^{-2} has accumulated since agricultural abandonment and where most C was stored as humified material with a turnover of 40 (O horizon) to 100+ (A horizon) years. Dynamics of SOM in mineral layers below the plow layer was much slower and ranged from 220 to 1760 years.

Another instance where C accumulation can be inferred from a disturbance event is in systems accumulating C following a fire. Trumbore and Harden (1997) applied a non-steady-state ^{14}C model to estimate SOM accumulation in boreal ecosystems in Manitoba, Canada that varied in soil drainage and fire history. This study also used different SOM models across the soil profile: Organic soil above the char layer was modeled using a non-steady-state post-bomb model, while deep organic and mineral soil below the char layer for most of the sites in this study were assumed to contain no bomb ^{14}C, and thus, long-term C dynamics during the 8000 years of soil formation were modeled using a pre-bomb non-steady-state model. High inputs coupled with slow decomposition resulted in C accumulation in surface organic layers following fire. Residence times above the char layer ranged from 6 to 250 years, while in deep organic and mineral layers, residence times ranged from 80 to 3000 years and were positively related to drainage, with the fastest turnover in well-drained sites.

In both of the previous examples of non-steady-state systems, the change in C inventory was estimated based on a known point at which the newly accumulated soil C pool was assumed to be zero. Non-steady-state models can also be applied to systems where changes in C inventory have been directly monitored over time. This approach was used to examine C dynamics in SOM from a fertilization experiment in Arctic tundra, where previous research found that nutrient addition increased C stocks in standing biomass, litter, and surface organic layers, but decreased C stocks in lower organic and mineral layers (Mack et al. 2004). Radiocarbon measurements and modeling were used to determine whether observed changes in SOM inventory were driven by altered inputs, decomposition rates, or both (Nowinski et al. 2008). The authors assumed steady-state conditions for the control plots, which were modeled either with a one- or two-pool model (fast- and slow-cycling pools), dependent upon soil layer. A non-steady-state model was applied to fertilized plots with the addition of a 'new' pool to the model to account for changes in plant community and litter quality shifts with fertilization. Increased plant inputs from fertilization resulted in C accumulation in surface soils (<5 cm), despite the fact that decomposition increased in some pools. In deeper organic (>5 cm) and mineral layers, nutrient addition decreased C pools through both decreased inputs and increased decomposition of SOM. Fertilization effects on residence times varied across SOM pools. In the deeper organic layer, fertilization did not affect dynamics of the slow-cycling pool, which ranged from 100 to 900 years. However, residence time of the faster-cycling pool declined from 25 to 75 years in control plots to 5–30 years in fertilized plots. Similarly, fertilization in the mineral layer did not affect the slow-cycling pool, which had a residence time of 2000–4000 years, but did decrease turnover of the faster pool from 150 to 600 years in unfertilized plots to 10–

20 years with fertilization. As illustrated by these studies, when information about changes in total C stocks is known, the application of non-steady-state models can be a useful method for understanding SOM dynamics in non-equilibrium systems.

6.4 Respiration and Other Ecosystem Carbon Losses

Ecosystems lose C back to the atmosphere and to the hydrosphere, with metabolic respiration of organisms being the largest pathway of these losses. Respiration losses originate from the activity of all organisms within an ecosystem, with a majority derived from microbial breakdown of SOM. This presents a challenge for understanding how C losses arising from multiple sources control overall rates of ecosystem C exchange with the atmosphere. Radiocarbon source partitioning is a powerful technique to explore how sources of respiration vary in their relative contribution to the overall respiratory flux from ecosystems. The technique of source partitioning is a common application of stable isotope measurements, but using ^{14}C as a source tracer is a relatively novel and distinct use compared to using it as an age tracer. In this section, we discuss the application of ^{14}C source partitioning to explore plant and microbial contributions to ecosystem and soil respiration, as well as partitioning microbial respiration among different-aged substrates or soil horizons. Radiocarbon source partitioning uses the isotopic value of total respiration (measured in situ or from laboratory incubations) and of the individual plant and soil sources contributing to total respiration (measured from isolated incubations or solid samples) in mixing models (Chap. 3) to calculate the proportional contribution of each individual source to total respiration. This methodology is used with ^{14}C measurements alone, or in combination with stable C isotope measurements as a powerful dual-isotope approach. Lastly, this section concludes by examining other C losses from ecosystems including the trace gas CH_4, and also inorganic and organic C dissolved in water. While these are relatively small C losses from ecosystems, they have proven to be important for understanding overall ecosystem C balance, and thus, the impact of ecosystems on climate change.

6.4.1 Radiocarbon of Respiration Sources

The fundamental underpinning of isotope partitioning is the separation of individual source pools with distinct isotopic values. Radiocarbon offers the benefits of large separation in ^{14}C values between plant (autotrophic) and microbial (heterotrophic) respiration, and among different heterotrophic source pools (Fig. 6.12). Because most of the C respired from plants is driven by recent photosynthesis, the ^{14}C value of autotrophic respiration, especially of aboveground plant parts, is generally similar to the atmospheric ^{14}C value in the year of measurement (Borken et al. 2006; Muhr and Borken 2009; Muhr et al. 2009). However, ^{14}C of root respiration

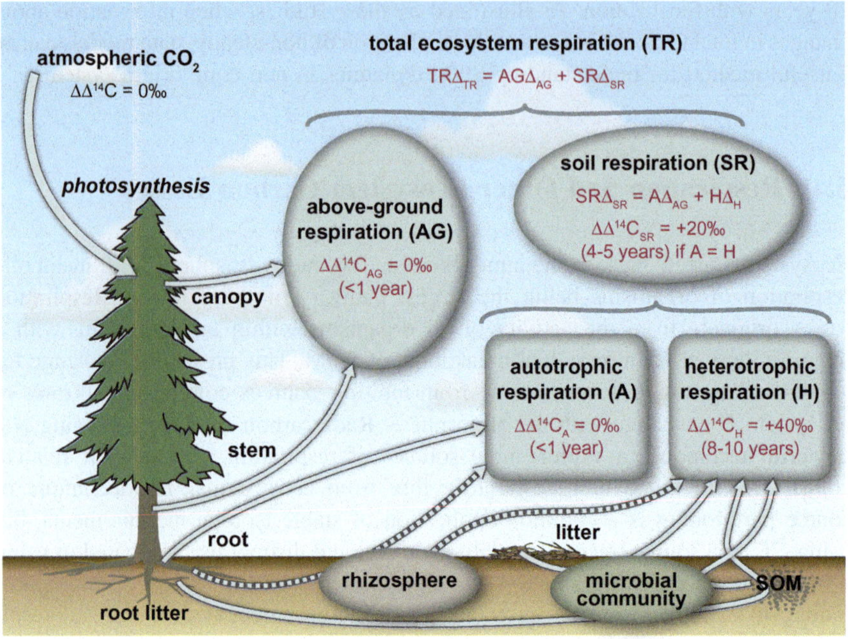

Fig. 6.12 The components of total ecosystem respiration are aboveground (AG) respiration and soil (SR) respiration, which is in turn made up of autotrophic (A) and heterotrophic (H) respiration. Autotrophic (plant) and heterotrophic (microbial) source pools have distinct $\Delta^{14}C$ values. Autotrophic respiration has $\Delta^{14}C$ values close to that of atmospheric CO_2, while heterotrophic respiration of the surface soil is generally more enriched in $\Delta^{14}C$ than the atmosphere. $\Delta\Delta^{14}C$ refers to the difference in $\Delta^{14}C$ as compared to the contemporary atmosphere, which is currently declining. The $\Delta\Delta$ value can be compared across different study years even though the absolute $\Delta^{14}C$ would differ depending on the year of measurement. Figure modified from Trumbore et al. (2006)

measured in isolation has sometimes been observed to be higher than atmospheric values (Hicks Pries et al. 2013; Czimczik et al. 2006; Schuur and Trumbore 2006), reflecting the use of stored C reserves greater than one year old for root growth and respiration (Czimczik et al. 2006; Gaudinski et al. 2009). Based on observed ^{14}C values, stored C reserves can have an estimated average age of 4–6 years, but these preliminary findings call for further investigation (Czimczik et al. 2006; Schuur and Trumbore 2006). Autotrophic respiration defined by isotope methods also typically includes fast-cycling soil C pools with residence times of hours to days such as that which fuels rhizosphere respiration. Operationally, rhizosphere (the soil zone around roots) C is often excluded from heterotrophic respiration as roots are typically removed from soil cores and ^{14}C measurements are made after one week to one month of soil incubation, by which time rhizosphere C is depleted and has little or no influence on the measured heterotrophic ^{14}C value (Czimczik et al. 2006; Schuur and Trumbore 2006; Muhr and Borken 2009; Muhr et al. 2009). Thus, while rhizosphere respiration is microbial respiration of recently fixed photosynthates such as root exudates and recently dead-root cells, it is usually included with the autotrophic source, together representing fast-cycling C.

In contrast to plant respiration, microbial respiration ^{14}C values are more variable depending on the mixture of soil C substrates consumed and the residence time of those substrates in the soil. Incubations of soil that are done to isolate microbial respiration ^{14}C values represent the metabolic activity of the full microbial community acting on the soil organic C pool simultaneously. Radiocarbon values from incubations that are greater than the atmospheric value in the year of measurement indicate that much of the soil C respired entered the ecosystem during or after the 1963 bomb peak, whereas incubation ^{14}C values less than the current atmosphere but greater than zero indicate a mixture of post- and pre-bomb soil C substrates. Finally, incubation ^{14}C values that are negative (using $\Delta^{14}C$ nomenclature) indicate a significant contribution of pre-bomb C that has undergone radioactive decay, although the small contribution of bomb-labeled substrate cannot be ruled out from these mixture measurements. Respiration of surface mineral and organic horizon soil incubations typically has post-bomb $\Delta^{14}C$ values ranging between 80 and 200 ‰ across a range of boreal, deciduous, and tropical forests (Borken et al. 2006; Czimczik et al. 2006; Schuur and Trumbore 2006; Muhr and Borken 2009; Nowinski et al. 2010; Posada and Schuur 2011). These values indicate that surface soil heterotrophic respiration is predominantly fueled by soil C pools with decadal-scale residence times across a range of ecosystems. In contrast, soil from deeper in the profile shows larger $\Delta^{14}C$ variability: 68 to 180 ‰ for boreal forest mineral soil horizons (Schuur and Trumbore 2006), 76 ‰ for a deciduous forest (Borken et al. 2006), and very negative values (−200 to −939 ‰) for deep permafrost soils (Dutta et al. 2006; Schuur et al. 2009; Nowinski et al. 2010). Variation in respiration ^{14}C from deeper soil incubations reflects the wide range of C age found in the soil profile, generally increasing in age with depth, but also depending on the particular mechanisms of soil C stabilization that are important for different ecosystems.

6.4.2 Partitioning Respiration In Situ

One of the long-standing challenges in ecosystem science has been to differentiate plant and microbial contributions to total soil or ecosystem respiration flux, and ^{14}C is an important tool with which to address this. Radiocarbon measurements of in situ soil respiration, measured as total belowground CO_2 losses to the atmosphere, reflect the mixed contribution of both plant root and microbial respiration sources, which differ in ^{14}C value as measured in isolated incubations described in the previous section. Partitioning the relative contribution of these two respiration sources is often accomplished using a two-source one-isotope mixing model (Chap. 3), which can derive a single solution for the relative contribution of each source to total soil respiration.

The relative contribution of plant and microbial respiration to total soil respiration can be highly variable among ecosystems and in response to changing ecological conditions. Several studies in black spruce-dominated boreal forests have used partitioning to explore the effects of disturbance by fire on soil respiration partitioning. Boreal forests have a large soil C pool that periodically burns and is, in part, released to the atmosphere. Time since fire is a strong control over the partitioning of total soil respiration—in young stands less than 20 % of total soil respiration was heterotrophic while in mature stands heterotrophic respiration contributed ~50 % to total soil respiration (Fig. 6.13) (Czimczik et al. 2006; Schuur and Trumbore 2006). This finding was opposite to the hypothesis that heterotrophic respiration would be highest in young stands as residues from the fire are decomposed. Radiocarbon partitioning has also been used to understand respiration responses to experimental manipulations such as soil moisture and snow cover. Rainfall exclusion experiments have shown that drier soil conditions decreased the contribution of heterotrophic respiration to soil respiration in a deciduous forest (Borken et al. 2006) and a Norway spruce forest (Muhr and Borken 2009), while root respiration contribution to soil respiration decreased but total root respiration flux remained constant. In a tundra ecosystem, root respiration contributed 40–80 % of growing season respiration under ambient snow conditions, but declined to 0–55 % with additional snow (Fig. 6.14) (Nowinski et al. 2010). Despite widely changing environmental and ecological conditions over a growing season, temporal variation in source contribution during this period has not been resolved in as fine detail (Cisneros-Dozal et al. 2006; Schuur and Trumbore 2006) until recently. By using ^{14}C and ^{13}C to better resolve source contributions, Hicks Pries et al. (2013) found autotrophic respiration peaked at the height of the growing season in tundra.

Separating plant and microbial contribution to total soil respiration gives analytically unique solutions because one isotope tracer can separate two source pools quantitatively. However, heterotrophic respiration often does not comprise one homogeneous source, but instead represents the mixed contribution from different-aged substrates or from different depths in the soil profile. The addition of more source pools calls for mixing models without unique solutions (Chap. 3), but there are approaches to address this. When determining the contribution from more than two sources, partitioning models can solve for a range of possible contributions from each source rather than solving for a single fixed contribution. For example, a study interested in ecosystem respiration of tundra undergoing permafrost thaw focused on the contribution of 'old' C arising from deep in the soil profile to this flux (Schuur et al. 2009). A three-source mixing model was used with ^{14}C to determine the contribution from plant respiration, surface soil, and deep soil to total ecosystem respiration. Across a permafrost thaw gradient where sites varied in the time since permafrost thaw initiated, respiration of old soil C contributed 7–23 % of total ecosystem respiration, increasing with time since thaw (Fig. 6.15). While the three-source, one-isotope modeling produced a range of possible solutions to the partitioning estimate, the negative $\Delta^{14}C$ values of old soil C respiration caused this source to be more constrained. A more recent study at the same site used

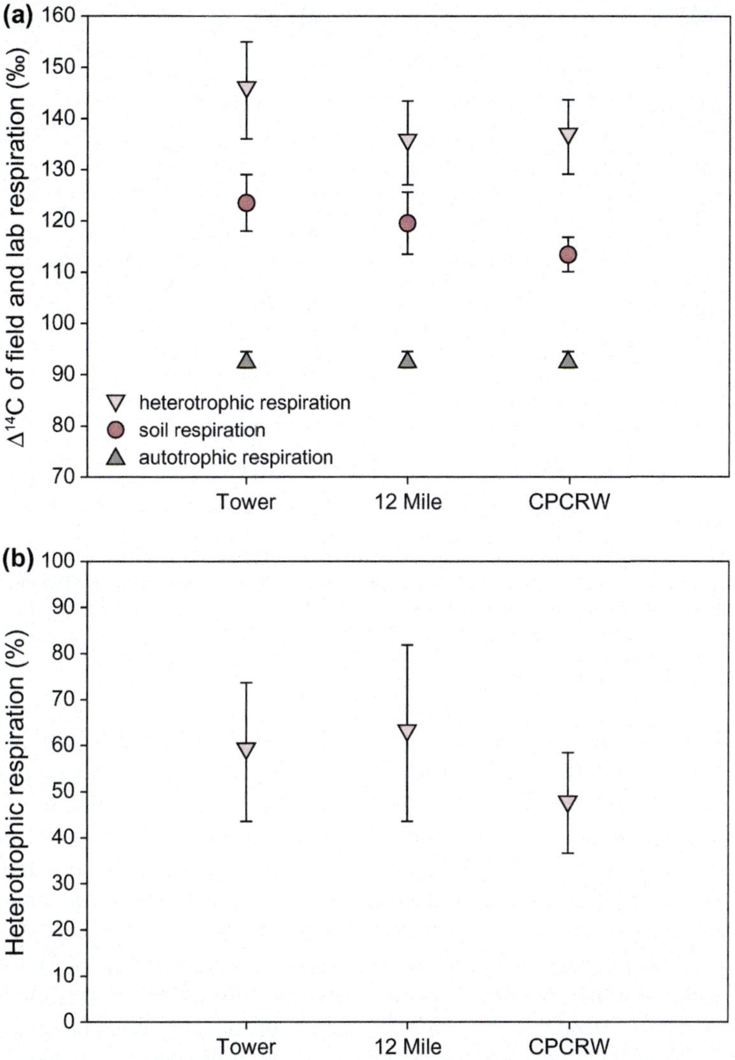

Fig. 6.13 (**a**) $\Delta^{14}C$ values of soil respiration (from field measurements), heterotrophic and autotrophic respiration (from incubations) and (**b**) the estimated contributions of heterotrophic respiration to total soil respiration in three mature boreal black spruce forests. Figure modified from Schuur and Trumbore (2006)

a four-source, two-isotope model (^{14}C and ^{13}C) and found a similar magnitude of increasing old soil respiration with deeper permafrost thaw (Hicks Pries et al. 2013). By using two isotopes, this study was able to resolve contributions from more sources and thus found permafrost thaw also increased the contribution of autotrophic respiration. Overall, these studies found that decades of thaw stimulated the

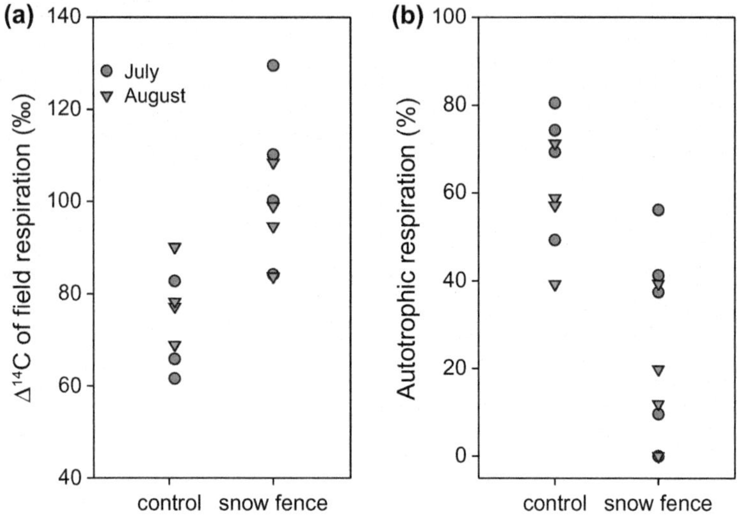

Fig. 6.14 In a tundra ecosystem, (**a**) higher $\Delta^{14}C$ values of field soil respiration in the snow fence treatment relative to the control were due to (**b**) declines in autotrophic contributions to total soil respiration. Autotrophic respiration contributed 40–80 % of growing season respiration under ambient snow conditions (control), but declined to 0–55 % in the snow fence treatment. Figure modified from Nowinski et al. (2010)

release of old permafrost soil C, making this tundra a possible C source. Further studies have similarly quantified contributions of old, deep C, which contributed 10–23 % of total ecosystem respiration in a British peatland (Hardie et al. 2009).

A single solution to source partitioning can also be found despite having multiple source pools by utilizing additional ecosystem information such as decomposition fluxes (Gaudinski et al. 2000), or by using other experimental techniques such as ^{14}C-labeled plant litter (Cisneros-Dozal et al. 2006). Gaudinski et al. (2000) used a series of equations that included isotope mass balance (Chap. 3), CO_2 flux, and a soil gas diffusion model to partition deciduous forest soil respiration into soil horizons and (separately) into different-aged sources. The different-aged sources were as follows: (1) recent C (photosynthate fixed in the past year); and (2) reservoir C (SOM > 1 year old), which was further split into leaf litter, root litter, humified organic material, and mineral-associated organics. Reservoir C contributed 41 % of soil respiration, and of the reservoir flux, 20 % came from humified organic material and 80 % came from recognizable plant litter (both leaf and root). In a temperate forest in Oak Ridge, Tennessee, researchers used an 'accidental' ^{14}C label to investigate contributions of root respiration, leaf litter decomposition, and soil decomposition to total soil respiration (Cisneros-Dozal et al. 2006). Normally, leaf litter respiration would have a ^{14}C value only slightly

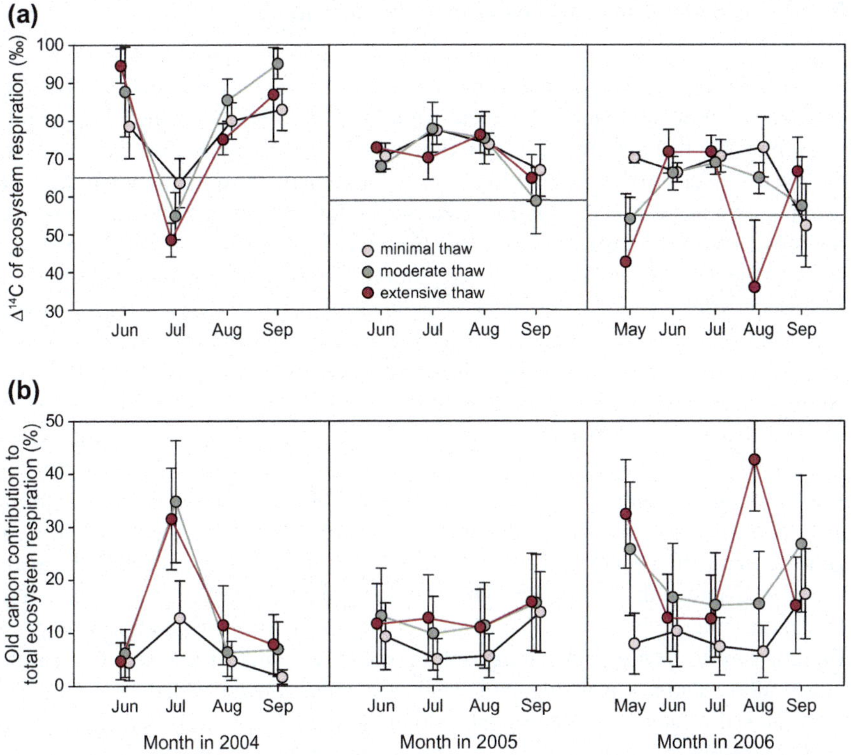

Fig. 6.15 (**a**) $\Delta^{14}C$ values of ecosystem respiration and (**b**) estimated contributions of old soil C to ecosystem respiration across a permafrost thaw gradient where sites varied in the time since permafrost thaw initiated. Contributions of old soil C to ecosystem respiration were greatest when the $\Delta^{14}C$ of ecosystem respiration were depleted. The respiration of old soil C contributed 7–23 % of total ecosystem respiration, increasing with time since thaw. Figure modified from Schuur et al. (2009)

older than root respiration, but without enough of a difference to partition the two pools separately. However, leaf litter that grew during a large, inadvertent release of ^{14}C-labeled CO_2 had two levels of $\Delta^{14}C$ values: either 200 ‰ or 1000 ‰, depending on the location where litter was collected. By replacing regular in situ leaf litter with either low- or high-labeled litter in new sites that were not influenced by the original ^{14}C release and utilizing a series of mass balance equations, litter contribution to soil respiration was found to increase from 1 % in dry conditions to 42 % in mesic conditions when the leaf moisture was high enough to promote microbial activity. By assimilating non-isotope information about ecosystem C fluxes, or by experimentally manipulating the isotope value of the source pool, further source partitioning can be made quantifiable beyond that possible with ^{14}C variation arising from natural radioactive decay or by the bomb label.

6.4.3 Partitioning Heterotrophic Respiration

In the previous section, soil respiration ^{14}C was measured in situ, but heterotrophic respiration from laboratory soil incubations can also be partitioned in a similar fashion except that autotrophic respiration is no longer a source since plant roots are removed. Heterotrophic respiration can be broken down into its components to examine which substrates within the heterogeneous mixture of SOM are supporting microbial activity. Heterotrophic respiration was partitioned in a boreal ecosystem to understand the potential of these ecosystems with large stores of organic soil C to feedback positively to climate change (Dioumaeva et al. 2002). Here, respiration of incubated boreal forest peat was partitioned into five potential sources—humified material, two size classes of roots, char, and wood fragments based on what could be physically separated from these organic soils. Respiration partitioning using ^{14}C showed that humified material contributed 30 % of the overall heterotrophic respiration flux for a feather moss peat even though this fraction was 75 % of the bulk soil C. In contrast, the two root pools contributed 50 % of the overall heterotrophic respiration flux even though these fractions were only 7 % of the bulk soil C. In another incubation study, respiration of Siberian permafrost soil was physically partitioned into four C fractions: DOC, mineral, humus, and organic matter fragments utilizing both ^{14}C and ^{13}C (Dutta et al. 2006). Utilizing both C isotopes is another way to handle partitioning among multiple sources because ^{13}C measurements allow another isotope mass balance equation to be added, which further constrains the range of source contributions (Chap. 3). In deep permafrost soils, DOC contributed substantially to respiration flux ranging from 30 to 86 % of the overall heterotrophic respiration flux, even though this pool only contained about 2 % of the bulk soil C. In both examples, partitioning of the heterotrophic respiration flux showed that C fractions that formed only a small part of the bulk soil C could contribute substantially to the overall flux, emphasizing again how ^{14}C measurements from incubations can be used to assess the fast turnover component of SOM.

Problems can arise in isotope partitioning, in particular if the isotopic value of total respiration falls outside the range of known source isotopic values. For ^{13}C, this can arise from issues with enzymatic fractionation, but for ^{14}C this mismatch indicates an unmeasured source that contributed to respiration but was not isolated and thus could not be included in the partitioning model. For example, another partitioning study in a Canadian boreal forest incubated surface soil to determine the ^{14}C of the heterotrophic source, but the measured ^{14}C value of in situ total soil respiration was even lower than the ^{14}C of the surface soil incubations; thus, the partitioning model could not produce a solution (Czimczik et al. 2006). Based on general knowledge of soil C accumulation, this suggested that deeper soil horizons that contained old C were contributing measurably to the soil respiration flux even though they were overlooked by the incubations. A follow-up study applied a ^{14}C partitioning model to soil incubation data in order to separate boreal forest soil respiration into five source pools: needles, dead roots, wood, moss, and amorphous

SOM. In this case, the ^{14}C value of the total heterotrophic respiration flux was greater than measured ^{14}C for all the sources, suggesting a microbial source of post-bomb organic C that the researchers were not able to physically isolate from the organic soil horizon (Czimczik and Trumbore 2007). Isotope partitioning, then, is limited in part by the ability to separate and measure important source pools from the heterogeneous mixture that is SOM. Comparing isotope values of individual source pools to the isotope value of the mixture is an important diagnostic as to the success of the separation techniques.

6.4.4 Other Ecosystem C Losses

Aside from respiration of CO_2 by organisms, C is lost from ecosystems by several other pathways that are important for ecosystem and landscape C balance. Disturbance by fire, microbial production of CH_4, and the transport of organic and inorganic C in water all have been studied with the isotopic approaches that have already been discussed in this chapter as well as approaches that are described in more detail in following sections. While a full discussion of these loss pathways is beyond the scope of this chapter, this brief overview highlights some isotope applications important for understanding these pathways.

Fire transforms ecosystem C pools and releases C directly back to the atmosphere, thus bypassing microbial degradation. The contribution of different ecosystem C pools to CO_2 released by a fire was estimated by measuring ^{13}C and ^{14}C isotopes both in CO_2 emitted from a burning boreal black spruce forest, and from solid plant and soil samples of 10 potential ecosystem C sources (Schuur et al. 2003). Isotope partitioning showed that 75 % of needles and fine branches and over 90 % of live surface and dead moss burned in the fire, whereas less than 30 % of cones, coarse branches, and wood burned. Using a 10 source mixing model with 2 isotopes produced relatively unconstrained ranges for potential source contributions, so additional constraints based on field observations of flammability, such as forcing surface soil to burn before deep soil, were added. These simple constraints were effective for allowing the isotope partitioning model to produce narrow ranges of combustion estimates for many of the pools, as well as a reasonable estimate of 2.5 kg C m^{-2} for overall C loss. This again highlights the utility of non-isotope ecosystem dynamics information that can be added to increase the effectiveness of source partitioning models.

Methane is an important C gas that can be released from ecosystems along with CO_2 as a result of microbial activity and the decomposition of organic matter in environments where oxygen is limited, such as wetlands and aquatic systems. Methane also has a global warming potential at least 25 times stronger than CO_2, thus plays an important role in determining the influence of C emissions from ecosystems to the atmosphere on climate change (IPCC 2007). The two main applications of ^{14}C measurements in terrestrial ecosystem research have been to detect the mean age and source of the substrates of CH_4 production using natural abundance ^{14}C, and to determine the metabolic pathway of CH_4 production and identify CH_4 consumers using ^{14}C pulse labeling. The former application is similar to what has been discussed in earlier sections of this chapter in regards to CO_2, while ^{14}C labeling will be discussed in greater detail in the following section and only in brief here.

Enriched ^{14}C values of CH_4 emissions measured in a range of ecosystems including peatlands, bogs, fens, and rice paddies across tropical, temperate, and boreal zones reflects the dominant use of recently fixed bomb-labeled soil C sources as a substrate for CH_4 production (Wahlen et al. 1989; Martens et al. 1992; Chanton et al. 1995; Nakagawa et al. 2002). However, ^{14}C depleted CH_4 emissions from lakes across the Arctic region demonstrate that old C sources, with an average of 27,000 ^{14}C years, can contribute to CH_4 production in some cases (Zimov et al. 1997; Walter et al. 2008). Siberian lake old CH_4 flux, often associated with emissions hotspots within lakes, could contribute an estimated 25 % to regional wintertime atmospheric CH_4 accumulation helping to drive the large amplitude of atmospheric CH_4 that occurs in northern latitudes. In addition to measurements of CH_4 flux to the atmosphere, natural abundance ^{14}C has also been used to examine the dynamics of CH_4 production within the peat profile. Several studies have shown that the ^{14}C age of CH_4 is consistently and markedly younger by 500–2000 years than the age of solid-phase peat found at the same depth (Aravena et al. 1993; Charman et al. 1994, 1999; Chanton et al. 1995; Chasar et al. 2000). Measurements of ^{14}C in dissolved organic and inorganic C suggest that CH_4 production in these peatlands is derived, in part, from recently fixed DOC that had been transported down the peat profile in pore water.

As with CO_2 fluxes detailed in the next section, the dynamics of CH_4 production and oxidation can be studied on timescales of minutes to days using ^{14}C pulse labeling. Radiocarbon labeling has been applied to track CH_4 production pathways and also as a biomarker to identify methane oxidizing bacterial communities. Radiocarbon-labeled methanogenic substrates such as acetate have been used to identify the contribution to CH_4 production from the two main CH_4 pathways: CO_2 reduction and acetate decarboxylation, in a range of wetland and lacustrine ecosystems (Glissman et al. 2004; Schulz and Conrad 1996; Kruger et al. 2001; Duddleston et al. 2002; Kotsyurbenko et al. 2004; Wand et al. 2006; Keller and Bridgham 2007). Other studies have labeled plants and plant litter (Megonigal et al. 1999; Juutinen et al. 2003) or tracked $^{14}CO_2$ applied to whole ecosystems (King and Reeburgh 2002; King et al. 2002; Christensen et al. 2003) to quantify the amount of CH_4 produced from recently fixed C, and to determine the time lags from C fixation to CH_4 production. Radiocarbon-labeled *Phragmites australis* litter applied in a fen mesocosm study showed that over a month-long incubation, 40 % of the CH_4 produced was derived from plant litter, while the remainder originated from SOM (Juutinen et al. 2003). Also, $^{14}CO_2$ applied to ecosystems and then emitted as $^{14}CH_4$ emissions from the soil demonstrated the rapid (<12 h) transfer of plant C to methanogens in the rhizosphere. In general, the application of ^{14}C methods to CH_4 has shown similar trends to CO_2 studies in that much of the return flux of C from ecosystems to the atmosphere is dominated by recent C, but that in some cases, in particular associated with disturbance events, old C can play an important role in determining ecosystem C balance.

Other than gaseous C loss to the atmosphere, ecosystems also experience loss of C as DOC, dissolved inorganic C (DIC) and particulate organic C (POC) that can leave ecosystems via leaching to groundwater and streams, or through erosion (Chap. 5). Hydrologic losses of C are small relative to respiration exchange with the atmosphere, but small additional C losses can be important when compared to the net difference between gross photosynthetic uptake and ecosystem respiration losses. Within the soil profile, DOC leaches from organic to mineral horizons where it can be either sequestered in the mineral soil and protected from microbial degradation, or leached out of the terrestrial ecosystem. Comparing the age of DOC to bulk SOM gives an indication of whether the mechanisms controlling dissolved and solid C are coupled or decoupled within the soil system. In a Scandinavian spruce forest, there was less than a 5 ‰ difference in $\Delta^{14}C$ values between DOC and bulk SOM in the B soil horizon, suggesting that there was a high rate of sorption and desorption of DOC within the mineral soil (Froberg et al. 2006). Therefore in that forested ecosystem, controls on DOC cycling were more likely to be governed by soil mineral properties than climate or biological factors. In contrast, the majority of studies find that DOC is of modern origin (post-bomb) and younger than the bulk SOM, implying that DOC and SOM cycling are decoupled and that the majority of DOC, for example 60–97 % in one study, is coming from recent photosynthate (Don and Schulze 2008). These findings have been replicated across a wide range of ecosystems including peatlands (Chanton et al. 2008; Clymo and Bryant 2008), grasslands (Don and Schulze 2008), and forests (Sanderman and Amundson 2008). Furthermore, DOC has been linked directly to respiration using ^{14}C as a source tracer. In a sedge-dominated peatland, the similarity of $^{14}CH_4$ and DIC to the ^{14}C of DOC suggested that DOC was the primary substrate for methanogens, whereas in a *Sphagnum* peatland and woody plant-dominated peatland, methanogens used a mixture of DOC and bulk SOM (Chanton et al. 2008). In peatlands, DOC cycling is likely controlled by the dominant vegetation type and its susceptibility to microbial degradation rather than physical processes within the soil profile.

After DOC is lost from terrestrial ecosystems, it enters streams and rivers where it can undergo further degradation or be transported to the ocean. Radiocarbon has been used to determine the sources of terrestrial DOC in freshwater streams and rivers, and to investigate whether water column microbes preferentially utilize terrestrial DOC. There is growing concern that climate change will increase the amount of old DOC entering rivers from terrestrial ecosystems, in particular for boreal rivers that drain permafrost-rich areas (Neff et al. 2006; Evans et al. 2007; Guo et al. 2007). In these landscapes, permafrost thaw could destabilize vast stores of SOM, altering regional C budgets and increasing the amount of organic C reaching northern oceans. A study of 3 major boreal rivers found that DOC was much younger in the Mackenzie and Yukon rivers than in the Sagavanirktok River with 72, 64, and 20 % of DOC coming from modern vegetation in each river, respectively. In all rivers, POC had a greater ^{14}C age than DOC, 4500–7500 years older in the Mackenzie and Yukon rivers, implying that unlike DOC, POC is derived from SOM stored in permafrost entering rivers through bank erosion (Guo et al. 2007). A contrasting study in Amazonian rivers found modern DOC with bomb ^{14}C across the entire watershed, but that the large evasion of supersaturated CO_2 from Amazonian rivers was derived from microbes preferentially respiring terrestrial organic C that was less than 5 years old (Mayorga et al. 2005). Again in contrast, old C detected in this study through DIC ^{14}C measurements was derived from inorganic C dissolution in watersheds with carbonate geology. In many ways, patterns of fire, CH_4 flux, and hydrologic C losses detected with isotopes tell largely similar stories to studies of CO_2 loss, but are important pathways for C loss from terrestrial ecosystems that cannot be overlooked when trying to understand the responses and impacts of changes in the global C cycle.

6.5 Low-Level Radiocarbon Labels

Up to this point in the chapter, we have been primarily discussing approaches where measured ^{14}C ratios were created as a consequence of either natural solar production in the stratosphere or thermonuclear weapons testing. However, application of low-level ^{14}C labels by researchers into specific study systems, a methodology based on traditional labeling techniques, is a relatively new and powerful approach that can track the fate of C as it moves through terrestrial ecosystems. Low-level ^{14}C labels, as defined here, contain radioactivity in amounts well below health and safety regulations; these experiments are now possible because of the enhanced precision of ^{14}C measured by AMS (Turteltaub and Vogel 2000). In general, a label is an isotopically distinct (^{14}C or ^{13}C, enriched or depleted) C substrate that can be introduced to the ecosystem (air, plant, or soil) in different forms including gas (e.g., CO_2), liquid (e.g., carbohydrate solutions), and solid (e.g., plant litter). Depending on how and where a label is applied, it can be used to follow the pathways and rates of C cycling across different timescales within an ecosystem (Fig. 6.1). In plants, potential pathways of study include C allocation to above- and belowground structures, carbohydrate storage pools, and respiratory processes. In soil, labeled materials can be used to track the transfer of C from plants to microbes, incorporation of C into SOM, and loss via microbial decomposition (respiration), volatilization, and leaching in dissolved C forms.

6.5.1 *Advantages of Enriched Low-Level ^{14}C Labels*

Isotopically enriched, low-level ^{14}C labels offer several advantages over traditional ^{14}C and ^{13}C labeling techniques (Table 6.1). In order to have sufficient ^{14}C content within a sample to measure by decay counting techniques, traditional enriched ^{14}C labels contain large amounts of radioactivity at levels (Megabecquerel; MBq) considered harmful to

Table 6.1 Advantages and disadvantages of isotope labeling techniques

Label	Advantages	Disadvantages
Traditional enriched ^{14}C label	Inexpensive analyses	Higher radioactivity for decay counting techniques Large sample sizes (g C) Strict health and safety regulations Limited to enclosures, juvenile, or small potted plants
Traditional enriched ^{13}C label	Safe (not regulated) Field applications Inexpensive analyses of samples (IRMS or laser)	Requires adding C to the system Less sensitive tracer Shorter-lived signal
Low-level enriched ^{14}C label	Safe (exempt quantities of radiation) Field applications Little C added to system Small sample sizes (μg-mg C) Sensitive tracer Longer-lived signal	Sample prep time for analyses by AMS Expensive analyses by AMS

organisms. For this reason, the majority of traditional ^{14}C labeling studies have been limited to juvenile, short-stature vegetation in greenhouses or growth chambers, with very few studies conducted under field conditions due to strict health and safety regulations (e.g., Friend et al. 1991). In contrast, a low-level ^{14}C label can be safely introduced in natural ecosystems without causing harmful exposure to radiation and thus not requiring health and safety permissions. For example, in the state of California, ^{14}C amounts below 0.037 Bq mL^{-1} (in air) and 296 Bq mL^{-1} (in liquid) are considered quantities exempt from regulation (California Department of Health Services 2005). Because the natural abundance of ^{14}C in the atmosphere is very low ($^{14}C/^{12}C$ $\sim 1 \times 10^{-12}$) or $\sim 4.7 \times 10^{-8}$ Bq mL^{-1}, a low-level ^{14}C label with an activity of $\sim 3.7 \times 10^{-5}$ Bq mL^{-1} air will increase the ^{14}C content to $\sim 1 \times 10^3$ times. This level is still easily detectable by AMS, while the atmospheric radioactivity produced will be three orders of magnitude below regulated levels.

A ^{13}C label measured by isotope ratio mass spectrometry (IRMS) or laser can also safely be used in natural ecosystems in any quantity and is significantly less expensive than a ^{14}C label measured by AMS. However, due to the high natural abundance in comparison with ^{14}C ($^{13}C/^{12}C$ ~ 0.01), it is a less sensitive tracer in that ^{13}C concentration can only be increased two orders of magnitude from 1 to 100 %. Therefore, increasing the signal strength of a ^{13}C label requires adding substantial quantities of C to the experimental system (elevated CO_2), or alternatively, labeling for an extended (days to years) period of time (continuous labeling), which limits the types of experiments that can be performed. The low-level ^{14}C label can be orders of magnitude greater than the ^{13}C label, allows for shorter labeling times, less C added to the system, and greater signal strength, and thus fills the experimental gaps where ^{13}C label is not useable. This is particularly advantageous for tracking the fate of C into small and longer-lived ecosystem C pools.

6.5.2 Pulse-Chase ^{14}C Labeling Studies

To date, the enriched low-level ^{14}C label method has been used in a pulse-chase labeling approach that is most appropriate for studying C cycling on timescales of hours to months. The 'pulse' is a one-time addition (although it can be repeated) of the labeled substrate to the experimental system. This is then followed by the 'chase' period, by which the concentration of the label is monitored over time as it moves into different ecosystem C pools. Assuming steady state, the residence time of the label in a specific C pool can be determined by the dilution of the label with new unlabeled C over time (Carbone et al. 2007). This experimental approach provides a snapshot of C-cycling dynamics at the time of the labeling and then can be applied repeatedly to capture temporal variability such as seasonal changes in plant allocation patterns.

Uniquely, a pulse-label approach is the only way to quantify the time lag between photosynthetic uptake of CO_2 from the atmosphere and respiration by plants, and how new assimilate is partitioned by plants into above- or belowground processes. A $^{14}CO_2$ label introduced in a large tent and photosynthesized by plants was used to trace the allocation of new assimilates into plant respiration, revealing unique information on the timing and contribution of recent versus older C to above- and belowground respiration sources (Fig. 6.16) (Carbone et al. 2007;

Fig. 6.16 Researchers assemble the low-level ^{14}C labeling tent over the black spruce trees before label application. The labeling study took place in 2004 in boreal forest near Thompson, Manitoba, Canada (*Photo credit* Mariah Carbone)

Carbone and Trumbore 2007). These studies demonstrated that a majority of newly assimilated C cycles through ecosystems quickly (Fig. 6.17). Time lags between assimilation and aboveground respiration were on the order of hours for all plant types. In contrast, the time required for translocation of the label belowground to the roots and to be used for belowground respiration varied from hours in mosses,

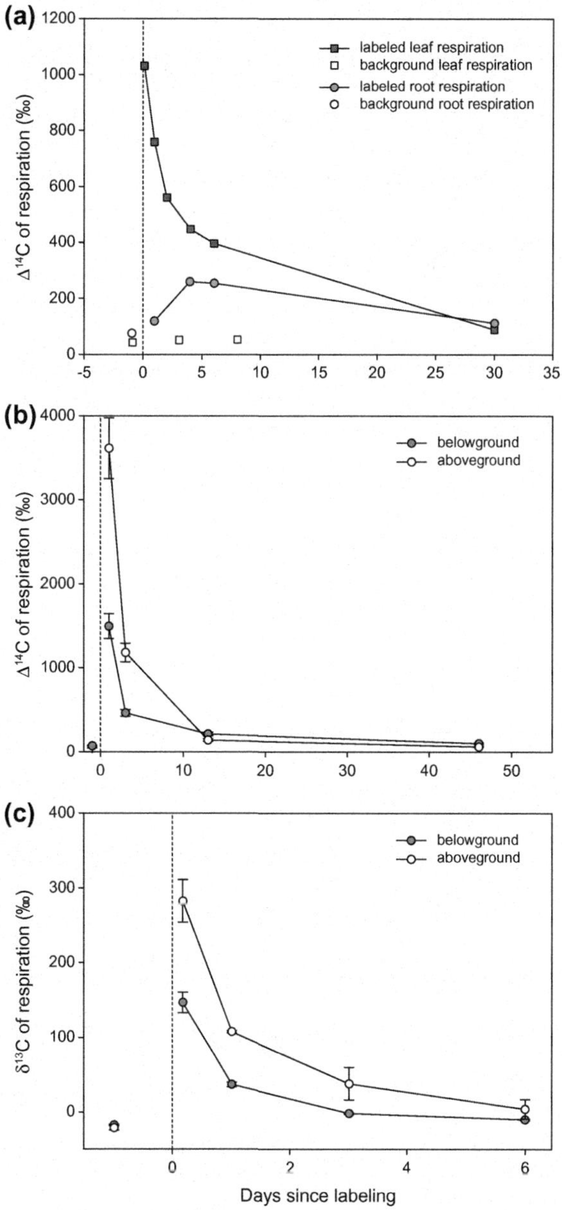

Fig. 6.17 (**a**) $\Delta^{14}C$ values of respiration for black spruce leaf and root respiration compared to background control, and (**b**) $\Delta^{14}C$ and (**c**) $\delta^{13}C$ values of respiration for grass leaf (aboveground) and root (belowground) respiration versus time since labeling. *Filled symbols* represent measurements taken after labeling; *open symbols* represent measurements taken before labeling and in unlabeled plots. Peak label concentrations are different between leaf and root respiration, reflecting differences in the timing and contribution of new assimilates to respiration. The higher-time-resolution $\delta^{13}C$ values (**c**) were used to quantify the short-term response (0–6 days), whereas the lower-time-resolution $\Delta^{14}C$ values (**a**) and (**b**) were used to quantify the long-term response (>6 days to many months) after labeling. Figure modified from Carbone et al. (2007)

grasses, and small shrubs to 4 days in trees. Allocation patterns of the label to above- and belowground respiratory processes also varied across plant types. Grasses respired 67 % of new assimilate belowground, whereas shrubs and trees respired only 20 % belowground. Notably, the small amount of label incorporation and longer residence time observed in black spruce root respiration supported ^{14}C root respiration studies described earlier that indicated a substantial portion of C fueling root respiration was derived from longer-lived stored carbohydrate pools (Czimczik et al. 2006; Schuur and Trumbore 2006).

Exploiting the benefits of both ^{13}C and ^{14}C, the application of a dual pulse-label can optimally quantify C cycling on multiple timescales (Carbone and Trumbore 2007). The ^{13}C label can be used for the short-term (hours to several days) response to take advantage of its greater ease and lower cost of analysis (i.e., IRMS or new portable laser technologies). Whereas the ^{14}C label measured by AMS, because of its greater signal strength, can be used to detect the small quantities of label in CO_2 respired several weeks to months following the label application. By applying such a dual label, Carbone and Trumbore (2007) increased the frequency of sampling during the initial chase period (0–6 days) with measurements of ^{13}C to better quantify the dynamics of the most rapidly cycling labeled C pools (Fig. 6.17). Then, the ^{14}C label was used to quantify longer-lived respiration sources over a month after labeling, as well as the remobilization of labeled storage C pools for new leaf respiration nine months after labeling (Fig. 6.18). This dual-label approach enabled researchers to identify three C pools used in plant metabolism in an intact

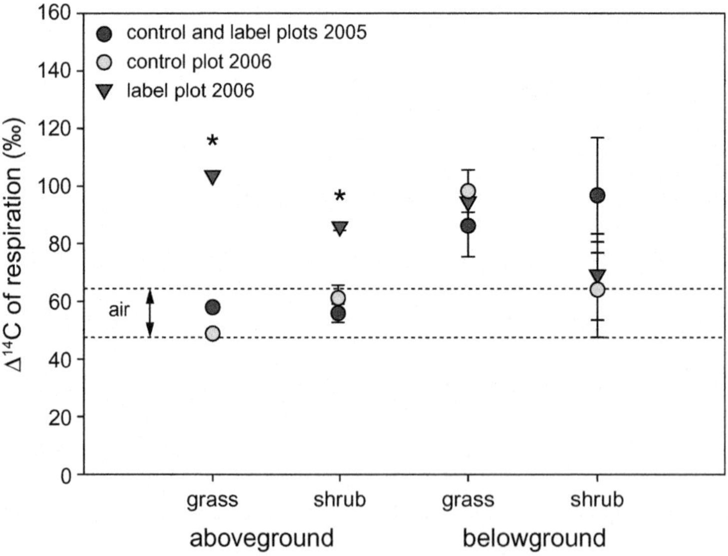

Fig. 6.18 $\Delta^{14}C$ values of respiration in grasses and shrubs nine months after labeling. Aboveground sources show substantial label contributions to respiration. Figure modified from Carbone and Trumbore (2007)

ecosystem: (1) a *fast* pool, composed of assimilation of the current day; (2) an *intermediate* pool, which integrated assimilation during the growing season with a turnover time of tens of days; and (3) a *storage* pool with a residence time of months to years, which was mobilized when necessary such as during initial leaf growth in the spring.

6.5.3 Enriched Isotope Background Study (EBIS)

Along the same theme as the intentionally applied low-level label experiments described in the previous section, an *unintentional* ecosystem-scale low-level ^{14}C label occurred in a temperate forest near Oak Ridge, Tennessee, into which hazardous waste incinerators released ^{14}C enriched CO_2 into the local atmosphere (Trumbore et al. 2002). Researchers have opportunistically used this low-level ^{14}C label to better understand soil C cycling in the Oak Ridge ecosystem and more recently expanded the application by harvesting labeled leaf and root litter from this site and applying it in other temperate forest ecosystems with different climates and soil textures (McFarlane et al. 2012).

The EBIS ^{14}C label has been especially useful for validating rates of decomposition processes and soil C dynamics. For example, litter decomposition determined by loss of ^{14}C-labeled litter over a growing season was about 30 % faster than rates obtained in litterbag studies at the site, highlighting the role of fragmentation and comminution by soil biota that can be excluded by traditional litterbag techniques (Hanson et al. 2005). In another study, soil C pools from labeled plots, experimentally separated by density, had distinct ^{14}C values suggesting that soil C pools with different cycling rates could be isolated with the laboratory fractionation scheme (Swanston et al. 2005). Interestingly, the dense fraction that was assumed to be slowest cycling also contained some ^{14}C label, thus demonstrating that this pool itself was likely not completely homogeneous but instead also contained a fast-cycling component that could not be separated. Similarly, investigations that followed the accidental label into roots found that both live-root turnover and dead-root decomposition take place on two different timescales: <1–2 years and $\sim$10 years, suggesting that both the live- and dead-root pools were each comprised of several distinct pools with different cycling rates (Riley et al. 2009). Other studies illuminated the role of storage C pools as the source of 55 % of new root growth (Gaudinski et al. 2009) and of early season root respiration (Cisneros-Dozal et al. 2006).

Perhaps most importantly, EBIS has enabled researchers to elucidate the relative contributions of roots and litter to soil respiration, soil C stocks, and microbial substrate use. Specifically, reciprocal transplants of labeled, high ^{14}C litter on plots with trees that were not exposed to the ^{14}C pulse (high ^{14}C litter, ambient ^{14}C roots) and vice versa (ambient ^{14}C litter, high ^{14}C roots) created conditions where the label could be traced to either litter or root sources (Fig. 6.19). This revealed that soil respiration flux and variability was dominated by litter decomposition that was

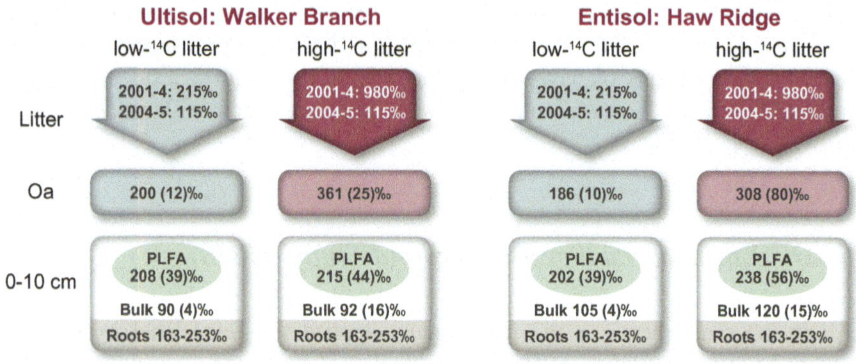

Fig. 6.19 $\Delta^{14}C$ values of labeled litter and ecosystem C pools in the Enriched Isotope Background Study experiment. Figure modified from Kramer et al. (2010)

strongly affected by moisture conditions, with root and rhizosphere respiration that were relatively constant over the growing season (Cisneros-Dozal et al. 2006). Soil microbes as a whole, as well as ectomycorrhizal fungi, obtained the majority of their C from roots, with <6 % and <2 % coming from leaf litter sources, respectively (Treseder et al. 2006; Kramer et al. 2010). By measuring multiple ecosystem C pools within this experimental manipulation of labeled litter, these studies highlighted the role of roots, rather than decaying leaves or previously existing soil C, in driving C-cycling processes of mineral soil. This suggests a decoupling of decomposition processes in the litter layer from that of the mineral soils and has led to new advances in soil respiration modeling of forest ecosystems (Parton et al. 2010).

6.5.4 *Free-Air Carbon Dioxide Enrichment Labels*

It is also possible to manipulate ^{14}C ratios in terrestrial ecosystems by the inverse approach—adding a ^{14}C-free C source—rather than new ^{14}C atoms to the study system. Depleted low-level ^{14}C labels have less ^{14}C than the background and thus require no compliance with radiation regulations. Similar to the ^{13}C labeling technique as described above, the potential dilution of background ^{14}C is limited in absolute magnitude, so a common depleted CO_2 label application is free-air CO_2 enrichment (FACE) studies, where total C levels are higher permitting a greater isotope dilution factor (Pataki et al. 2003). FACE experiments were designed to study the effects of elevated CO_2 concentrations on ecosystem function, with a particular focus on C cycling. Under FACE fumigation, CO_2 concentrations are generally raised $\sim$200 ppm above ambient levels by fumigation with a fossil CO_2 source ($\Delta^{14}C = -1000$ ‰), resulting in an elevated CO_2 atmosphere which is very depleted in ^{14}C (and also ^{13}C) compared to the background atmosphere. Thus, C fixed by photosynthesis into plant tissue and incorporated into SOM in the elevated CO_2 plots since the initiation of FACE is isotopically distinct from older, pre-experiment soil C. FACE experiments have been done in many ecosystems over the past two decades, including large-statured forest ecosystems, which are otherwise difficult to label. The continuous nature of the fumigation label and the longevity of the experiments make FACE experiments a unique opportunity to focus on decadally-cycling C pools (Fig. 6.1). Many of the FACE experiments have been actively labeling by CO_2 fumigation for more than a decade, meaning that pre-FACE C is >10 years old. Effectively, FACE fumigation results in ^{14}C and ^{13}C-labeled soil C pools of known age that can be resolved using ^{14}C mass balance/isotopic mixing models described earlier in this chapter. Thus, the label offers the unique opportunity to compare global change factors, such as warming, on soil C pools of different ages (e.g., Andrews et al. 2000; Hopkins et al. 2012).

Many studies have taken advantage of ^{13}C label in FACE (e.g., Matamala et al. 2003; Lichter et al. 2008; Pregitzer et al. 2008) to examine C cycling belowground; however, the addition of ^{14}C measurements offers significant additional advantages. In FACE, the primary difference between the two isotopes is the magnitude of the label ($\sim$300 ‰ in ^{14}C and $\sim$10 ‰ in ^{13}C; Table 6.2), with the larger ^{14}C label allowing more precise estimates of FACE and pre-FACE C pools. The greater label magnitude also means that variation in target CO_2 concentration has a much larger effect on resulting isotopic values of photosynthate. Cotton plants grown under elevated CO_2 had greater variance in ^{14}C as compared to variance in ^{13}C, relative to the variance of those isotopes in the ambient control (Leavitt et al. 1994). Nevertheless, the magnitude of the label is large enough to overcome this variance, and partitioning of the FACE C contribution to respiration turns out to be relatively insensitive to variance in the ^{14}C value of the FACE atmosphere (Hopkins et al. 2012).

Another advantage to using the ^{14}C signal in FACE experiments is the ability to study the effect of elevated CO_2 on soil C cycling using this isotope tracer much in the same way as has been described throughout this chapter (Torn et al. 2013; Leavitt et al. 1994; Harrison 2004; Hopkins et al. 2012). While the ^{13}C label in FACE can be used to track biological processes in the elevated CO_2 plot, it is not

Table 6.2 Approximate isotope values for C pools with different turnover times and exposed to a C label in the form of a free-air CO_2 enrichment (FACE) experiment

Pool	Age	$\delta^{13}C$ (‰)	$\Delta^{14}C$ (‰)
Photosynthate (FACE control)	Hours–weeks	−27	45 (for calender year 2008)
Photosynthate (FACE elevated)	Hours–weeks	−39	−309
Young soil C (FACE elevated)	1–5 years	−34	−300
Older soil C	15–100s years	−26.5	60–145 (mean age 5–20 years)

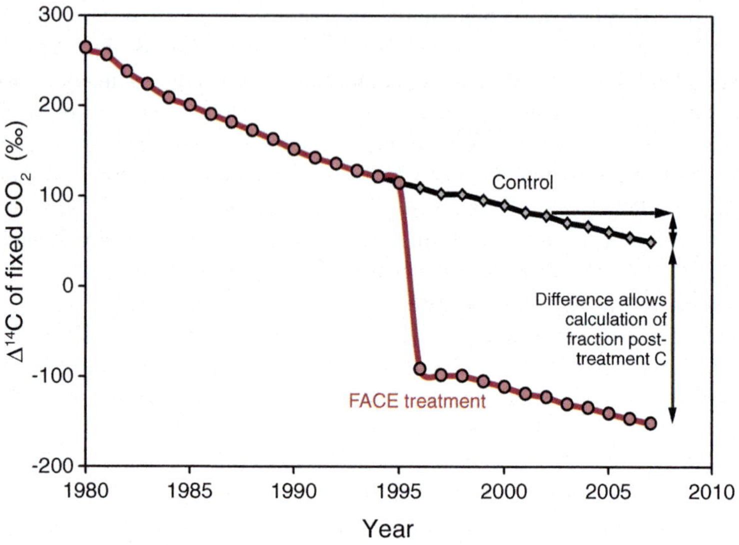

Fig. 6.20 $\Delta^{14}C$ values of fixed C (photosynthate) over the duration of a free-air CO_2 enrichment (FACE) experiment: control plots $\Delta^{14}C$ (*gray*) and elevated CO_2 plots $\Delta^{14}C$ (*pink*). Figure modified from Hopkins et al. (2012)

present in the control plots in significant levels with which to study the effects of CO_2 concentration on ecosystem C cycling (but see Billings and Ziegler 2008). In contrast, there is ^{14}C label in both ambient and elevated CO_2 treatments due to the time-varying ^{14}C content of the atmosphere (and thus new photosynthate) from the decline of atmospheric bomb-derived ^{14}C, and in the elevated CO_2 plot, the same signal exists with an additional −300 ‰ offset (Fig. 6.20). Using both bomb ^{14}C label and FACE ^{14}C label together, changes in the C source of heterotrophic respiration can be detected in response to CO_2 enrichment (Hopkins et al. 2012).

Further differences in the response of ^{14}C and ^{13}C in FACE to biochemical processes means that measurement of both isotopes gives unique and complementary information. For example, biochemical processes often cause isotope fractionation, which is recorded in the value of ^{13}C, but not in ^{14}C, which is commonly corrected for mass-dependent fractionation (Chaps. 3 and 9). This can be important in situations where mass-dependent fractionation might be indicative of a C cycle process, such as shifts in substrate use by microbes, or in temperature manipulation experiments as a result of the well-known effect of temperature on isotope fractionation. For example, there was a 3–5 ‰ fractionation between 4 and 22 °C in ^{13}C of CO_2 respired by soil incubations from a southern pine forest enriched with CO_2, which would not have been detected by measuring ^{14}C alone (Andrews et al. 2000). In such situations, it is preferable to use ^{14}C instead to partition age of respiration sources, while ^{13}C may be used as an indicator of substrate status of the microbial community. In a follow-up experiment at the same site, the same temperature-dependent fractionation effect was observed in respired $^{13}CO_2$, while the $^{14}CO_2$ flux was used to determine that the respiration of pre-FACE and post-FACE C pools was stimulated by a similar amount with warming (Hopkins et al. 2012). Finally, use of both C isotopes also gives an extra constraint for C cycle modeling and allows partitioning of an additional C source to respiration with isotopic mass balance as described earlier in this chapter.

6.5.5 Additional ^{14}C Label Application Techniques

Adding low-level ^{14}C-labeled compounds directly to soil, either in aqueous solution (such as dissolved sucrose) or as a solid (such as labeled charcoal) is perhaps the most closely related in approach to traditional ^{14}C labeling experiments of the past. In fact, the turnover of specific C compounds in soil has been studied by applying traditional, high-level ^{14}C-labeled compounds directly to soil (van Veen and Paul 1981), but now can presently be accomplished at much lower levels with AMS, greatly simplifying experimental procedures. In fact, natural materials produced during the bomb peak of the mid-twentieth century are enriched enough in ^{14}C relative to the background environment to be used as a label. Sucrose from sugar cane grown from 1965 to 1971, presently serving as a global ^{14}C standard (ANU), has been applied directly to soils to investigate the potential of 'priming,' where addition of a labile C substrate leads to increased decomposition of slower-cycling SOM (Fontaine et al. 2007). The priming effect can only be truly quantified via the addition of C isotope-labeled compounds that are distinct in both ^{14}C and ^{13}C from native soil C. In this way, the isotope mass balance of ^{13}C can be used to determine the contribution of the added substrate to respiration, and ^{14}C can be used to determine the source of excess (primed) soil C-derived CO_2 released due to addition of sucrose (Fig. 6.21) (Kuzyakov and Bol 2006). ANU sucrose added to soil incubations from a southern pine-elevated CO_2 experiment increased the contribution of pre-FACE (>10 year C) to CO_2 respired, along with overall stimulation of respiration by the sucrose addition (Hopkins et al. 2014). These results, however, contrasted with an incubation of soils from the Sierra Nevada mountains of California where neither stimulation of respiration nor a change in isotopes of respired CO_2 was detected with the addition of ANU sucrose (Koarashi et al. 2012). These differences in priming response likely stem from divergent limitations to microbes in these contrasting soil systems—respiration from soils in the Sierra Nevada of California increased with water addition, while pine soils from the southeastern USA did not, illustrating that microbial respiration in the former were water limited, whereas the latter incubations were substrate limited.

Pulse labels of liquid or solid substrates applied directly to the soil can also be used to examine uptake and usage rates of specific decomposer communities. One

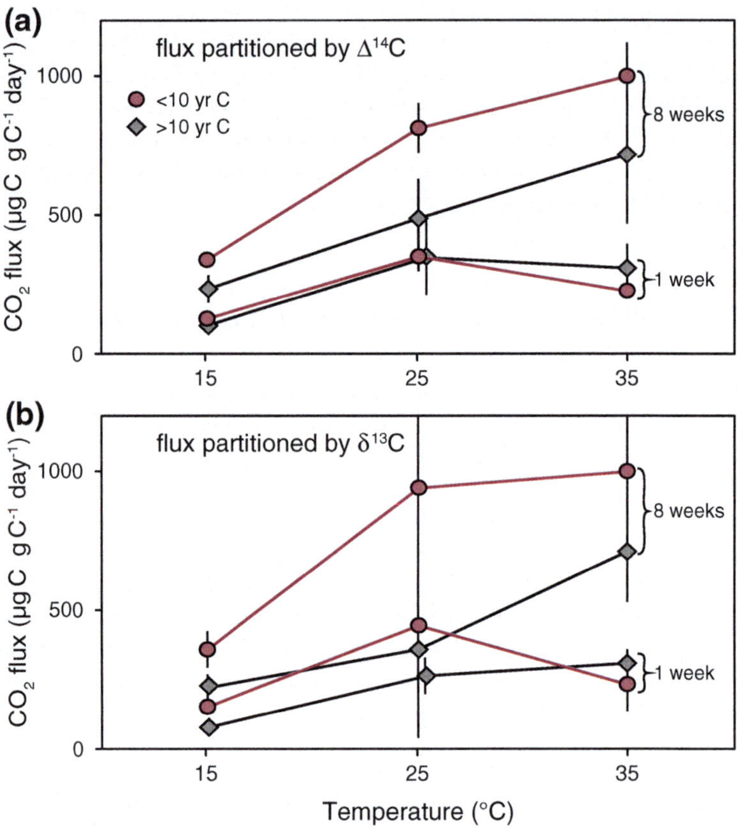

Fig. 6.21 Comparing (**a**) ^{14}C partitioning and (**b**) ^{13}C partitioning of respired C age in FACE-elevated plots. The FACE signal was used to partition the >10-year C (pre-FACE) and <10-year C (FACE C) respired during incubation at three temperatures. Patterns are similar with the two isotopes, but ^{14}C partitioning has much smaller error bars

test of the potential of this approach was the injection of a ^{14}C-labeled glycine solution into a boreal forest soil to determine how fast and over what distances the labeled C was taken up, transported, and assimilated by fungi (Czimczik et al. 2005). To avoid disturbance to the intact soil ecosystem, measurements of ^{14}C in both aboveground fungal sporocarps and respiration from those sporocarps were made to detect the translocated label. Labeled glycine was metabolized by the fungi and respired from their sporocarps within 5 h, indicating rapid uptake and transport of glycine C over several meters through fungal rhizomorphs. Methodologically, this study showed that only a small amount ($\sim$0.0001 g C) of ^{14}C-labeled glycine was necessary, thus minimizing perturbations to competitive relationships among plants and soil microorganisms, as well as any possible 'priming' effects associated with adding a labile substrate to the soil.

6.5.6 Compound-Specific ^{14}C Labeling

Fungal uptake of ^{14}C-glycine and subsequent measurement of respiration from sporocarps is a unique study where a labeled substrate was followed through the assimilation process. In general, addition of ^{14}C-labeled compounds followed by measurement of $^{14}CO_2$ allows the determination of compound turnover rates, but generally do not provide more specific information about metabolic pathway or decomposer organism. This is because decomposition is the activity of a diverse microbial community and because processes are often occurring belowground within the spatially heterogeneous soil matrix. The role of specific microbial functional types in degrading certain compounds may be detected by tracing labeled C directly into their biomass. The most commonly used biomarker is the phospholipid fatty acid (PLFA), a class of polar lipids that comprise the membrane of microbial cells, and whose unique structure can be used to identify particular microbial functional groups (i.e., bacteria, fungi). Since PLFAs break down in the soil after cell death, this isolates metabolically active microbes. Extraction of PLFAs, followed by ^{14}C analysis, can be used to investigate decomposition pathways in soil by enabling identification of microbes and their C source.

One of the first studies of ^{14}C-PLFA yielded ^{14}C values close to that of the atmosphere, illustrating that most microbial C is derived from recent photosynthate with relatively fast turnover times (Rethemeyer et al. 2004). This technique requires a large difference in ^{14}C between C sources in order to detect differences because the error estimated from replicate analyses was 40 ‰ (Kramer et al. 2010); thus, the technique is most powerful in conjunction with the addition of a ^{14}C label. In the EBIS experiment described earlier, ^{14}C-PLFA showed that roots were most likely the predominant source of microbial C rather than decomposing litter or SOM, confirming results from laboratory soil incubations (Kramer et al. 2010). Under natural abundance ^{14}C conditions, the technique is most useful for detecting degradation of ancient or industrial, fossil fuel-derived ^{14}C-free materials. For example, ancient C ($\Delta^{14}C = -990$ ‰) in black shale was shown to be biodegradable by aerobic prokaryotes by culturing microorganisms on a kerogen substrate in order to get sufficient PLFA material for ^{14}C analysis (Petsch et al. 2001). Similarly, ^{14}C-PLFA values of −800 to −900 ‰ demonstrated microbial degradation of petroleum-derived hydrocarbons in the contaminated soils of a bioremediation site compared to the modern values (40 to −150 ‰) found at the control site (Cowie et al. 2010). Thus, detecting the activity of particular microbes using compound-specific ^{14}C measurement can be an important technique where and when the experimental system or the scientific question allows for it.

Compound-specific ^{14}C analysis has become increasingly feasible due to advances in AMS and analytical techniques such as small sample sizes (Chap. 9). While the measurement of PLFA biomarkers has been the most widely used technique to date, it is becoming increasingly possible to measure and interpret ^{14}C in other compounds of terrestrial origin in ocean sediments and DOC (Chap. 5). Measurements of ^{14}C have been performed on isolated compounds in environmental samples such as lignin phenols (biomarker derived from vascular plants, Ingalls et al. 2010), polycyclic aromatic hydrocarbons (a toxic product of incomplete combustion of organic matter, e.g., Kumata et al. 2006), and *n*-alkanes (a biomarker derived from leaf waxes, e.g., Uchikawa et al. 2008). One notable finding with respect to the terrestrial C cycle using these techniques is that not all organic C lost in the dissolved form from thawing permafrost soils is remineralized to CO_2. Large differences detected in the ^{14}C of n-alkanes suggested that younger, peat-derived C remained in suspension in the water column while old, mineral-stabilized C settled out into the sediment layer in a high-latitude environment (Vonk et al. 2010).

6.6 Conclusions and Future Directions

Using ^{14}C as a source tracer and as a measurement of age has provided unique and exciting insight into many aspects of terrestrial ecosystem C dynamics. All of the approaches outlined above have potential new applications both by applying these techniques in new ecosystems, but also by combining techniques together for even more powerful outcomes. As we have seen, applying replicated ^{14}C measurements to ecosystems along environmental gradients, or with experimental manipulations has pushed this analytical tool far beyond its archaeological and geochemical origin. Furthermore, studies using ^{13}C and ^{14}C have often been pursued independently because different driving factors act upon these isotopes. Applying a dual-isotope approach offers another path forward, especially when trying to determine ecosystem sources when there are many, and in combination with other non-isotope datasets. Making use of inadvertent labels such as bomb ^{14}C and FACE studies or purposeful low-level labels allows researchers to focus on multiple timescales relevant to ecosystem C dynamics. Indeed, as bomb ^{14}C levels in the atmosphere steadily decline, the attenuation of this global label will make the application of low-level ^{14}C labels to in situ ecosystems increasingly desirable for studying C cycling in the future. Importantly, while the bomb ^{14}C label attenuates through time, the continued addition of ^{14}C-free C into the atmosphere will continue to alter atmospheric isotope ratios in whole new ways. The implication of this continued alteration of atmospheric ^{14}C has yet to be fully incorporated into our future study designs and will likely offer new opportunities not even yet dreamed of. Finally, continual advances in AMS and analytical techniques are allowing for new applications of ^{14}C in terrestrial ecosystems. Specifically, any compound that can be isolated from the environment might someday be measurable for ^{14}C content with the right experimental conditions and design. In sum, the many bright new paths forward, revealed by the research described here using ^{14}C to discern the dynamics of terrestrial ecosystems, await the next generation of ^{14}C researchers.

References

Andersen, G.L., and K. Krzywinski. 2007. Longevity and growth of Acacia tortilis; insights from 14C content and anatomy of wood. *BMC Ecology* 7: 4.

Andrews, J.A., R. Matamala, K.M. Westover, and W.H. Schlesinger. 2000. Temperature effects on the diversity of soil heterotrophs and the delta C-13 of soil-respired CO_2. *Soil Biology and Biochemistry* 32: 699–706.

Aravena, R., B.G. Warner, D.J. Charman, L.R. Belyea, S.P. Mathur, and H. Dinel. 1993. Carbon isotopic composition of deep carbon gases in an ombrogenous peatland, northwestern Ontario, Canada. *Radiocarbon* 35: 271–276.

Baisden, W.T., R. Amundson, A.C. Cook, and D.L. Brenner. 2002. Turnover and storage of C and N in five density fractions from California annual grassland surface soils. *Global Biogeochemical Cycles* 16: 64.

Barber, K.E., and P.G. Langdon. 2007. What drives the peat-based palaeoclimate record? A critical test using multi-proxy climate records from northern Britain. *Quaternary Science Reviews* 26: 3318–3327.

Barber, K.E., F.M. Chambers, and D. Maddy. 2003. Holocene palaeoclimates from peat stratigraphy: macrofossil proxy climate records from three oceanic raised bogs in England and Ireland. *Quaternary Science Reviews* 22: 521–539.

Billings, S.A., and S.E. Ziegler. 2008. Altered patterns of soil carbon substrate usage and heterotrophic respiration in a pine forest with elevated CO_2 and N fertilization. *Global Change Biology* 14: 1025–1036.

Biondi, F., S.D.J. Strachan, S. Mensing, and G. Piovesan. 2007. Radiocarbon analysis confirms the annual nature of Sagebrush growth rings. *Radiocarbon* 49: 1231–1240.

Blaauw, M. 2010. Methods and code for 'classical' age-modelling of radiocarbon sequences. *Quaternary Geochronology* 5: 512–518.

Blaauw, M., and J.A. Christen. 2005. Radiocarbon peat chronologies and environmental change. *Journal of the Royal Statistical Society Series C-Applied Statistics* 54: 805–816.

Borken, W., K. Savage, E.A. Davidson, and S.E. Trumbore. 2006. Effects of experimental drought on soil respiration and radiocarbon efflux from a temperate forest soil. *Global Change Biology* 12: 177–193.

California Department of Health Services. 2005. California Code of Regulations. Title 17. Division 1. Chapter 5. Subchapter 4. Group 2. Article 9. par. 30237. Schedule C.

Camargo, P.B., R.D.P. Salomao, S. Trumbore, and L.A. Martinelli. 1994. How old are large Brazil Nut trees (*Bertholletia excelsa*) in the Amazon? *Scientia Agricola* 51: 389–391.

Carbone, M.S., and S.E. Trumbore. 2007. Contribution of new photosynthetic assimilates to respiration by perennial grasses and shrubs: residence times and allocation patterns. *New Phytologist* 176: 124–135.

Carbone, M.S., C.I. Czimczik, K.E. McDuffee, and S.E. Trumbore. 2007. Allocation and residence time of photosynthetic products in a boreal forest using a low-level (14)C pulse-chase labeling technique. *Global Change Biology* 13: 466–477.

Castanha, C., S. Trumbore, and R. Amundson. 2008. Methods of separating soil carbon pools affect the chemistry and turnover time of isolated fractions. *Radiocarbon* 50: 83–97.

Chambers, J.Q., N. Higuchi, and J.P. Schimel. 1998. Ancient trees in Amazonia. *Nature* 391: 135–136.

Chanton, J.P., P.H. Glaser, L.S. Chasar, D.J. Burdige, M.E. Hines, D. I. Siegel, L.B. Tremblay, and W.T. Cooper. 2008. Radiocarbon evidence for the importance of surface vegetation on fermentation and methanogenesis in contrasting types of boreal peatlands. *Global Biogeochemical Cycles* 22.

Chanton, J.P., J.E. Bauer, P.A. Glaser, D.I. Siegel, C.A. Kelley, S.C. Tyler, E.H. Romanowicz, and A.H. Lazrus. 1995. Radiocarbon evidence for the substrates supporting methane formation within northern Minnesota peatlands. *Geochimica et Cosmochimica Acta* 59: 3663–3668.

Charman, D.J., R. Aravena, and B.G. Warner. 1994. Carbon dynamics in a forested peatland in north-eastern Ontario, Canada. *Journal of Ecology* 82: 55–62.

Charman, D.J., R. Aravena, C.L. Bryant, and D.D. Harkness. 1999. Carbon isotopes in peat, DOC, CO_2, and CH_4 in a Holocene peatland on Dartmoor, southwest England. *Geology* 27: 539–542.

Chasar, L.S., J.P. Chanton, P.H. Glaser, D.I. Siegel, and J.S. Rivers. 2000. Radiocarbon and stable carbon isotopic evidence for transport and transformation of dissolved organic carbon, dissolved inorganic carbon, and CH_4 in a northern Minnesota peatland. *Global Biogeochemical Cycles* 14: 1095–1108.

Christensen, T., N. Panikov, M. Mastepanov, A. Joabsson, A. Stewart, M. Oquist, M. Sommerkorn, S. Reynaud, and B. Svensson. 2003. Biotic controls on CO_2 and CH_4 exchange in wetlands—a closed environment study. *Biogeochemistry* 64: 337–354.

Cisneros-Dozal, L.M., S. Trumbore, and P.J. Hanson. 2006. Partitioning sources of soil-respired CO_2 and their seasonal variation using a unique radiocarbon tracer. *Global Change Biology* 12: 194–204.

Clymo, R.S. 1984. The limits to peat bog growth. *Philosophical Transactions of the Royal Society of London Series B-Biological Sciences* 303: 605–654.

Clymo, R., and C. Bryant. 2008. Diffusion and mass flow of dissolved carbon dioxide, methane, and dissolved organic carbon in a 7 m deep raised peat bog. *Geochimica Et Cosmochimica Acta* 72: 2048–2066.

Cowie, B.R., B.M. Greenberg, and G.F. Slater. 2010. Determination of microbial carbon sources and cycling during remediation of petroleum hydrocarbon impacted soil using natural abundance C-14 analysis of PLFA. *Environmental Science and Technology* 44: 2322–2327.

Czimczik, C.I., and S.E. Trumbore. 2007. Short-term controls on the age of microbial carbon sources in boreal forest soils. *Journal of Geophysical Research-Biogeosciences* 112: 8.

Czimczik, C.I., K.K. Treseder, M.S. Carbone, and S.E. Trumbore. 2005. Radiocarbon—a low-impact tool to study nutrient transport by soil fungi under field conditions. *New Phytologist* 166: 595–600.

Czimczik, C.I., S.E. Trumbore, M.S. Carbone, and G.C. Winston. 2006. Changing sources of soil respiration with time since fire in a boreal forest. *Global Change Biology* 12: 957–971.

Dioumaeva, I., S. Trumbore, E.A.G. Schuur, M.L. Goulden, M. Litvak, and A.I. Hirsch. 2002. Decomposition of peat from upland boreal forest: Temperature dependence and sources of respired carbon. *Journal of Geophysical Research-Atmospheres* 108.

Don, A., and E.D. Schulze. 2008. Controls on fluxes and export of dissolved organic carbon in grasslands with contrasting soil types. *Biogeochemistry* 91: 117–131.

Duddleston, K.N., M.A. Kinney, R.P. Kiene, and M.E. Hines. 2002. Anaerobic microbial biogeochemistry in a northern bog: acetate as a dominant metabolic end product. *Global Biogeochemical Cycles* 16.

Dutta, K., E.A.G. Schuur, J.C. Neff, and S.A. Zimov. 2006. Potential carbon release from permafrost soils of northeastern Siberia. *Global Change Biology* 12: 2336–2351.

Ellis, C.J., and L. Rochefort. 2006. Long-term sensitivity of a high Arctic wetland to Holocene climate change. *Journal of Ecology* 94: 441–454.

Evans, C., C. Freeman, L. Cork, D. Thomas, B. Reynolds, M. Billett, M. Garnett, and D. Norris. 2007. Evidence against recent climate-induced destabilisation of soil carbon from C-14 analysis of riverine dissolved organic matter. *Geophysical Research Letters* 34.

Fichtler, E., D.A. Clark, and M. Worbes. 2003. Age and long-term growth of trees in an old-growth tropical rain forest, based on analyses of tree rings and C-14. *Biotropica* 35: 306–317.

Fissore, C., C.P. Giardina, R.K. Kolka, C.C. Trettin, G.M. King, M.F. Jurgensen, C.D. Barton, and S.D. McDowell. 2008. Temperature and vegetation effects on soil organic carbon quality along a forested mean annual temperature gradient in North America. *Global Change Biology* 14: 193–205.

Fissore, C., C.P. Giardina, C.W. Swanston, G.M. King, and R.K. Kolka. 2009. Variable temperature sensitivity of soil organic carbon in North American forests. *Global Change Biology* 15: 2295–2310.

Fontaine, S., S. Barot, P. Barre, N. Bdioui, B. Mary, and C. Rumpel. 2007. Stability of organic carbon in deep soil layers controlled by fresh carbon supply. *Nature* 450:277–U210.

Frank, D., T. Depriest, K. McLauchlan, and A. Risch. 2011. Topographic and ungulate regulation of soil C turnover in a temperate grassland ecosystem. *Global Change Biology* 17: 495–504.

Friend, A.L., G. Scarasciamugnozza, J.G. Isebrands, and P.E. Heilman. 1991. Quantification of 2 year old hybrid poplar root systems—morphology, biomass, and C-14 distribution. *Tree Physiology* 8: 109–119.

Froberg, M., D. Berggren, B. Bergkvist, C. Bryant, and J. Mulder. 2006. Concentration and fluxes of dissolved organic carbon (DOC) in three Norway spruce stands along a climatic gradient in Sweden. *Biogeochemistry* 77: 1–23.

Gaudinski, J.B., S.E. Trumbore, E.A. Davidson, and S.H. Zheng. 2000. Soil carbon cycling in a temperate forest: radiocarbon-based estimates of residence times, sequestration rates and partitioning of fluxes. *Biogeochemistry* 51: 33–69.

Gaudinski, J.B., S.E. Trumbore, E.A. Davidson, A.C. Cook, D. Markewitz, and D.D. Richter. 2001. The age of fine-root carbon in three forests of the eastern United States measured by radiocarbon. *Oecologia* 129: 420–429.

Gaudinski, J.B., M.S. Torn, W.J. Riley, C. Swanston, S.E. Trumbore, J.D. Joslin, H. Majdi, T.E. Dawson, and P.J. Hanson. 2009. Use of stored carbon reserves in growth of temperate tree roots and leaf buds: analyses using radiocarbon measurements and modeling. *Global Change Biology* 15: 992–1014.

Glissman, K., K.-J. Chin, P. Casper, and R. Conrad. 2004. Methanogenic pathway and archaeal community structure in the sediment of eutrophic Lake Dagow: effect of temperature. *Microbial Biology* 48: 389–399.

Gorham, E. 1991. Northern peatlands—role in the carbon-cycle and probable responses to climatic warming. *Ecological Applications* 1: 182–195.

Guo, L.D., C.L. Ping, and R.W. Macdonald. 2007. Mobilization pathways of organic carbon from permafrost to arctic rivers in a changing climate. *Geophysical Research Letters* 34.

Hanson, P.J., C.W. Swanston, C.T. Garten, D.E. Todd, and S.E. Trumbore. 2005. Reconciling change in Oi-horizon carbon-14 with mass loss for an oak forest. *Soil Science Society of America Journal* 69: 1492–1502.

Hardie, S.M.L., M.H. Garnett, A.E. Fallick, N.J. Ostle, and A.P. Rowland. 2009. Bomb-14C analysis of ecosystem respiration reveals that peatland vegetation facilitates release of old carbon. *Geoderma* 153: 393–401.

Harrison, K.G. 2004. Soil carbon CO_2 fertilization factor: the measure of an ecosystem's capacity to increase soil carbon storage in response to elevated CO_2 levels. *Geochemistry Geophysics Geosystems* 5.

Hicks Pries, C., E. Schuur, and G. Crummer. 2013. Thawing permafrost increases old soil and autotrophic respiration in tundra: partitioning ecosystem respiration using $\delta^{13}C$ and $\Delta^{14}C$. *Global Change Biology* 19: 649–661.

Hobbie, E.A., N.S. Weber, J.M. Trappe, and G.J. van Klinken. 2002. Using radiocarbon to determine the mycorrhizal status of fungi. *New Phytologist* 156: 129–136.

Hopkins, F.M., M.S. Torn, and S.E. Trumbore. 2012. Warming accelerates decomposition of decades-old carbon in forest soils. *Proceedings of the National Academy of Sciences of the United States of America* 109: E1753–E1761.

Hopkins, F.M., T.R. Filley, G. Gleixner, M. Lange, S.M. Top, and S.E. Trumbore. 2014. Increased belowground carbon inputs and warming promote loss of soil organic carbon through complementary microbial responses. *Soil Biology and Biochemistry* 76: 57–69.

Horvitz, C.C., and L. Sternberg. 1999. C-14 dating of tree falls on Barro Colorado Island (Panama): a new method to study tropical rain forest gap dynamics. *Journal of Tropical Ecology* 15: 723–735.

Ingalls, A.E., E.E. Ellis, G.M. Santos, K.E. McDuffee, L. Truxal, R.G. Keil, and E.R.M. Druffel. 2010. HPLC purification of higher plant-derived lignin phenols for compound specific radiocarbon analysis. *Analytical Chemistry* 82: 8931–8938.

Jenkinson, D.S., D.E. Adams, and A. Wild. 1991. Model estimates of CO_2 emissions from soil in response to global warming. *Nature* 351: 304–306.

Joslin, J.D., J.B. Gaudinski, M.S. Torn, W.J. Riley, and P.J. Hanson. 2006. Fine-root turnover patterns and their relationship to root diameter and soil depth in a C-14-labeled hardwood forest. *New Phytologist* 172: 523–535.

Juutinen, S., T. Larmola, R. Remus, E. Mirus, W. Merbach, J. Silvola, and J. Augustin. 2003. The contribution of *Phragmites australis* litter to methane (CH_4) emission in planted and non-planted fen microcosms. *Biology and Fertility of Soils* 38: 10–14.

Keller, J.K., and S.D. Bridgham. 2007. Pathways of anaerobic carbon cycling across an ombrotrophic-minerotrophic peatland gradient. *Limnology and Oceanography* 52: 96–107.

King, J., and W. Reeburgh. 2002. A pulse-labeling experiment to determine the contribution of recent plant photosynthates to net methane emission in arctic wet sedge tundra. *Soil Biology and Biochemistry* 34: 173–180.

King, J.Y., W.S. Reeburgh, K.K. Thieler, G.W. Kling, W.M. Loya, L.C. Johnson, and K. J. Nadelhoffer. 2002. Pulse-labeling studies of carbon cycling in Arctic tundra ecosystems: the contribution of photosynthates to methane emission. *Global Biogeochemical Cycles* 16: 10.

Koarashi, J., W.C. Hockaday, C.A. Masiello, and S.E. Trumbore. 2012. Dynamics of decadally cycling carbon in subsurface soils. *Journal of Geophysical Research* 117: G03033.

Kotsyurbenko, O.R., K.J. Chin, M.V. Glagolev, S. Stubner, M.V. Simankova, A.N. Nozhevnikova, and R. Conrad. 2004. Acetoclastic and hydrogenotrophic methane production and methanogenic populations in an acidic West-Siberian peat bog. *Environmental Microbiology* 6: 1159–1173.

Kramer, C., S. Trumbore, M. Froeberg, L.M.C. Dozal, D. Zhang, X. Xu, G.M. Santos, and P. J. Hanson. 2010. Recent (<4 year old) leaf litter is not a major source of microbial carbon in a temperate forest mineral soil. *Soil Biology and Biochemistry* 42: 1028–1037.

Kruger, M., P. Frenzel, and R. Conrad. 2001. Microbial processes influencing methane emission from rice fields. *Global Change Biology* 7: 49–63.

Kuhry, P. 1994. The role of fire in the development of sphagnum-dominated peatlands in western boreal Canada. *Journal of Ecology* 82: 899–910.

Kuhry, P. 2008. Palsa and peat plateau development in the Hudson Bay Lowlands, Canada: timing, pathways and causes. *Boreas* 37: 316–327.

Kumata, H., M. Uchida, E. Sakuma, T. Uchida, K. Fujiwara, M. Tsuzuki, M. Yoneda, and Y. Shibata. 2006. Compound class specific C-14 analysis of polycyclic aromatic hydrocarbons associated with PM10 and PM1.1 aerosols from residential areas of suburban Tokyo. *Environmental Science and Technology* 40: 3474–3480.

Kurokawa, H., T. Yoshida, T. Nakamura, J.H. Lai, and T. Nakashizuka. 2003. The age of tropical rain-forest canopy species, Borneo ironwood (*Eusideroxylon zwageri*), determined by C-14 dating. *Journal of Tropical Ecology* 19: 1–7.

Kuzyakov, Y., and R. Bol. 2006. Sources and mechanisms of priming effect induced in two grassland soils amended with slurry and sugar. *Soil Biology and Biochemistry* 38: 747–758.

Leavitt, S.W., E.A. Paul, B.A. Kimball, G.R. Hendrey, J.R. Mauney, R. Rauschkolb, H. Rogers, K.F. Lewin, J. Nagy, P.J. Pinter, and H.B. Johnson. 1994. Carbon-isotope dynamics of free-air CO_2-enriched cotton and soils. *Agricultural and Forest Meteorology* 70: 87–101.

Leifeld, J., M. Zimmermann, J. Fuhrer, and F. Conen. 2009. Storage and turnover of carbon in grassland soils along an elevation gradient in the Swiss Alps. *Global Change Biology* 15: 668–679.

Lichter, J., S.A. Billings, S.E. Ziegler, D. Gaindh, R. Ryals, A.C. Finzi, R.B. Jackson, E.A. Stemmler, and W.H. Schlesinger. 2008. Soil carbon sequestration in a pine forest after 9 years of atmospheric CO_2 enrichment. *Global Change Biology* 14: 2910–2922.

Mack, M.C., E.A.G. Schuur, M.S. Bret-Harte, G.R. Shaver, and F.S. Chapin. 2004. Ecosystem carbon storage in arctic tundra reduced by long-term nutrient fertilization. *Nature* 431: 440–443.

Mack, M.C., M.S. Bret-Harte, T.N. Hollingsworth, R.R. Jandt, E.A.G. Schuur, G.R. Shaver, and D.L. Verbyla. 2011. Carbon loss from an unprecedented Arctic tundra wildfire. *Nature* 475: 489–492.

Martens, C.S., C.A. Kelley, and J.P. Chanton. 1992. Carbon and hydrogen isotopic characterization of methane from wetlands and lakes of the Yukon-Kuskokwim Delta, western Alaska. *Journal of Geophysical Research* 97D: 16689–16701.

Martinez-Ramos, M., and E.R. Alvarez-Buylla. 1998. How old are tropical rain forest trees? *Trends in Plant Science* 3: 400–405.

Matamala, R., M. Gonzalez-Meler, J. Jastrow, R. Norby, and W. Schlesinger. 2003. Impacts of fine root turnover on forest NPP and soil C sequestration potential. *Science* 302: 1385–1387.

Mayorga, E., A.K. Aufdenkampe, C.A. Masiello, A.V. Krusche, J.I. Hedges, P.D. Quay, J.E. Richey, and T.A. Brown. 2005. Young organic matter as a source of carbon dioxide outgassing from Amazonian rivers. *Nature* 436: 538–541.

McFarlane, K.J., M.S. Torn, P.J. Hanson, R.C. Porras, C.W. Swanston, M.A. Callaham Jr., and T. P. Guilderson. 2012. Comparison of soil organic matter dynamics at five temperate deciduous forests with physical fractionation and radiocarbon measurements. *Biogeochemistry*. doi:10.1007/s10533-012-9740-1.

Megonigal, J.P., S.C. Whalen, D.T. Tissue, B.D. Bovard, D.B. Albert, and A.S. Allen. 1999. A plant-soil-atmosphere microcosm for tracing radiocarbon from photosynthesis through methanogenesis. *Soil Science Society of America Journal* 63: 665–671.

Mozeto, A.A., P. Fritz, M.Z. Moreira, E. Vetter, R. Aravena, E. Salati, and R.J. Drimmie. 1988. Growth-rates of natural Amazonian forest trees based on radiocarbon measurements. *Radiocarbon* 30: 1–6.

Muhr, J., and W. Borken. 2009. Delayed recovery of soil respiration after wetting of dry soil further reduces C losses from a Norway spruce forest soil. *Journal of Geophysical Research-Biogeosciences* 114.

Muhr, J., W. Borken, and E. Matzner. 2009. Effects of soil frost on soil respiration and its radiocarbon signature in a Norway spruce forest soil. *Global Change Biology* 15: 782–793.

Nakagawa, F., N. Yoshida, A. Sugimoto, E. Wada, T. Yoshioka, S. Ueda, and P. Vijarnsorn. 2002. Stable isotope and radiocarbon compositions of methane emitted from tropical rice paddies and swamps in Southern Thailand. *Biogeochemistry* 61: 1–19.

Neff, J.C., J.C. Finlay, S.A. Zimov, S.P. Davydov, J.J. Carrasco, E.A. G. Schuur, and A.I. Davydova. 2006. Seasonal changes in the age and structure of dissolved organic carbon in Siberian rivers and streams. *Geophysical Research Letters* 33.

Nowinski, N.S., S.E. Trumbore, E.A.G. Schuur, M.C. Mack, and G.R. Shaver. 2008. Nutrient addition prompts rapid destabilization of organic matter in an arctic tundra ecosystem. *Ecosystems* 11: 16–25.

Nowinski, N.S., L. Taneva, S.E. Trumbore, and J.M. Welker. 2010. Decomposition of old organic matter as a result of deeper active layers in a snow depth manipulation experiment. *Oecologia* 163: 785–792.

Olk, D.C., and E.G. Gregorich. 2006. Overview of the symposium proceedings, "Meaningful pools in determining soil carbon and nitrogen dynamics". *Soil Science Society of America Journal* 70: 967–974.

Osterkamp, T.E., M.T. Jorgenson, E.A.G. Schuur, Y.L. Shur, M.Z. Kanevskiy, J.G. Vogel, and V. E. Tumskoy. 2009. Physical and ecological changes associated with warming permafrost and thermokarst in interior Alaska. *Permafrost and Periglacial Processes* 20: 235–256.

Parton, W.J., P.J. Hanson, C. Swanston, M. Torn, S.E. Trumbore, W. Riley, and R. Kelly. 2010. ForCent model development and testing using the Enriched Background Isotope Study (EBIS) experiment.

Parton, W.J., D.S. Schimel, C.V. Cole, and D.S. Ojima. 1987. Analysis of factors controlling soil organic matter levels in Great Plains grasslands. *Soil Science Society of America Journal* 51: 1173–1179.

Pataki, D.E., D.S. Ellsworth, R.D. Evans, M. Gonzalez-Meler, J. King, S.W. Leavitt, G.H. Lin, R. Matamala, E. Pendall, R. Siegwolf, C. Van Kessel, and J.R. Ehleringer. 2003. Tracing changes in ecosystem function under elevated carbon dioxide conditions. *BioScience* 53: 805–818.

Paul, E.A., H.P. Collins, and S.W. Leavitt. 2001. Dynamics of resistant soil carbon of midwestern agricultural soils measured by naturally occurring C-14 abundance. *Geoderma* 104: 239–256.

Paul, E., S. Morris, R. Conant, and A. Plante. 2006. Does the acid hydrolysis-incubation method measure meaningful soil organic carbon pools? *Soil Science Society of America Journal* 70: 1023–1035.

Petsch, S.T., T.I. Eglinton, and K.J. Edwards. 2001. C-14-dead living biomass: evidence for microbial assimilation of ancient organic carbon during share weathering. *Science* 292: 1127–1131.

Posada, J.M., and E.A.G. Schuur. 2011. Relationships among precipitation regime, nutrient availability, and carbon turnover in tropical rain forests. *Oecologia* 165: 783–795.

Pregitzer, K.S., A.J. Burton, J.S. King, and D.R. Zak. 2008. Soil respiration, root biomass, and root turnover following long-term exposure of northern forests to elevated atmospheric CO_2 and tropospheric O_3. *New Phytologist* 180: 153–161.

Rethemeyer, J., C. Kramer, G. Gleixner, G.L.B. Wiesenberg, L. Schwark, N. Andersen, M. J. Nadeau, and P.M. Grootes. 2004. Complexity of soil organic matter: AMS C-14 analysis of soil lipid fractions and individual compounds. *Radiocarbon* 46: 465–473.

Riley, W.J., J.B. Gaudinski, M.S. Torn, J.D. Joslin, and P.J. Hanson. 2009. Fine-root mortality rates in a temperate forest: estimates using radiocarbon data and numerical modeling. *New Phytologist* 184: 387–398.

Sanderman, J., and R. Amundson. 2008. A comparative study of dissolved organic carbon transport and stabilization in California forest and grassland soils. *Biogeochemistry* 309–327.

Schell, D.M. 1983. C-13 and C-14 abundances in Alaskan aquatic organisms—delayed production from peat in Arctic food webs. *Science* 219: 1068–1071.

Schulz, S., and R. Conrad. 1996. Influence of temperature on pathways to methane production in the permanently cold profundal sediment of Lake Constance. *FEMS Microbiology Ecology* 20: 1–14.

Schuur, E.A.G., and S.E. Trumbore. 2006. Partitioning sources of soil respiration in boreal black spruce forest using radiocarbon. *Global Change Biology* 12: 165–176.

Schuur, E.A.G., S.E. Trumbore, M.C. Mack, and J.W. Harden. 2003. Isotopic composition of carbon dioxide from a boreal forest fire: inferring carbon loss from measurements and modeling. *Global Biogeochemical Cycles* 17: 1.

Schuur, E.A.G., J.G. Vogel, K.G. Crummer, H. Lee, J.O. Sickman, and T.E. Osterkamp. 2009. The effect of permafrost thaw on old carbon release and net carbon exchange from tundra. *Nature* 459: 556–559.

Smith, L.C., G.M. MacDonald, A.A. Velichko, D.W. Beilman, O.K. Borisova, K.E. Frey, K.V. Kremenetski, and Y. Sheng. 2004. Siberian peatlands a net carbon sink and global methane source since the early Holocene. *Science* 303: 353–356.

Strand, A., S. Pritchard, M. McCormack, M. Davis, and R. Oren. 2008. Irreconcilable differences: fine-root life spans and soil carbon persistence. *Science* 319: 456–458.

Swanston, C.W., M.S. Torn, P.J. Hanson, J.R. Southon, C.T. Garten, E.M. Hanlon, and L. Ganio. 2005. Initial characterization of processes of soil carbon stabilization using forest stand-level radiocarbon enrichment. *Geoderma* 128: 52–62.

Tamm, C.O., and H.G. Ostlund. 1960. Radiocarbon dating of soil humus. *Nature* 185: 706–707.

Tierney, G.L., and T.J. Fahey. 2002. Fine root turnover in a northern hardwood forest: a direct comparison of the radiocarbon and minirhizotron methods. *Canadian Journal of Forest Research-Revue Canadienne De Recherche Forestiere* 32: 1692–1697.

Tolonen, K., and J. Turunen. 1996. Accumulation rates of carbon in mires in Finland and implications for climate change. *Holocene* 6: 171–178.

Torn, M.S., C.W. Swanston, C. Castanha, and S.E. Trumbore. 2009. Storage and turnover of organic matter in soil. In *Biophysico-chemical processes involving natural nonliving organic matter in environmental systems*, 219–272. Wiley.

Torn, M.S., S.E. Trumbore, O.A. Chadwick, P.M. Vitousek, and D.M. Hendricks. 1997. Mineral control of soil organic carbon storage and turnover. *Nature* 389: 170–173.

Torn, M.S., A.G. Lapenis, A. Timofeev, M.L. Fischer, B.V. Babikov, and J.W. Harden. 2002. Organic carbon and carbon isotopes in modern and 100-year-old-soil archives of the Russian steppe. *Global Change Biology* 8: 941–953.

Torn, M.S., P.M. Vitousek, and S.E. Trumbore. 2005. The influence of nutrient availability on soil organic matter turnover estimated by incubations and radiocarbon modeling. *Ecosystems* 8: 352–372.

Torn, M.S., M. Kleber, E.S. Zavaleta, B. Zhu, C.B. Field, and S.E. Trumbore. 2013. A dual isotope approach to isolate soil carbon pools of different turnover times. *Biogeosciences* 10: 8067–8081.

Townsend, A.R., P.M. Vitousek, and S.E. Trumbore. 1995. Soil organic-matter dynamics along gradients in temperature and land-use on the island of Hawaii. *Ecology* 76: 721–733.

Treseder, K.K., M.S. Torn, and C.A. Masiello. 2006. An ecosystem-scale radiocarbon tracer to test use of litter carbon by ectomycorrhizal fungi. *Soil Biology and Biochemistry* 38: 1077–1082.

Trumbore, S.E. 1993. Comparison of carbon dynamics in tropical and temperate soils using radiocarbon measurements. *Global Biogeochemical Cycles* 7: 275–290.

Trumbore, S. 2006. Carbon respired by terrestrial ecosystems—recent progress and challenges. *Global Change Biology* 12: 141–153.

Trumbore, S. 2009. Radiocarbon and soil carbon dynamics. *Annual Review of Earth and Planetary Sciences* 37: 47–66.

Trumbore, S., and J. Gaudinski. 2003. The secret lives of roots. *Science* 302: 1344–1345.

Trumbore, S.E., and J.W. Harden. 1997. Accumulation and turnover of carbon in organic and mineral soils of the BOREAS northern study area. *Journal of Geophysical Research-Atmospheres* 102: 28817–28830.

Trumbore, S.E., and S.H. Zheng. 1996. Comparison of fractionation methods for soil organic matter C-14 analysis. *Radiocarbon* 38: 219–229.

Trumbore, S.E., J.S. Vogel, and J.R. Southon. 1989. AMS 14C measurements of fractionated soil organic matter: an approach to deciphering the soil carbon cycle. *Radiocarbon* 31: 644–654.

Trumbore, S.E., E.A. Davidson, P.B. Decamargo, D.C. Nepstad, and L.A. Martinelli. 1995. Belowground cycling of carbon in forests and pastures and eastern Amazonia. *Global Biogeochemical Cycles* 9: 515–528.

Trumbore, S.E., O.A. Chadwick, and R. Amundson. 1996. Rapid exchange between soil carbon and atmospheric carbon dioxide driven by temperature change. *Science* 272: 393–396.

Trumbore, S.E., J.L. Bubier, J.W. Harden, and P.M. Crill. 1999. Carbon cycling in boreal wetlands: a comparison of three approaches. *Journal of Geophysical Research-Atmospheres* 104: 27673–27682.

Trumbore, S., J.B. Gaudinski, P.J. Hanson, and J.R. Southon. 2002. A whole-ecosystem carbon-14 label in a temperate forest. *EOS* 83(265): 267–268.

Trumbore, S., E.S. Da Costa, D.C. Nepstad, P.B. De Camargo, L. Martinelli, D. Ray, T. Restom, and W. Silver. 2006. Dynamics of fine root carbon in Amazonian tropical ecosystems and the contribution of roots to soil respiration. *Global Change Biology* 12: 217–229.

Turteltaub, K.W., and J.S. Vogel. 2000. Bioanalytical applications of accelerator mass spectrometry for pharmaceutical research. *Current Pharmaceutical Design* 6: 991–1007.

Uchikawa, J., B.N. Popp, J.E. Schoonmaker, and L. Xu. 2008. Direct application of compound-specific radiocarbon analysis of leaf waxes to establish lacustrine sediment chronology. *Journal of Paleolimnology* 39: 43–60.

van Veen, J.A., and E.A. Paul. 1981. Organic-carbon dynamics in grassland soils.1. Background information and computer-simulation. *Canadian Journal of Soil Science* 61: 185–201.

Vieira, S., S. Trumbore, P.B. Camargo, D. Selhorst, J.Q. Chambers, N. Higuchi, and L.A. Martinelli. 2005. Slow growth rates of Amazonian trees: consequences for carbon cycling. *Proceedings of the National Academy of Sciences of the United States of America* 102: 18502–18507.

von Lutzowa, M., I. Kogel-Knabner, K. Ekschmittb, H. Flessa, G. Guggenberger, E. Matzner, and B. Marschner. 2007. SOM fractionation methods: relevance to functional pools and to stabilization mechanisms. *Soil Biology and Biochemistry* 39: 2183–2207.

Vonk, J.E., L. Sanchez-Garcia, I. Semiletov, O. Dudarev, T. Eglinton, A. Andersson, and O. Gustafsson. 2010. Molecular and radiocarbon constraints on sources and degradation of terrestrial organic carbon along the Kolyma paleoriver transect, East Siberian Sea. *Biogeosciences* 7: 3153–3166.

Wahlen, M., N. Tanaka, R. Henry, B. Deck, J. Zeglen, J.S. Vogel, J. Southon, A. Shemesh, R. Fairbanks, and W. Broecker. 1989. C-14 in methane sources and in atmospheric methane—the contribution from fossil carbon. *Science* 245: 286–290.

Walter, K.M., J.P. Chanton, F.S. Chapin, E.A.G. Schuur, and S.A. Zimov. 2008. Methane production and bubble emissions from arctic lakes: isotopic implications for source pathways and ages. *Journal of Geophysical Research-Biogeosciences* 113.

Wand, U., V.A. Samarkin, H.M. Nitzsche, and H.W. Hubberten. 2006. Biogeochemistry of methane in the permanently ice-covered Lake Untersee, central Dronning Maud Land, East Antarctica. *Limnology and Oceanography* 51: 1180–1194.

Wattel-Koekkoek, E.J.W., P. Buurman, J. van der Plicht, E. Wattel, and N. van Breemen. 2003. Mean residence time of soil organic matter associated with kaolinite and smectite. *European Journal of Soil Science* 54: 269–278.

Worbes, M., and W.J. Junk. 1999. How old are tropical trees? The persistence of a myth. *Iawa Journal* 20: 255–260.

Zimov, S.A., Y.V. Voropaev, I.P. Semiletov, S.P. Davidov, S.F. Prosiannikov, F.S. Chapin III, M. C. Chapin, S. Trumbore, and S. Tyler. 1997. North Siberian lakes: a methane source fueled by Pleistocene carbon. *Science* 277: 800–802.

Chapter 7
Paleoclimatology

J.R. Southon, R. De Pol-Holz and E.R.M. Druffel

7.1 Introduction

The Earth has undergone natural changes in climate that are manifest by glacial–interglacial (IG) periods during the last 2.5 million years. During the past 800,000 years (800 kyr), these glacial–IG cycles have occurred approximately every 100 kyr. These glacial–IG cycles are influenced by three astronomical components. The first two, tilt of Earth's spin axis and eccentricity of Earth's orbit around the Sun, change the intensity of the seasons. The third, precession or wobble of Earth's spin axis, affects the interaction between the tilt and eccentricity. When the amount of sunlight incident at high northern latitudes in summer is greatest, and when there is an immense amount of ice tied up on land, massive melting occurs that ends the glaciation abruptly. The last time this occurred was approximately 14 kya, which marked the start of the Holocene.

Variation of the amount of solar radiation reaching Earth cannot explain the magnitude of climate change. Instead, climate change is accompanied by changes in the global C cycle on land, in the ocean, and in the atmosphere. With the measurements of trace gas concentrations in ancient glacial ice layers, it was revealed that atmospheric CO_2 concentration was one-third lower (195 ppm) during the last glacial maximum (LGM) 20 kya than that during the present IG (12 kya to 200 years ago) (280 ppm) (Fig. 7.1a). Our ability to predict future climate depends on resolving changes in past climate with changes in the C cycle.

This chapter begins by presenting the basic components of paleoclimate science, particularly for the last glacial–IG cycle, and outlines how radiocarbon (^{14}C) has

J.R. Southon · E.R.M. Druffel (✉)
Department of Earth System Science, University of California, Irvine, CA, USA

R. De Pol-Holz
Department of Oceanography, University of Concepcion, Concepcion, Chile

E.A.G. Schuur et al. (eds.), *Radiocarbon and Climate Change*,
DOI 10.1007/978-3-319-25643-6_7

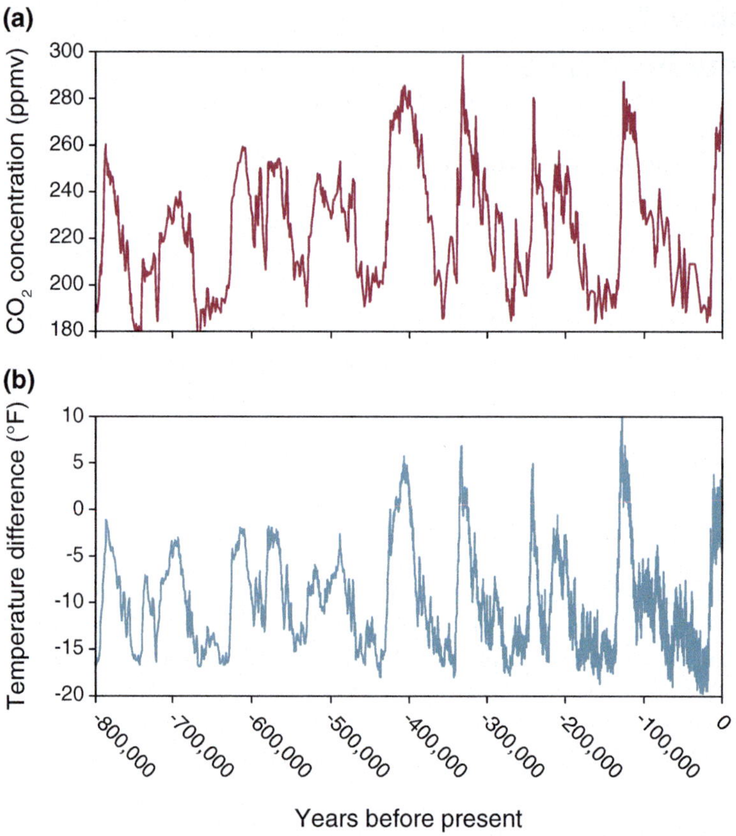

Fig. 7.1 Time histories for the past 400 kyr for **a** pCO_2 as recorded in Vostok ice cores and **b** temperature change. Figure modified from Petit et al. (1999)

helped to provide a timescale for the last of these climate changes. Early ^{14}C dates of felled trees buried within glacial moraines from North America were the first evidence that allowed geologists to put a time frame on the final advance of the last glaciation, known as the Younger Dryas (12–13 kya). The second part of the chapter delves into a recent conundrum in paleo-oceanographic research that reveals evidence of old water in the intermediate and deep ocean during the "mystery interval" from 17.5 to 14.5 kya. This old water would provide evidence of large shifts in the C cycle associated with climate change. A description of ocean water mixing and distributions of inorganic and organic C in the world ocean is provided in Chap. 5 and offers background information for this chapter. The last section describes how ^{14}C is formed in the atmosphere, the processes that determine the shape of calibration curves, and the datasets that are used to construct them. This section fills in the details behind the Chap. 3 section that outlines how ^{14}C dates are calibrated. In summary, this chapter focuses on using the full 50 kyr timescale of ^{14}C to understand the dynamics of C reservoirs on Earth during the last glacial–IG period.

7.2 Paleoclimate—The Basics

7.2.1 *Proxies of Past Climate Change*

Historical records of past climate that have been directly measured, such as temperature records on land and at sea, exist for only the past ~150 years. To obtain older records of past climate, we rely on a variety of proxies, mainly in sediment, ice, corals, and speleothems (cave deposits). These proxies reveal information regarding the main aspects of past climate: ice volume on Earth, temperature, aridity, atmospheric gas content, and ocean chemistry. Proxy materials by their nature are layered and so preserve temporal changes in the climate system that left their imprint that we can measure today. To obtain an in-depth discussion of these proxies and their uses for revealing past climate changes, the reader is referred to Broecker's *The Glacial World According to Wally* (1992) and Broecker and Denton (1990).

Temperature records are obtained from land proxies (pollen in lake sediments, gas content of water in ancient aquifers, tree ring widths) and ocean proxies (e.g., $\delta^{18}O$ in corals and planktonic foraminifera in ocean sediments). During the past deglaciation (15–9 kyr), sea level rose about 120 m because of the melting of large ice sheets that had accumulated on land during the cold glacial period. Evidence is found for the presence of large ice sheets in glacial moraines that mark their edges, ancient shorelines that mark past sea level, and decreases in the total volume of water in the world ocean recorded in $^{18}O/^{16}O$ ($\delta^{18}O$) ratios in the calcium carbonate of benthic (deep sea, bottom dwelling) foraminifera obtained from marine sediment cores. Over the past 800 kyr, the volume of ice sheets during the glacial and IG periods has waxed and waned every ~100 kyr. Synthesis studies that include hundreds of deep-sea sediment cores around the world revealed that the shorter period signals of changes in the Earth's tilt (41 kyr) and precession (23 kyr) were also visible in the $^{18}O/^{16}O$ ratios in benthic foraminifera, as expected for astronomical control of climate (Imbrie 1982). Astronomical controls may have been the trigger in changing climate, but are not the whole story since there needed to be feedbacks from the C cycle to get the observed level of climate change and melting of the ice sheets.

One of the most important proxy records of past temperature is the $\delta^{18}O$ values recorded in layers of glacial ice cores sampled from Greenland and Antarctica. Fluctuations of $\delta^{18}O$ record changes in the local (latitudinal) temperature of the ice layer at the time it was laid down because higher air temperature increases the proportion of the heavier isotope in precipitation (Fig. 7.1b). These records revealed that the temperature during the LGM (20–26.5 kya) was on average 10 °C lower than it was in our present IG period and that the lowering of temperature occurred gradually over tens of thousands of years prior to the LGM, but rose rapidly within a few thousand years to the current conditions. The saw-toothed pattern of temperature that defined glacial/IG cycles is also present in the concentration of CO_2 in air (measured as pCO_2) record obtained from air bubbles frozen within ice core layers (Fig. 7.1a). These air bubbles contain a record of pCO_2 from the past atmosphere, which varied due to redistribution of C among the atmosphere, land, and oceans. Superimposed on the overall ~10 °C drop in temperature during the glacial period were smaller (1–3 °C), more abrupt fluctuations of temperature that happened on 1.5 kyr periods, called Dansgaard–Oeschger (D-O) cycles (Fig. 7.1b). These D-O events are believed to have been caused by changes in ocean circulation and associated C storage shifts (see Sect. 7.3).

Changes in aridity, or the amount of precipitation, during these same time periods has been more difficult to reconstruct as compared to temperature. One proxy for precipitation change—fluctuations in the sizes of closed basin lakes—shows no consistent, global pattern of change between glacial and IG periods. In contrast, windblown dust accumulation recorded in loess soils (continental dust layers), ocean sediments, and ice cores has indicated that rates of dust accumulation were 10–20 times higher during the glacial compared to IG periods. This may have been because there was less vegetation on land, pointing toward less precipitation, or that the wind speed was higher during the glacial periods.

Records of the concentration of the greenhouse gas CO_2 in air are obtained from ice cores and show that pCO_2 was 195 ppm during the glacial periods, 1/3 lower than Holocene values of 280 ppm (Fig. 7.1a). Causes as to why the pCO_2 was lower during the glacials are primarily found in the ocean. Because the ocean contains 60 times the amount of C that is in the atmosphere, the amount of CO_2 in the atmosphere is dependent on ocean chemistry. Colder surface waters during the glacial periods absorbed more CO_2. However, there was 1/3 less organic matter on land during the glacials, which adds to the amount of excess C that would need to have been stored in the ocean during the glacials.

The biologic pump, the fixation of CO_2 by photosynthesis in the mixed ocean layer and transfer of organic matter to the ocean's interior, was found to have been stronger during the glacials and provides a way to store C in the deep ocean. Records of $\delta^{13}C$ values in benthic foraminifera show that a significant amount of terrestrial C from land was indeed stored in ocean water during the glacials.

7.2.2 Meridional Overturning Circulation

The deep ocean conveyor introduced in Chap. 5 represents surface to deep mixing and transport of water throughout the world's oceans (Fig. 7.2). Cold and salty surface water sinks in the northern North Atlantic, and this North Atlantic Deep Water (NADW) flows to the South Atlantic, becoming entrained into the deep eastward current that surrounds Antarctica. It then flows north into the deep Indian and Pacific oceans before rising again where surface currents reconnect back to the source in the North Atlantic (Fig. 7.2). This conveyor is called meridional overturning circulation (MOC) and is driven by a combination of sinking of dense NADW, Southern Ocean wind stress that imparts eastward movement, and the small differences in water density that exist between NADW and the less dense waters further down the conveyor's path. There is evidence that this circulation system has changed during the glacial–IG cycles, which also provides a mechanism for changes in the C cycle observed during past times.

The first main mode of the MOC that characterizes current IG conditions is a strong northern source of NADW (Fig. 7.3 "warm"). The second main mode that operated during the glacial periods was characterized by reduced flow and shallower depth of formation of NADW. This mode is referred to as the glacial "on" mode (interstadials) (Fig. 7.3 "cold") that abruptly stopped at the glacial "off" mode (stadials) (Fig. 7.3 "off") where no NADW was formed. The NADW that forms during the IG periods is salty and deep in contrast to NADW formed during the glacial "on" mode periods that was fresher and did not descend as deeply into the ocean, leaving the deep ocean more isolated and poorly mixed, and older with respect to ^{14}C. Sea ice covered the Antarctic Ocean year-round in the "cold" glacial periods and likely contributed to these changes in the MOC. In the glacial "off" mode, no NADW was produced; instead, it was replaced by Glacial North Atlantic Intermediate Water. In the glacial "off" mode, large amounts of ice in the northern North Atlantic may have contributed to the lack of NADW formation. Widespread ice was detected by the presence of small grains of sand and rocks in ocean

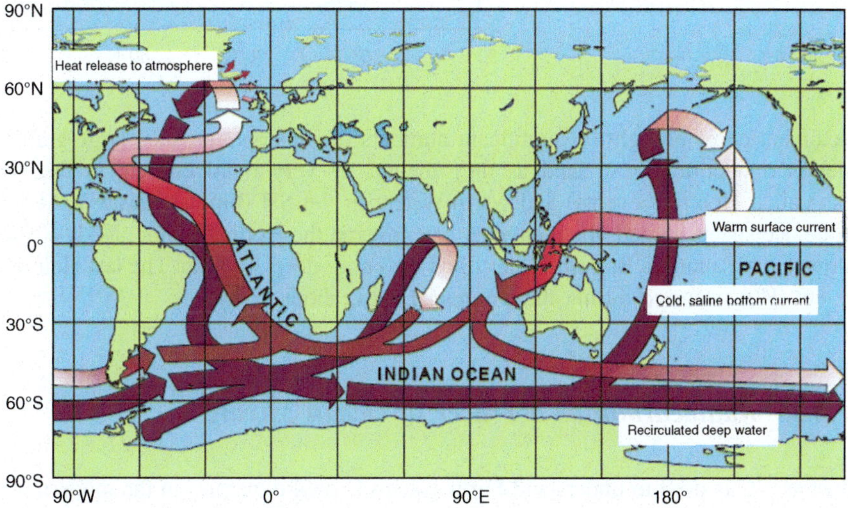

Fig. 7.2 Schematic of the meridional overturning circulation (MOC) system that shows the direction of flow of deep water (*purple*) and surface return flow (*red*). Figure taken from Windows to the World Web site, image courtesy CLIVAR, after W. Broecker, modified by E. Maier-Reimer

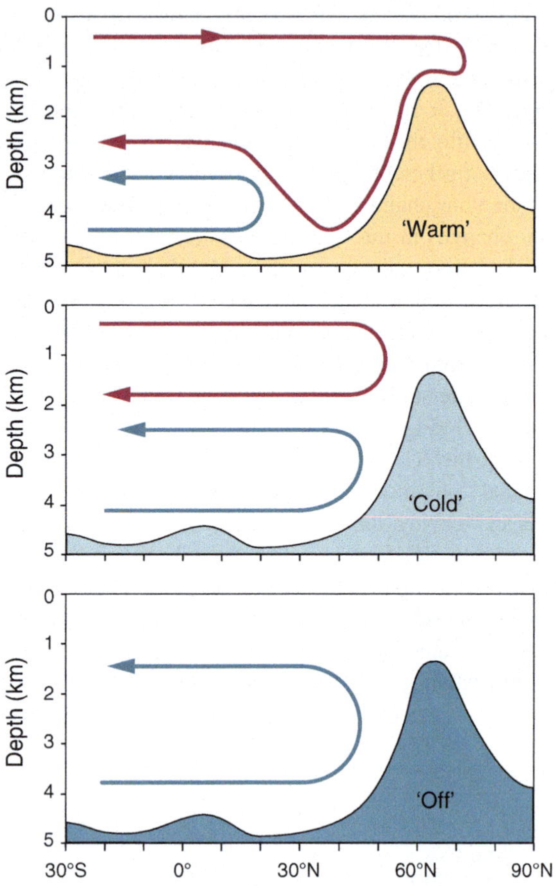

Fig. 7.3 Schematic of the three modes of ocean circulation that prevailed during different times of the last glacial period. Shown is a section along the Atlantic; the rise in bottom topography symbolizes the shallow sill between Greenland and Scotland. North Atlantic overturning is shown by the *red line* and Antarctic Bottom Water by the *blue line*. Figure modified from Rahmstorf (2002)

sediment cores brought by significant numbers of icebergs that shed large-grained debris transported from land as they melted by very rapid cooling in the NH (so-called Heinrich events) and warm waters in the SH (bipolar seasaw). At this time, ice sheets surged, the meltwater cap covered the North Atlantic, NADW shut down, and Antarctic Bottom Water filled the entire deep Atlantic. The last Heinrich event (H0) occurred during the Younger Dryas, about 12 kya.

7.2.3 *Radiocarbon as a Tracer of Ocean Mixing*

Just as ^{14}C is used to understand Earth system C pool dynamics in the modern era, it has also been applied to understanding C pool dynamics of the past in order to interpret paleoclimate.

Most of the ^{14}C (90 %) decays in the deep ocean because this is where most of the dissolved organic carbon (DIC) resides. As shown in Chap. 5, ^{14}C is higher in the atmosphere, lower in the surface ocean, and lowest in the deep ocean because of the time it takes newly produced ^{14}C in the atmosphere to mix throughout the ocean.

Foraminifera are tiny organisms in the ocean with calcium carbonate shells that are preserved in ocean sediments and can be analyzed for ^{14}C content. Some species of foraminifera live only at the bottom of the ocean (benthic), while other species live in the water column near the surface (planktonic), and thus, the shells of these species reflect the ^{14}C of the water where they lived. These shells can be separated by species from layers within the sediment column, and the ^{14}C age difference between benthic and planktonic foraminifera species from the same layer in a sediment column estimates the time since the deep water was at the surface. This is known as the apparent ventilation age, which refers to the last time that the water in the deep ocean was able to exchange C with the atmosphere (Adkins et al. 1998). Considering the differences in ocean mixing that occurred in the glacial period, if the MOC is either reduced (glacial "on" mode) or completely stopped (glacial "off" mode), the atmosphere and the surface ocean $\Delta^{14}C$ values increase. This occurs because newly produced ^{14}C in the atmosphere is diluted in a smaller C pool in the surface ocean, and ^{14}C in the deep ocean decreases because there is reduced input of ^{14}C from the surface based on less mixing, while radioactive decay continues to remove ^{14}C from the deep ocean based on the half-life (see Sects. 7.3 and 7.4). These changes in ^{14}C of different water masses within the ocean are then recorded through time by the foraminifera deposited in sediment layers. Age differences between benthic and planktonic foraminifera measured in individual layers that then change through time (depth) in sediment cores are used as a proxy for changes in the ocean circulation that are thought to have occurred over glacial–IG cycles (other tracers include Cd/Ca from foraminifera, Pa/Th of sediments, $\delta^{13}C$ and $\delta^{18}O$ from foraminifera; see Broecker and Denton (1990)).

Apart from the changes in ocean circulation, ocean ^{14}C can also be influenced by changes in the atmospheric ^{14}C production rate as has happened over the past 50 kyr. This influence of changing production needs to be disentangled if ocean ^{14}C measurements are to be interpreted as a fingerprint of ocean circulation changes. A recent change in atmospheric ^{14}C production via nuclear weapons testing (see Chap. 4) showed that a doubling of atmospheric ^{14}C ($\Delta^{14}C$ up to 1000 ‰) was attenuated as it mixed into the surface ocean. This caused a $\Delta^{14}C$ increase of ~150 ‰ with peak values occurring 10 years later than peak values in the atmosphere (see Chap. 5, Fig. 5.1). Deep ocean $\Delta^{14}C$ changes in response to a perturbation like this are diluted by the huge, deep ocean DIC pool and the slow rate of exchange of C with the surface ocean and the atmosphere and so are attenuated even further. Indeed, bomb ^{14}C has penetrated only to 1200 m in the Pacific in 60 years.

Changes in ocean mixing have a large influence on $\Delta^{14}C$ values of the Earth system, and thus, measurements of past C pools (or their proxies) can be used to interpret shifts that occurred between glacial and IG periods. For example, atmospheric ^{14}C increases with decreased ocean mixing as fewer ^{14}C atoms are dissolved into the ocean, even if cosmic ray production of ^{14}C were to remain constant.

7.3 Case Study: The Old Water Mystery During the Last Glacial–Deglacial Transition

Measurements of corals and of foraminifera in ocean sediments as proxies reveal that atmospheric $\Delta^{14}C$ values were higher 21 kya LGM) than today, with $\Delta^{14}C$ values being almost 500 ‰ higher than the preindustrial atmosphere (Fig. 7.4). The transition from the glacial to the present IG period was characterized by an abrupt reduction in atmospheric $\Delta^{14}C$ that occurred in several stages. One reduction that occurred between 17 and 14.5 kya has received particular attention due to its large magnitude (190 ‰ difference in $\Delta^{14}C$) and that it coincided with the first step in the glacial–IG increase in atmospheric CO_2 concentration. Untangling the cause of this ^{14}C decline can be useful as a tracer for understanding the shifts in Earth system C dynamics and pools that occurred during this important climate regime shift.

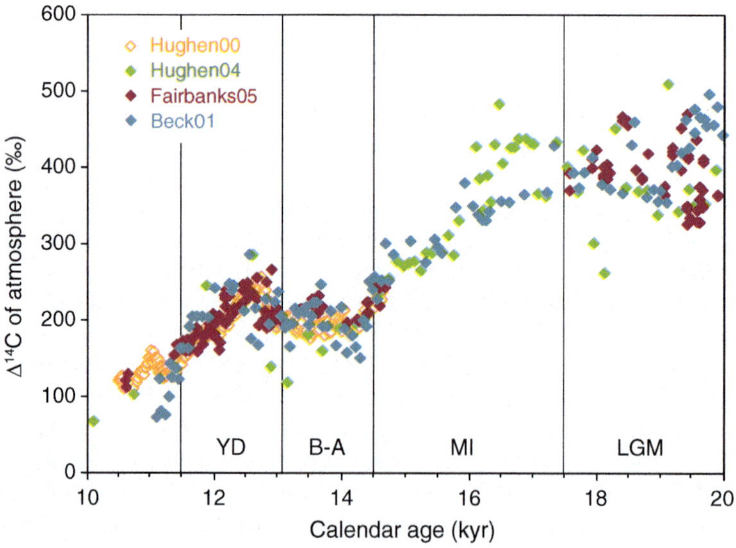

Fig. 7.4 Reconstructions of the atmospheric ^{14}C for the period 10–20 kya (Hughen et al. 2000, 2004a). Fairbanks et al.'s (2005) data come from Th-dated corals, and Beck et al.'s (2001) data are from a Bahamian stalagmite. Figure modified from Broecker and Barker (2007)

7.3.1 The 190 ‰ Drop in Atmospheric Radiocarbon

The magnitude and timing of the 190 ‰ decline in atmospheric $\Delta^{14}C$ are well documented using several independent proxies. Several climate archives using warm water corals and stalagmites show this ^{14}C decline independently and have been absolutely time-dated using uranium/thorium (U-Th) dating, putting a relatively precise age on this event (Fairbanks et al. 2005). These proxies correlate well with independent ^{14}C measurements made from foraminifera from sediment cores in an anoxic ocean basin, after a fixed and constant ^{14}C age difference between the surface ocean and the atmosphere (reservoir age) was applied (Hughen et al. 2006). Sediment layers from these cores cannot be age-dated precisely using U-Th methods, but instead are correlated with known events recorded in yet other long-time cores such as ice cores from Greenland. For changes in atmospheric ^{14}C to be indicative primarily of changes in ocean mixing, it was also important to assess any changes in cosmic ray production of ^{14}C. Cosmic ray production did not appear to concurrently decrease during this time interval, based on the evidence from the accumulation of ^{10}Be (another cosmic ray-produced isotope) in Greenland ice (see Sect. 7.4). Putting this 190 ‰ decline of atmospheric $\Delta^{14}C$ into the context of the global C cycle, the atmosphere would need to be diluted by a release of 5000 Pg of ^{14}C-free CO_2, which could occur if large amounts of old CO_2 were released from the deep ocean that had been depleted in ^{14}C by radioactive decay. This amount is equivalent to 15 % of the entire current ocean DIC pool; thus, such a net release would signify major changes in deep ocean mixing as has been proposed to occur during transitions between glacial and IG periods.

7.3.2 Source of Old Carbon Dioxide Emissions

The mechanism for old C emissions to the atmosphere relates to the reduced MOC (glacial "on" mode) thought to be operating during the last glacial period. This mode reduced contact between the deep ocean and the surface of the Southern Ocean, which is where main contact between the deep ocean and the atmosphere exists today. This promoted the buildup of DIC in the deep ocean because decomposition of organic matter incoming from the surface ocean continued, while degassing back to the atmosphere was reduced. As the contact with the deep ocean was impeded, the atmospheric ^{14}C levels increased and at the same time, the ^{14}C produced in the high atmosphere was not transferred to the ocean depths. This mechanism was suggested by the finding of extremely salty bottom water in the South Atlantic Ocean that yielded a glacial age salinity larger than calculated from samples elsewhere or calculated from the buildup of excess continental ice (Adkins et al. 2002) (Fig. 7.3 "cold"). The density of this salty bottom water, likely created as a result of wintertime expansion of sea ice, would have prevented this water mass from mixing with the rest of the ocean, and the ice would prevent contact with the atmosphere. In that case, the ^{14}C of that water mass would have been depleted by radioactive decay producing a reservoir of DIC that was ^{14}C-poor (Fig. 7.3 "cold"). Alternatively, the sudden addition of ^{14}C-free CH_4 stored as hydrates in ocean-margin sediments is unlikely to have been the source of the old C because there is no prominent peak seen in the ice core CH_4 record, and the amount of CH_4 oxidized would have doubled the ocean's surface pCO_2 and there is no geochemical evidence that this happened. Likewise, an increase in subaerial volcanic activity during deglaciations is unlikely because there was not a large increase in activity during the "mystery interval."

If the old CO_2 emissions came from an isolated, deep reservoir in the Southern Ocean, its existence should have been recorded as a significant age difference between coexisting pairs of planktonic and benthic foraminifera in sediment layers. The expected 5000 Pg C emission that was needed to cause the drop in atmospheric $\Delta^{14}C$ has direct consequences for the size of the old water pool being sought. Age differences between benthic and planktonic foraminifera in several sediment cores retrieved from the upper 2000 m of the ocean do not show any significant variation during the LGM compared to core surface measurements (modern sediment) that yielded an average difference of 1500 years (Broecker et al. 2004). Earlier studies showed 2000-year age differences in Pacific waters at a depth of 3210 m (Shackleton et al. 1988). These measurements constrain the possible size of a ^{14}C-depleted deep ocean pool since the total average depth of the ocean is 3600 m. As the volume occupied by this ^{14}C-depleted abyssal reservoir is reduced, the required age difference rises rapidly. For example, if the isolated abyssal reservoir

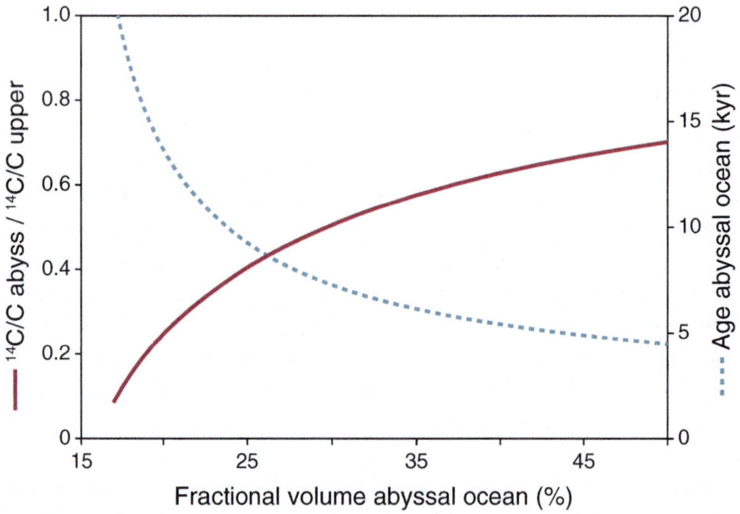

Fig. 7.5 $^{14}C/C$ ratio for the proposed isolated LGM abyssal reservoir to that for the remainder of the ocean and its calculated age with respect to the surface ocean, assuming that the average age difference for upper ocean C (above 2 km) at the LGM was about 1500. Figure modified from Broecker and Barker (2007)

occupied only 30 % of the ocean's volume, the required age difference would need to be increased from 4400 to 6500 ^{14}C years and even more if it was smaller (Fig. 7.5). It is technically challenging to detect these foraminifera age differences in deep ocean settings due to slow sedimentation rates, possible reworking of sediments by organisms, and the corrosive environment of the deep ocean that dissolves foraminifera shell. As such, no cores to date have provided direct evidence that an old, deepwater reservoir existed during the LGM.

7.3.3 Sedimentary Evidence of the Old CO_2 Reservoir

Although there was some evidence from reservoir ages off New Zealand that very old water could have been present at intermediate depths of the Southern Ocean (Sikes et al. 2000), surprisingly there has been a great deal of excitement in light of the recent finding of extremely old waters at intermediate depths off Baja California in the eastern North Pacific Ocean (Marchitto et al. 2007). Coinciding with the deglacial onset of atmospheric CO_2 increase and the decrease of its ^{14}C content, the benthic foraminiferal ^{14}C content off Baja California dropped dramatically by

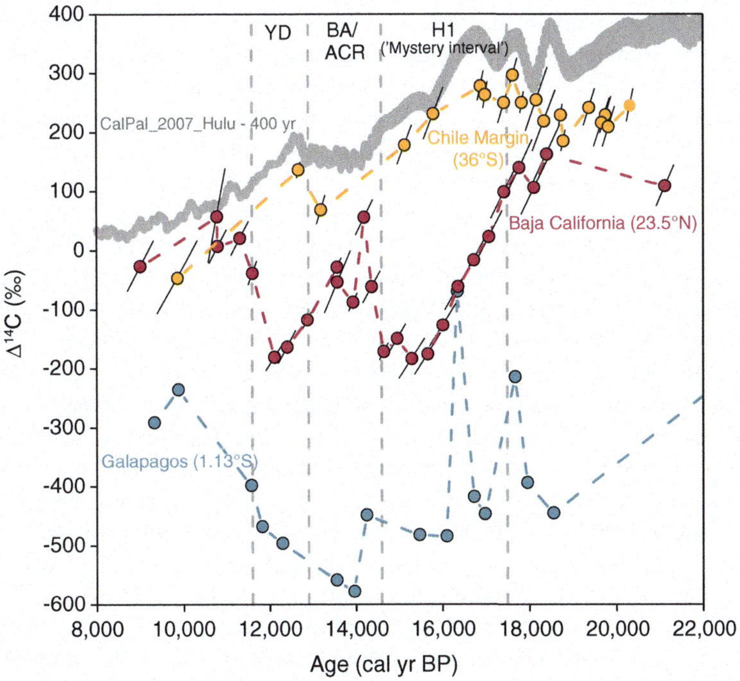

Fig. 7.6 Radiocarbon content of intermediate waters of the Pacific Ocean. Figure modified from De Pol-Holz et al. (2010)

Fig. 7.7 Schematic indicating the proposed mechanism of the ITCZ and the Southern Hemisphere westerly winds having moved southward during the last glacial period. Figure modified from Toggweiler (2009); original by J.L. Toggweiler, Carnegie Mellon University

400 ‰ (Fig. 7.6). This represents an age offset of more than 4000 years between the intermediate waters that bathed the core site at that time and the contemporaneous atmosphere. Similar and even larger offsets have been observed in benthic foraminifera from intermediate water sediment cores off Galapagos and the Arabian Sea (Stott et al. 2009; Bryan et al. 2010). Benthic and planktonic age differences as large as 8000 years are documented for the Galapagos, while at the Arabian Sea, the age difference mimics the record of Baja California (Fig. 7.6).

It has been hypothesized that this very old intermediate water is the result of increased ventilation in the Southern Ocean and the mixing of the very old abyssal water with the surface ocean in regions where active intermediate water mass formation occurs (Anderson et al. 2009). A clear confirmation of this process was the finding of increased wind-driven, upwelling diatom fluxes in different regions of the Southern Ocean during Heinrich H1 (17 kya), a glacial “off” mode. Because isotopic equilibration with the atmosphere takes decades to complete, the wintertime deepening of the pycnocline (depth of the greatest change in water density) and positive wind-driven downward mixing would “contaminate” recently ventilated waters with very low ^{14}C water while having high oxygen content. A likely mechanism by which all of these factors could be functioning in the proposed synchronicity would be the southward displacement of the Intertropical Convergence Zone (ITCZ) during the dramatic decrease of North Atlantic overturning during Heinrich H1. This southward displacement would bring the core of the westerly wind belt to be aligned with the Drake Passage and favor a higher rate of Ekman transport (wind-driven) near Antarctica. Micro- and macronutrient fertilization of the surface ocean would then favor diatom growth and the export of

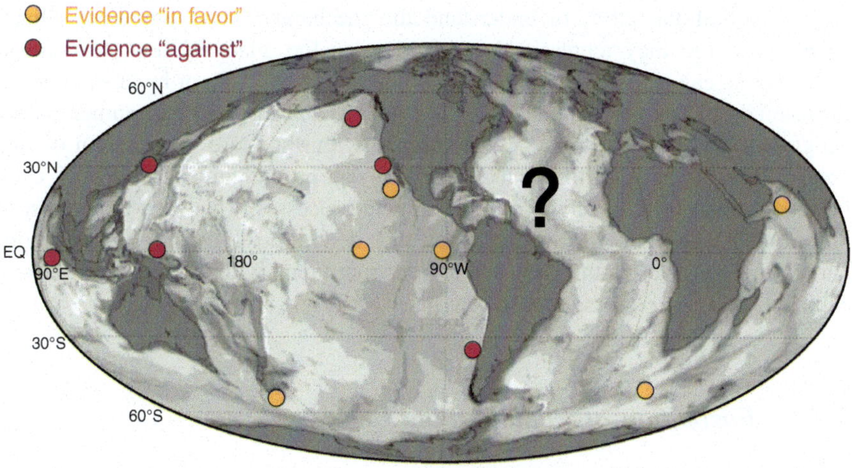

Fig. 7.8 A summary of where old water has been found [Arabian Sea, Bryan et al. (2010); Southern Ocean, Skinner et al. (2010); Galapagos, Stott et al. (2009); equatorial mid-Pacific, Keigwin, unpublished data; southwest Pacific, Sikes et al. (2000); Baja CA, Marchitto et al. (2007)] and where it has not been found [Chile margin, De Pol-Holz et al. (2010); Santa Barbara basin, Magana et al. (2010); Gulf of Alaska, Galbraith et al. (2007); Kamchatka, J. Southon, S. Gorbarenko, unpublished data; Sumatra, De Pol-Holz et al. unpublished data; New Guinea, Broecker et al. (2007)]

biogenic opal to the ocean floor (Fig. 7.7). Recent results from intermediate waters off Chile at a site closer to the active areas of formation of intermediate waters show no evidence of abyssal ^{14}C-depleted C during the deglaciation (Fig. 7.6).

Notwithstanding possible caveats in the Chilean margin records, the absence of direct evidence for the old water reservoir, that is, the lack of extremely large differences in planktonic and benthic foraminiferal ages in the deep Pacific, Gulf of Alaska, Sea of Okhotsk, New Zealand, and off Sumatra (De Pol-Holz et al. 2010) (Fig. 7.8) suggests that there was no pool of very old water at the bottom of the ocean during the last glacial. Yet the controversy is not completely resolved.

7.4 Calibration of Radiocarbon During the Last 50,000 Years

The reconstruction of atmospheric $\Delta^{14}C$ and consequent calibration of the ^{14}C timescale has been a goal of the scientific community for several decades. This need arose after the discovery that production of ^{14}C atoms in the atmosphere was not constant, which was an underlying assumption of ^{14}C dating originally formulated by Libby. Determining past changes in atmospheric ^{14}C allows ^{14}C dates to be converted into actual calibrated calendar dates. Calibrated ^{14}C chronologies are a critical part of the effort to understand the mechanisms and timing of climate change, because they represent the link between the relatively few paleoclimate archives that can be dated absolutely (radiometrically by U-Th or by layer counting in materials that deposit annual layers or bands), and the far more numerous and geographically diverse set of ^{14}C-dated records. Furthermore, comparison of calibrated ^{14}C records from different C reservoirs can provide information on C cycle turnover rates and their relationship to climate processes. This section is aimed at providing background on the processes that determine the shape of calibration curves and the datasets that are used to construct them, with the details on how to calibrate a ^{14}C date provided in Chap. 3.

7.4.1 Formation of Radiocarbon

Radiocarbon is made when incoming galactic cosmic rays (mostly protons and alpha particles, at gigaelectronvolt—GeV—energies) interact with oxygen and nitrogen nuclei in the atmosphere. These energetic collisions shatter the nuclei, producing a shower of secondary neutrons, protons, and mesons along the track of the incoming primary particle, and these secondaries in turn interact with other nuclei and produce yet more fragments (Fig. 7.9). The charged particles in these showers interact with electrons and nuclei and are brought to a halt in relatively

Fig. 7.9 A rendition of how cosmic rays interacting with atoms in the atmosphere might look if the particles were actually visible (NASA)

short distances, but the uncharged neutrons travel a much greater distance, colliding with many atmospheric nuclei before losing excess energy. A few secondary ("knock-on") neutrons are produced with sufficient energy to initiate the so-called spallation reactions, which produce a variety of other cosmogenic radioisotopes as well as ^{14}C (^{10}Be from N and O, ^{26}Al and ^{36}Cl from Ar, etc.). Most are produced at much lower energies (Goldhagen et al. 2002), but since the $^{14}N(n, p)^{14}C$ reaction can be initiated even by thermal neutrons that have lost essentially all of their initial excess kinetic energy in repeated collisions, almost all atmospheric neutrons ultimately contribute to ^{14}C production regardless of their initial energy.

These reactions occur as cosmic rays interact with the atmosphere, but they are not distributed equally within it. The major component of Earth's geomagnetic field approximates a dipole magnet, and the field strength at low latitudes is sufficient to deflect most incoming cosmic rays, thus preventing them from reaching the atmosphere. However, at the Earth's poles, the incoming primary particles are traveling along the magnetic field lines and are not deflected, so that the primary and secondary fluxes, and therefore the ^{14}C production, are greatest at high geomagnetic latitudes. Because the primary cosmic rays lose energy in their collisions with nuclei, lower energy primaries are slowed down and stopped at high altitudes and so the primary and secondary fluxes decrease exponentially with depth into the atmosphere, with a characteristic absorption length of 150–200 g cm^{-2} (15–20 % of the total atmospheric thickness). As a result, about 2/3 of all ^{14}C production takes place in the stratosphere, and the production per unit area is highest at high latitudes.

The two major reasons why calibration is required are temporal changes in the C cycle that can redistribute ^{14}C within the C cycle as discussed earlier in this chapter and changes in ^{14}C production due to variations in the flux of galactic cosmic rays reaching the Earth's atmosphere. The solar system as a whole is shielded against galactic cosmic rays by the magnetic field imbedded in the solar wind of charged particles streaming out from the Sun, and the Earth itself is further shielded by the geomagnetic field; there is evidence that both the heliomagnetic and the geomagnetic shielding have varied over time.

7.4.2 Records of the Geomagnetic Field

The strength of the Earth's geomagnetic field is determined by the dynamics of the Earth's molten iron core, and past variations are recorded in lavas and in marine and terrestrial sediments. Figure 7.10 shows a time series for the virtual axial dipole moment, a measure of the changing geomagnetic field derived by combining (stacking) measurements of remnant magnetism in deep-sea sediment cores, together with the corresponding calculated ^{14}C production rates (Laj et al. 2002). Over the past few thousand years, the geomagnetic field was higher than at any other time over the past ~50 kyr, so that ^{14}C production during the recent past was lower than in glacial time; furthermore, there were two periods ~41 and 35 kya when the field went almost to zero, leading to ^{14}C production peaks. These events, named the Laschamp

(41 kya) and Mono Lake events (35 kya) after the terrestrial localities where they were first detected, were accompanied by large changes in the geomagnetic field direction (azimuth and declination), and it is assumed that they represent aborted magnetic reversals—periods where the entire geomagnetic field began to switch poles but eventually settled back into the previous configuration.

Figure 7.11 shows $\Delta^{14}C$ data from the latest IntCal13 calibration curve (Reimer et al. 2013). There is a clear resemblance between these data and the Laj et al. (2002) ^{14}C production values (Fig. 7.10), with generally higher values before ~20 kya, peaks ~34 and 41 kya, and a slow fall after 20 kya that is due at least in part to the decay of the excess ^{14}C accumulated in the C cycle during the period of high production. However, simple C cycle models (Muscheler et al. 2004) show that the increased $\Delta^{14}C$ in the glacial is too large to be explained by production increases alone.

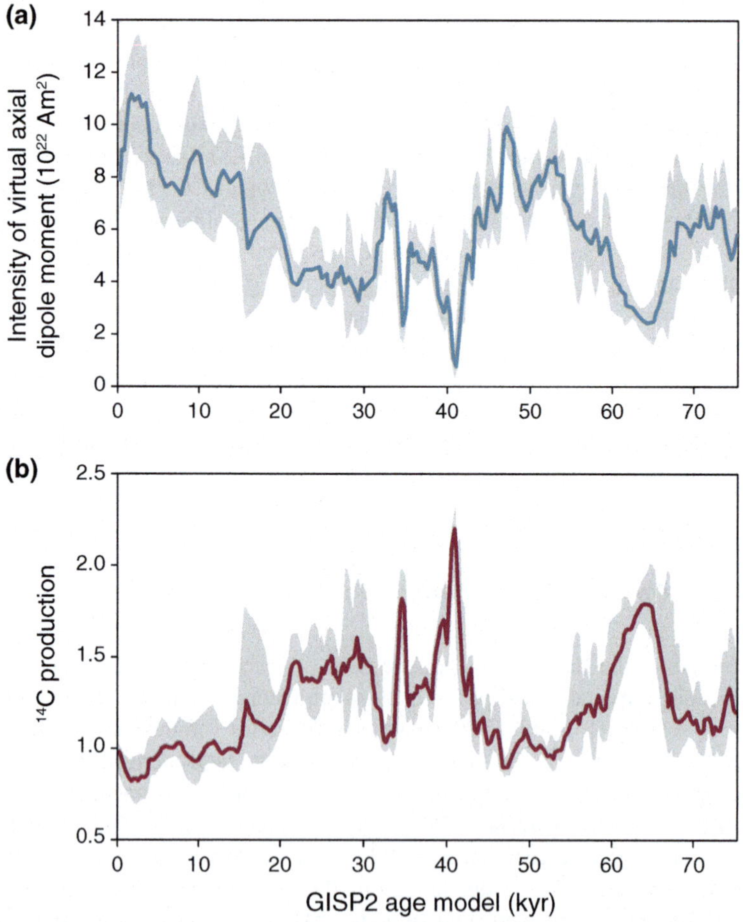

Fig. 7.10 Long-term changes in ^{14}C production. Estimates of (**a**) the geomagnetic dipole and (**b**) the calculated corresponding ^{14}C production based on paleomagnetic signals in sediments (NAPIS-75 stack: Laj et al. 2002). Figure modified from Laj et al. (2002)

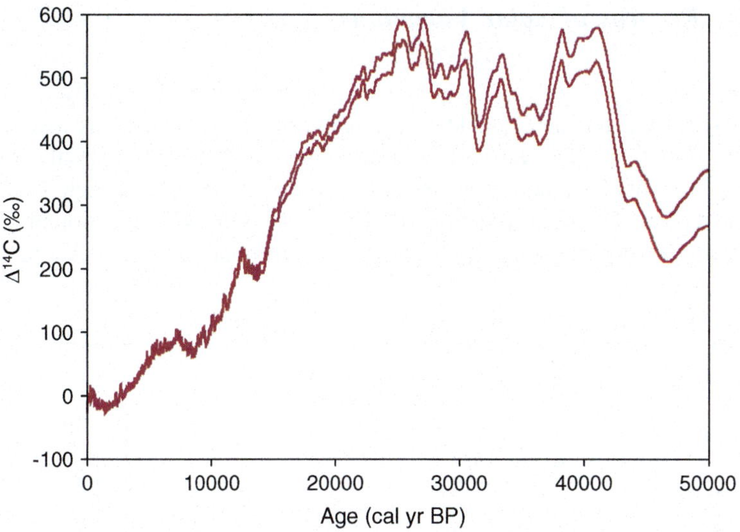

Fig. 7.11 $\Delta^{14}C$ over the past 50,000 years (IntCal13; Reimer et al. 2013) shows an overall resemblance to the ^{14}C production estimates in Fig. 7.10, but the $\Delta^{14}C$ variations are too large to be explained by production changes alone

A second independent measure of ^{14}C production can be derived from measurements of ^{10}Be in polar ice cores (Finkel and Nishiizumi 1997; Muscheler et al. 2005). Like ^{14}C, ^{10}Be is produced by cosmic ray neutrons, but ice core ^{10}Be is thought to be a more precise signal of the production of isotopes in the atmosphere by cosmic rays compared to ^{14}C because there is no ^{10}Be equivalent of the C cycle to damp the production signal. ^{10}Be is scavenged by aerosols and directly deposited in precipitation or by dry deposition. Surprisingly, when fluxes of ^{10}Be are computed from the measured ^{10}Be concentrations using layer-counted Greenland ice core timescales, the results show very little variation on multi-kyr timescales apart from those due to glacial/IG precipitation changes (Muscheler et al. 2005). This apparent contradiction with the geomagnetic results could be explained if polar ^{10}Be found in ice cores is overwhelmingly a result of direct deposition of ^{10}Be produced locally in the polar stratosphere, rather than a globally integrated signal of ^{10}Be production. The local production at the poles would be less sensitive to the geomagnetic field compared to global production as a whole, but would instead vary according to the shielding of the entire solar system by the Sun's magnetic field (see Sect. 7.4.3). However, atmospheric transport models suggest that stratospheric mixing is sufficiently rapid that while polar enhancement does exist, a total dominance of polar production in high-latitude ^{10}Be deposition does not occur. Hence, the questions of the relative importance of local versus global ^{10}Be input in polar ice, and the extent to which polar ^{10}Be is a proxy for solar versus geomagnetic shielding, remain open.

7.4.3 Records of Solar Variations

Short-term (~100–200 year) variation in ^{14}C production is usually ascribed to changes in solar magnetic shielding against cosmic rays, since it is assumed that changes in the geomagnetic field are relatively slow. This is not strictly true. The geomagnetic field strength has decreased by 10 % since Friedrich Gauss first measured it in 1835, corresponding to a 5 % increase in ^{14}C production in less than 200 years. Nevertheless, as Fig. 7.12a shows, centennial scale increases in $\Delta^{14}C$ in

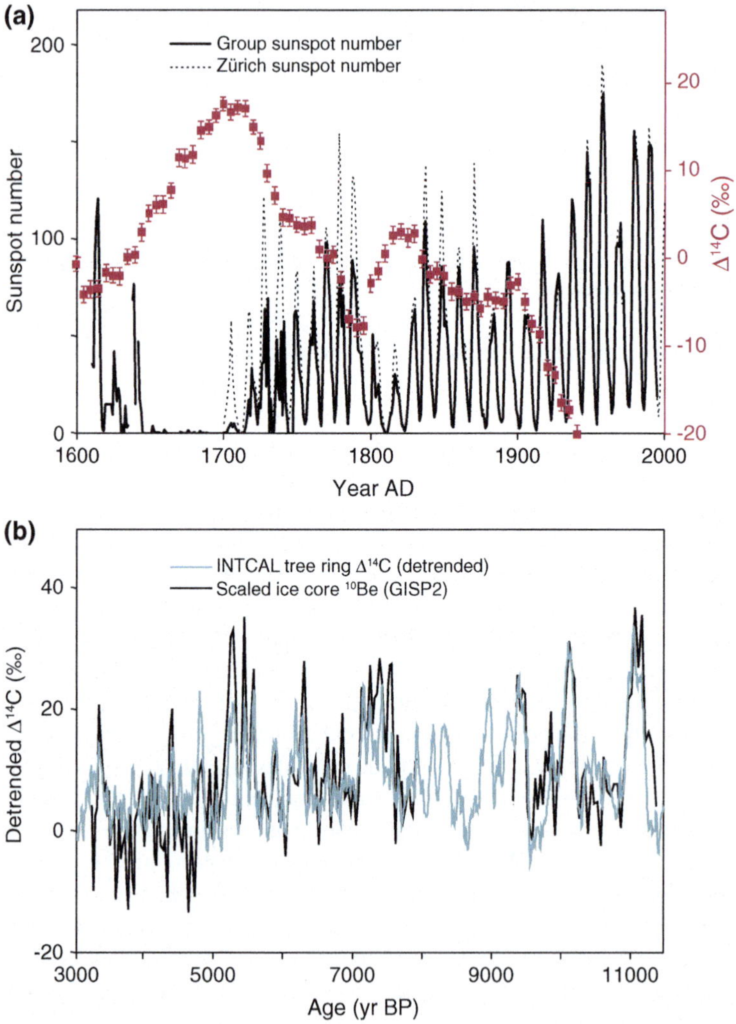

Fig. 7.12 **a** $\Delta^{14}C$ variations over the latter part of the Little Ice Age are strongly anticorrelated with sunspot numbers [figure modified from Vaquero (2007)]; **b** $\Delta^{14}C$ in tree rings and ^{10}Be concentration in polar ice cores show similar centennial-scale variations through the Holocene

tree rings from the Little Ice Age are closely correlated with strong reductions in sunspot frequency (Eddy 1976; Vaquero 2007), suggesting that the $\Delta^{14}C$ increases are linked to solar changes, presumably through reductions in heliomagnetic shielding. Furthermore, the $\Delta^{14}C$ variations lag changes in ^{10}Be deposition in polar ice cores by about 20 years, and this ^{10}Be-^{14}C lag is just what models predict for the damping effect of the C cycle on ^{14}C production changes. There are similar strong correlations between ^{10}Be and ^{14}C throughout the Holocene (Fig. 7.12b), and at least some of these events can be identified with periods of cooling around the North Atlantic, most recently during the Little Ice Age.

The ^{10}Be-^{14}C lag increases to about 70 years in the early Holocene, whereas models predict a constant offset, but this probably reflects small cumulative errors in the layer-counting ice core timescales. The correlation between the two isotopes is so strong that the heliomagnetic/production explanation seems robust. It is therefore tempting to seek a solar explanation for the associated centennial-scale cooling events, and reduced solar output at ultraviolet frequencies leading to changes in stratospheric ozone heating and changes in atmospheric circulation have been suggested as the cause (Shindell et al. 2001). However, the likely variations in irradiance are too small to have significantly affected climate, so that if solar irradiance changes were indeed the cause, there is an unknown climate amplifier at work, possibly involving changes in the MOC of the oceans.

7.4.4 Calibration Records: Tree Rings

Given these known changes in ^{14}C production over time (plus possible effects of C cycle variations), a calibration curve for ^{14}C is required. This has been derived by measuring the ^{14}C content of known-age samples that have been independently dated. Tree rings form an ideal atmospheric ^{14}C calibration archive because they sample the atmospheric CO_2 pool directly, they are formed annually in many parts of the world, and they can be counted directly to assess the year of growth. Furthermore, long time sequences can be developed by matching ring patterns in multiple individual trees, both live and dead, to extend these atmospheric records thousands of years into the past. The first multimillennial, dendrochronological tree ring series was completed by Wesley Ferguson at the University of Arizona using bristlecone pine (*Pinus longaeva*) collected from the White Mountains in Eastern California. This was the earliest long atmospheric ^{14}C record using bristlecone pine (Suess and Linick 1990). Individual bristlecone pine trees can live for more than 5000 years. However, the longest atmospheric ^{14}C sequences currently available are now from Central Europe and are based on dead trees unearthed and matched to chronologies of living trees. Irish and German oak chronologies extend back 9 and 11 kyr, respectively, and have been linked to European pine chronologies that go back another 3 kyr (Reimer et al. 2013). Since oaks and pines typically live for just a few centuries, the development of long records required finding and dating a great number of individual trees whose life spans overlapped through time. This has taken many decades, starting in the 1940s and 1950s, and has involved an enormous amount of work extracting buried logs from riverbanks, quarries, and construction sites.

Critical aspects of tree ring reconstructions include selection of trees that were exposed to stress, and replication. Trees growing in ideal environments may show little interannual variation in ring width (growth) even when a growth factor such as precipitation has year-to-year variation, because the changes may not approach the tolerance limits of that species. For the location of many trees, though, temperature and moisture conditions are seldom optimal, so distinctive ring width patterns emerge. These patterns of ring width are used to cross-date individual trees with one another with the idea that high growth years for one individual correspond to good environmental conditions for growth that likely affect other individuals similarly. However, the situation is complicated by the likelihood of purely local stresses on a particular tree leading to anomalous ring widths or even missing or false rings. Therefore, replication of the records using multiple trees that cover any particular time period is essential for developing robust chronologies. Studies of European chronologies from the early Holocene and the Younger Dryas suggest that the dominant stress was temperature, rather than moisture as it was in the later Holocene. Since temperature changes are often more widespread than moisture variation, it is easier to correlate ring width patterns between regional chronologies that cover extended areas, which partially compensates for the scarcity of trees that prevailed in Central Europe under the harsh conditions of the Younger Dryas and the general difficulties of finding and recovering dead trees from so long ago.

Measurements of ^{14}C from tree rings whose calendar age is known by direct counting are universally considered the gold standard as a record of changing atmospheric ^{14}C used for calibrating ^{14}C dates to actual calendar dates. The dendrochronological records have usually been made with sufficient replication to remove local noise present in any individual tree and are precise and accurate to within one year. However, the oldest portion of any regional sequence is usually based on relatively sparse data with little replication because by definition it consists of the very oldest trees that have been found. As an example, Fig. 7.13 shows how the tentative extension of the master tree ring series from Central Europe through the mid–Younger Dryas from 12.2 to 12.6 kya is based on just a few trees. Likewise, the ^{14}C coverage of the oldest sections of the chronology is often initially sparse. Both of these problems typically decrease over time as ring sequences from more trees are added to the chronology and as more ^{14}C dates are carried out, but it is prudent to remember that the earliest few hundred years of any chronology are a work in progress and may be subject to change.

Currently, the Northern Hemisphere atmospheric ^{14}C record is determined back to 12.6 kya using the European tree ring data (Reimer et al. 2013). Single-year ^{14}C measurements are available back to 440 year BP, and most of the record is covered at decadal resolution, though the data around 12 kya are notably thin (Fig. 7.14). A recent tentative reconstruction back to 14 kya (Hua et al. 2009) uses decadal ^{14}C data from Southern Hemisphere tree ring records to span a short gap between the master Northern Hemisphere sequence that extends back into the Younger Dryas

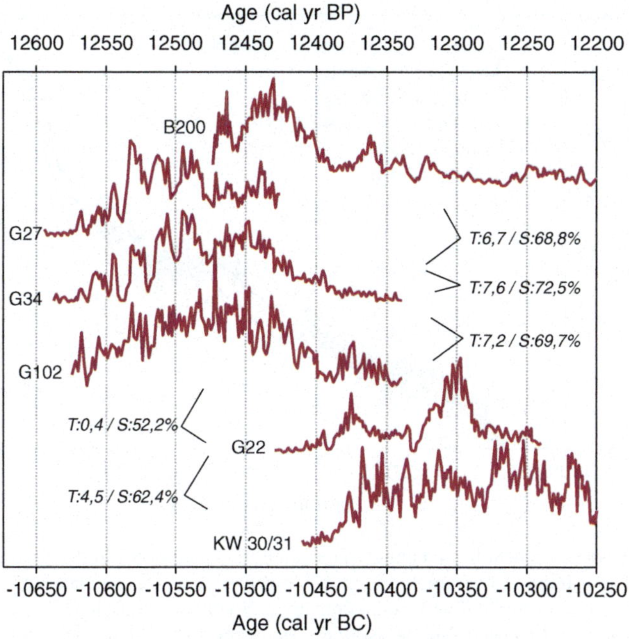

Fig. 7.13 Tree ring width data for the mid- to late-Younger Dryas period, showing the wiggle-matches used to extend the record from 12,200 BP to 12,600 BP. In this interval, the number of trees that make up the time series is very small

and a 1300-year "floating" pine chronology from the preceding Allerød warm period (Kromer et al. 2004). A North–South interhemispheric offset of 40 ^{14}C years, based on the average value for the past 2000 years, was used in the wiggle-matching. This interhemispheric offset is simply an estimate, but the same value gives reasonable results for comparisons of North–South records in the early Holocene, and it seems unlikely that the calendar ages based on the preliminary comparison will shift by more than a few decades.

Figure 7.14 shows the Younger Dryas portion of the extended tree ring series, which resolves several prior chronological problems. The drop in tree ring ^{14}C ages (Δ^{14}C increase) at ~12.7 kya now agrees with the timing and intensity of a large peak in cosmogenic ^{10}Be in the Greenland GISP2 ice core, consistent with the view that both represent production rate (cosmic ray intensity) changes (Muscheler et al. 2008). The timing also agrees with the onset of the Younger Dryas as recorded by climate proxies in varved European lake sediments (e.g., Brauer et al. 2008), and those in turn are consistent with ^{14}C records from other (unvarved) European paleoclimate sequences which show that a sharp decrease in ^{14}C ages occurred right at the Allerød/Younger Dryas transition.

Prior to 14 kya, a few floating tree ring sequences exist, notably from New Zealand kauri (Turney et al. 2007). These trees are often very large, living

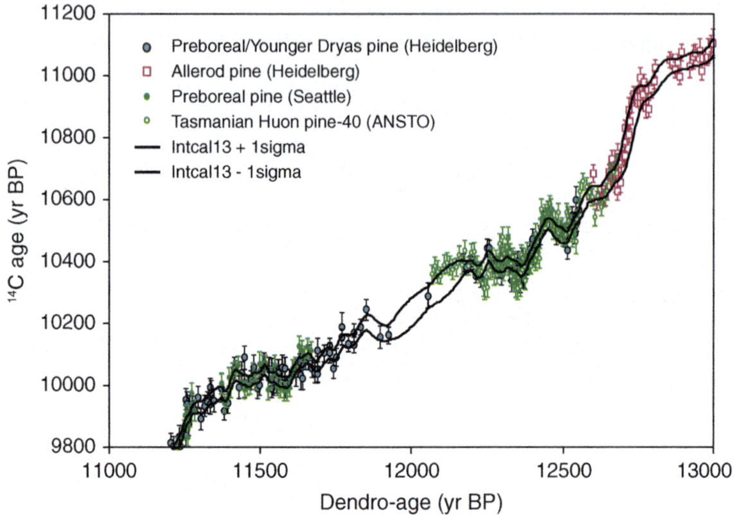

Fig. 7.14 Radiocarbon data from Younger Dryas tree rings measured in the Seattle, Heidelberg, and ANSTO (Australia) laboratories are shown together with the estimated 67 % confidence interval (±1 sigma) for the IntCal13 calibration curve. Southern Hemisphere data have been used to bridge a short gap between the master Northern Hemisphere dataset extending to 12,600 BP and another 1300 years of "floating" data. Note that the interval around 12,000 BP currently has few radiocarbon measurements, though the dendrochronology extends well beyond this gap

1000 years or more, and the records extend well back into Marine Isotope Stage 3 (60–30 kya) including many trees that fall beyond the ^{14}C dating limit. Unfortunately, the trees recovered so far are clumped together in groups of similar age separated by long gaps, and prospects for a major extension of the continuous tree ring record seem remote.

7.4.5 Radiocarbon Tree Ring Data: What Are the Limitations?

When using the tree ring data to obtain calendar dates from ^{14}C measurements, it is important not to lose sight of possible limitations of this calibration technique. Interlaboratory bias in the high-precision measurement of ^{14}C in tree rings is a major concern, and interlaboratory comparisons are routinely carried out to test this. These typically show that any interlaboratory offsets are of the order of a decade or two at most; however, because ^{14}C ages are increasingly measured with precisions of ±20 years or better, these are not necessarily trivial errors.

Regional ^{14}C variation is another potential problem for determining a calendar date at a particular location. A 2200-year Southern Hemisphere curve has been derived from tree ring data (Hogg et al. 2013), and North–South comparisons have shown a variable offset in ^{14}C ranging from almost zero to about 80 ^{14}C years over this period. This is thought to arise from the effects of strong air–sea gas exchange over the Southern Ocean, where ^{14}C-depleted deep ocean water outcrops at the surface (Chap. 5). This effect of Southern Ocean degassing of old water and the resulting interhemispheric offset probably represents the largest spatial variation in atmospheric ^{14}C in the prebomb era, and it is likely that North–South offsets remained within the same range over the entire Holocene. Within the Southern Hemisphere itself, atmospheric transport models predict postbomb variations in ^{14}C equivalent to just a few decades between the tropics and the Southern Ocean (Rodgers et al. 2011); hence, prebomb offsets, which would have been smaller, were probably insignificant except at high latitudes.

In parts of the tropics, trees are influenced at different times within the year by air masses from the Northern and Southern hemispheres, due to shifts of the ITCZ. This effectively "moves" trees from one hemisphere to another for parts of the year and could plausibly introduce shifts of a decade or so in prebomb ^{14}C records. Postbomb atmospheric $\Delta^{14}C$ data show that this process does occur, but few precise prebomb ^{14}C tree ring records from tropical locations are available to confirm the magnitude of the effect. At midlatitudes, the injection into the troposphere of ^{14}C from the stratosphere, where most ^{14}C production occurs, peaks in the spring as warming breaks down the stratification at the troposphere–stratosphere boundary. The possibility of regional offsets due to variations in the timing of peak plant growth relative to this spring injection has been raised, and comparisons of Anatolian data with the German oak records (Kromer et al. 2010) suggest that a time-varying bias of a decade or two has occurred over at least some intervals.

To summarize, currently available analytical precision and calibration tools permit researchers to use the dendrochronological ^{14}C data to produce precise, calibrated dates that in favorable cases may span intervals of just a few decades. However, these high-precision results may require careful scrutiny and should not be accepted uncritically, because several mechanisms exist that can potentially introduce decadal biases.

7.4.6 Calibration Beyond Tree Rings—Varved Lake Sediments

Until recently, the "atmospheric" calibration records covering the period older than 14 kya were actually based on marine data corrected for marine–terrestrial ^{14}C offsets, because no suitable records that sampled the atmospheric ^{14}C pool existed until Bronk Ramsey and colleagues published the Japanese Lake Suigetsu macrofossil ^{14}C record in 2012 (Ramsey et al. 2012) (Fig. 7.15). Suigetsu is a varved lake on the western shore of Honshu that satisfies many of the criteria for an ideal site. Since the lake is anoxic, the sediments are not bioturbated, and surrounding hills have helped protect the sedimentary varves from wind-induced disturbance. The area of the local

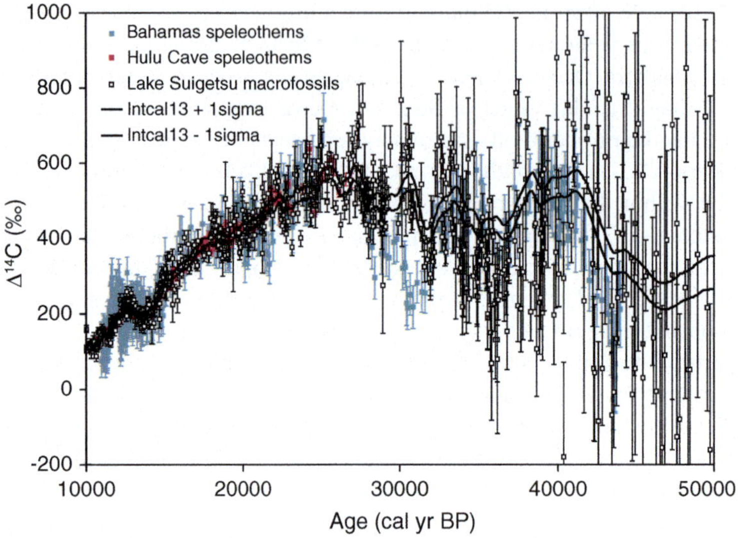

Fig. 7.15 Terrestrial (macrofossil plus speleothem) calibration data for the period 10–50 kyr BP plus the IntCal13 calibration curve

catchment is very small, which reduces the likelihood of reworked macrofossils entering the lake, and the only significant river feeding the lake enters via a shallow channel from Lake Mikata that acts as a catch basin for flood debris.

An initial attempt to develop a Suigetsu record in the 1990s was done with a limited budget that only allowed for extraction of a single core, and core recovery was incomplete. A new, large-scale effort undertaken in 2006 (http://www.suigetsu.org) produced multiple cores with overlapping stratigraphy, and a timescale for the period back to 40 kyr was developed using semiautomated varve counting techniques. The new Suigetsu data are the best currently available approximation to a "pure" terrestrial ^{14}C record, though reworking of macrofossils is a concern in spite of the limited catchment area. Also, the possibility of systematic over- or undercounting of varves must be considered.

7.4.7 Other Atmospheric Calibration Records—Speleothems

A pioneering study by Beck et al. (2001) produced the first large-scale record of ^{14}C dates on speleothem calcite, spanning 10–50 kya, from a U-Th-dated stalagmite recovered from a submerged "blue hole" cave in the Bahamas. A second record from the same cave was published by Hoffmann et al. (2010). The detrital content of the calcite is relatively high, so that corrections to the U-Th ages of 1500–2000 years are required to account for detrital Th present when the speleothem initially formed. Comparisons of ^{14}C results from the Younger Dryas section of the speleothems with tree ring data showed that the ^{14}C ages also required an adjustment of ~1500 years to correct for incorporation of geological carbonate in the groundwater from which the speleothem carbonate precipitated (the so-called dead carbon fraction or DCF correction). These large corrections to both the calendar and ^{14}C ages introduce significant uncertainty.

A second long speleothem record is from Hulu Cave near Nanjing, China. The Hulu speleothems exhibit high U concentrations and very low detrital Th levels, which allow very precise U-Th dating. A calibration record from the Hulu H82 speleothem (Southon et al. 2012), based on closely interleaved samples for U-Th and ^{14}C, covers the period 10–27 kya at centennial resolution or better to ~20 kya, dropping to a few samples per kyr near the bottom of the speleothem where growth rates are lower. The DCF correction for the period of overlap with the tree ring ^{14}C data is just 460 years, and since it was unchanged through the major climate shifts into and out of the Younger Dryas, there are grounds for hope that it may also have remained constant through stadial–interstadial shifts during the glacial.

Unlike the Suigetsu macrofossils, neither of these records samples the atmospheric ^{14}C pool directly. The required DCF corrections introduce additional uncertainty into the speleothem ^{14}C data, and furthermore, those corrections are based on the untested assumption that the DCF remained constant back to ~50 kya. However, within the limits imposed by analytical uncertainties and detrital Th corrections, the speleothem U-Th ages do provide additional constraints for the Suigetsu varve count, which is subject to cumulative counting errors estimated at ~6 % at 40 kya, and beyond 40 kya is simply extrapolated assuming constant sedimentation rate. To the extent to which the three records give consistent results, each strengthens the others (Fig. 7.15); the speleothem results (and other calibration data) were used by Bronk Ramsey and colleagues to refine the varve counting results and generate the final Suigetsu timescale.

7.4.8 Marine Calibration Records

As stated above, marine results were used exclusively for calibration beyond the tree ring limit until recently, because insufficient terrestrial data existed. Apart from their obvious uses for calibrating marine dates and for investigating past ocean circulation and C cycle changes, they are still used to constrain the atmospheric calibration curve, because the relevant datasets are from low-latitude locations that are thought to be least sensitive to possible changes in marine–terrestrial ^{14}C offsets (marine reservoir ages). For that application, the ^{14}C ages have been corrected to equivalent atmospheric values using recent (prebomb) reservoir ages from the same locations (see Chap. 5) with a large uncertainty of ±200 years assigned to those corrections (for a reservoir age database, see http://intcal.qub.ac.uk/marine). Three types of records have provided ^{14}C data: reef-building corals that can be independently dated by U-Th; planktonic foraminifera from varved marine sediments;

and foraminifera from sedimentary records that can be tied to other chronologies by wiggle-matching sediment climate proxies.

The major problem for all marine-based data is that marine reservoir ages may have varied over time in a spatially inhomogeneous fashion, even at low latitudes; if so, an average or "best" marine curve may not exist. Some temporal changes probably did occur, at least on a global scale. Ignoring all other changes, CO_2 exchange between the ocean and a glacial atmosphere with pCO_2 of 200 ppm would drop to 5/7 of the exchange rate for a 280 ppm Holocene atmosphere, requiring a glacial ocean–atmosphere ^{14}C offset 7/5 of the Holocene value to balance the effects of ^{14}C decay in the deep ocean. Because Holocene reservoir ages are $\sim$400 years for much of the warm surface ocean, glacial-IG changes of $\sim$160 years in the offset could be expected from this cause alone.

In fact, the actual magnitude and even the sign of any long-term reservoir age changes are uncertain, because changes in wind speeds and ice cover that may have profoundly affected air–sea gas exchange, particularly at high latitudes, are difficult to quantify. Additionally, the global-scale changes in ocean circulation discussed earlier in this chapter plus the higher ^{14}C production due to reduced and variable geomagnetic field intensities during the glacial would also have altered marine–terrestrial ^{14}C offsets. There is also some evidence for superimposed regional changes. Several studies have documented very large glacial reservoir age shifts at high latitudes (e.g., Skinner et al. 2010). In addition, the presence of very ^{14}C-depleted waters at shallow intermediate depths in the eastern Pacific and the Arabian Sea during deglaciation (Fig. 7.6) raises the possibility that surface reservoir ages in these and perhaps other low-latitude locations may also have varied if the old water mass outcropped at the surface.

The use of ^{14}C in climate and C cycle models has been suggested as a way to predict glacial reservoir ages and hence fill in the gaps in the data, but given our current limited understanding of glacial ocean circulation and of the changes that occurred during deglaciation (Sect. 7.3), model predictions must be treated cautiously. Thus, there is a continuing need to measure more long-term calibration records, even from marine or other C pools that do not directly sample atmospheric ^{14}C. If several disparate calibration datasets with their own strengths and weaknesses agree over a given interval, the combined data probably give an accurate representation of how ^{14}C varied over that period, and a larger selection of records may make it easier to recognize outliers.

7.4.9 Uranium-Thorium-Dated Corals

The Fairbanks et al. (2005) compilation from Barbados and the southwest Pacific is the largest set of U-Th-dated corals currently available (Fig. 7.16). It represents the results of decades of shallow water drilling in search of ancient paleoreef sites drowned by rising sea levels and/or buried under later coral growth. Other smaller coral datasets [see Reimer et al. (2013) for details] are all from the Pacific. As well as difficulties of access, major problems with coral samples include the opportunities for secondary calcite contamination caused by the very large surface area that corals present at both macroscopic (pores) and microscopic (needle-like aragonite crystals) scales. In addition, some atolls "float" on a submerged pool of fresh or brackish water that can bring younger carbonate into buried reefs. Since most reef-building corals grow within a few tens of meters of the ocean surface, many reefs from Marine Isotope Stage 3 were exposed to the atmosphere and therefore vulnerable to contamination during peak sea-level depression at the LGM. However, since the corals can be directly dated by U-Th, they are an invaluable resource because they not only yield calibration data directly, but also provide spot checks on other less directly dated marine archives.

7.4.10 Marine Calibration Records—Sedimentary Records

Planktonic foraminifera in a core from the Cariaco basin off Venezuela were dated to derive a deglacial calibration record spanning 10.5–14.7 kya (Fig. 7.16, Hughen et al. 2004b). Cariaco sediments deposited during deglaciation are laminated, and varve counting provided a floating calendar chronology. A ^{14}C wiggle-match with the tree ring calibration record fixed the floating record in time, and the reservoir-corrected data show good agreement with the extended tree ring record (Fig. 7.16) apart from a short interval over the Allerød/Younger Dryas boundary (12.7–13 kya) when reservoir ages within the basin (or perhaps regionally) appear to have been extremely low. That section of the record was removed from the calibration database. The lesson is that replication of disparate records is critically important for building a robust calibration dataset that is free of local variations.

Three other marine sediment records of foraminifera ^{14}C are presented in Fig. 7.16: data from a longer (unvarved) Cariaco record (Hughen et al. 2006) and similar records from the Iberian margin in the North Atlantic and the Pakistani margin in the Arabian Sea (Bard et al. 2013). None of these have their own timescales, but climate proxies in the sediments ($\delta^{18}O$, $\delta^{15}N$, sediment color, etc.) show strong correlations with $\delta^{18}O$ in layer-counted Greenland ice cores and in U-Th-dated tropical and subtropical speleothems. Earlier realizations of these datasets used timescales derived from correlations with Greenland $\delta^{18}O$, but current versions are correlated with $\delta^{18}O$ in Hulu Cave speleothems. Thus, the newer timescales are ultimately based on U-Th dates, and the revised datasets show better consistency with the directly U-Th-dated coral results.

Reservoir-corrected Cariaco data are systematically younger than the terrestrial and speleothem results over the period 16–17.5 kya: Either the Cariaco chronology is skewed in this interval by an erroneous correlation with the Hulu Cave $\delta^{18}O$ data, or the local Cariaco reservoir age may have been very low during this time (the Heinrich 1 stadial) as it was in the early Younger Dryas. No published coral data are

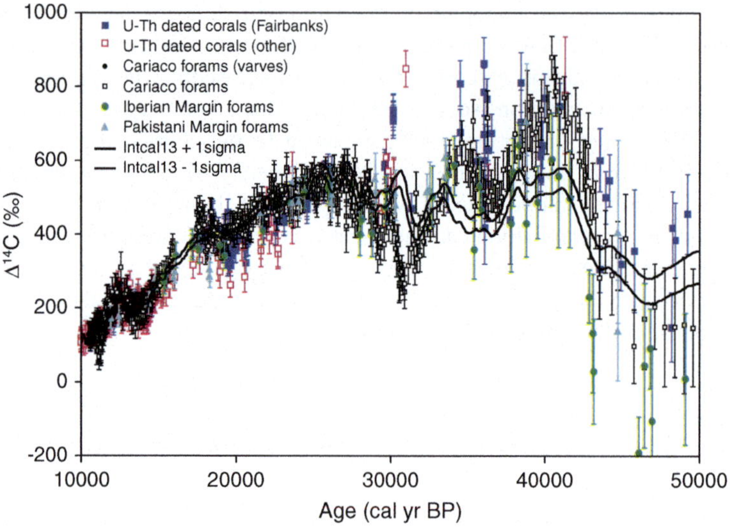

Fig. 7.16 Marine (coral plus foraminifera) calibration data for the period 10–50 kya plus the IntCal13 calibration curve. The marine data have been reservoir-corrected for comparison with the atmospheric IntCal13 curve, but clear systematic differences exist beyond 30 kya

available in this time range for cross-checks, and the Cariaco data in this interval have been omitted from the calibration curves. The eastern North Atlantic cores used for the Iberian margin record are located far offshore from coastal upwelling zones where reservoir ages may have increased during Heinrich stadials and other cold intervals, and the reservoir-corrected data appear consistent with the atmospheric record back to ~30 kya. The Pakistani margin results are also generally in agreement with the atmospheric data.

Prior to 30 kya, there are large significant systematic differences between the reservoir-corrected marine data and the atmospheric curve, including periods where the adoption of completely unphysical negative reservoir ages would be required to bring the marine and terrestrial data into agreement, i.e., the differences are not due to reservoir age changes. The patterns of disagreement resemble those that would arise if the independent chronologies for some of the records are in error, as opposed to systematic problems with the ^{14}C measurements themselves. However, all we can confidently say at present is that the marine and terrestrial data are in serious disagreement and resolution of these problems will probably require development of additional detailed ^{14}C records to validate or correct the existing datasets. For the moment, the older section of the recently published Marine13 calibration curve (Reimer et al. 2013) is based on the atmospheric IntCal13 curve with a marine reservoir offset applied, on the grounds that the coral results for that period show very large scatter, and the detailed Suigetsu varve chronology, which largely defines IntCal13 in that interval, is more likely to be correct than the timescales for the sedimentary marine ^{14}C data.

7.5 Calibration Computer Codes and Their Datasets

A selection of the available calibration codes can be found on the Web site of the journal Radiocarbon (http://www.radiocarbon.org). Others are available from individual laboratories, again reachable from Radiocarbon or via Google, and those Web sites provide detailed operating instructions and/or links to online manuals. All of these codes use some or all of the datasets described above and they typically give very similar results. However, for the oldest parts of the curves where results from the individual calibration datasets disagree (Figs. 7.15 and 7.16), the calibrated ages obtained may be significantly different depending on which data are used by each software package. The Fairbanks et al.'s (2005) curve relies solely on their own compilation of U-Th-dated coral data. IntCal13 (Reimer et al. 2013) and the CALPAL curve (http://www.calpal.de) include the additional coral results shown in Fig. 7.16 plus the sedimentary records discussed above. The new IntCal13 curve also includes the speleothem records from China and the Bahamas and, most importantly, the Suigetsu macrofossil dataset. CALPAL offers a "build-your-own" feature for picking a selection of the available datasets, which is interesting but dangerous, in the sense that the onus is on users employing that option to be careful to document which datasets are used. The Bayesian analysis codes BCal and OxCal, which integrate stratigraphic information with the ^{14}C data, both use the IntCal dataset.

7.6 Conclusions and Future Directions

The use of ^{14}C as a tracer in paleoclimate proxy systems is a powerful tool for revealing past climate change, and ^{14}C dating is essential for placing paleoclimate records on a common timescale. The quest for unveiling the mechanisms behind the "mystery interval" $\Delta^{14}C$ decline has put our previous knowledge of the Earth as a system to a test and it seems that we are failing. Future refinements of the ^{14}C cosmogenic production rates, deglacial volcanism, or just simply the finding of an extraordinary deep Pacific site could illuminate the future path of paleoclimate research. Radiocarbon calibration is the glue that joins ^{14}C-dated paleoclimate records with ice cores and other absolutely dated archives. The recent IntCal13 calibration curve is the first version to incorporate a ~50-kyr record based on materials that directly sampled the atmospheric ^{14}C pool, but additional measurements to better constrain the older part of the curve and further development of regional and marine calibrations will be required to take full advantage of the analytical capabilities that are now available for ^{14}C.

References

Adkins, J.F., H. Cheng, E.A. Boyle, E.R.M. Druffel, and R.L. Edwards. 1998. Deep-sea coral evidence for rapid change in ventilation of the deep north Atlantic 15,400 years ago. *Science* 280: 725–728.

Adkins, J.F., K. McIntyre, and D.P. Schrag. 2002. The salinity, temperature, and delta O-18 of the glacial deep ocean. *Science* 298: 1769–1773.

Anderson, R.F., S. Ali, L.I. Bradtmiller, S.H.H. Nielsen, M.Q. Fleisher, B.E. Anderson, and L.H. Burckle. 2009. Wind-driven upwelling in the Southern Ocean and the deglacial rise in atmospheric CO_2. *Science* 323: 1443–1448.

Bard, E., G. Menot, F. Rostek, L. Licari, P. Boening, R.L. Edwards, H. Cheng, Y. Wang, and T. J. Heaton. 2013. Radiocarbon calibration/comparison records based on marine sediments from the Pakistan and Iberian margins. *Radiocarbon* 55: 1999–2019.

Beck, J.W., D.A. Richards, R.L. Edwards, B.W. Silverman, P.L. Smart, D.J. Donahue, S. Hererra-Osterheld, G.S. Burr, L. Calsoyas, A.J.T. Jull, and D. Biddulph. 2001. Extremely large variations of atmospheric C-14 concentration during the last glacial period. *Science* 292: 2453–2458.

Brauer, A., G.H. Haug, P. Dulski, D.M. Sigman, and J.F.W. Negendank. 2008. An abrupt wind shift in Western Europe at the onset of the Younger Dryas cold period. *Nature Geoscience* 1: 520–523.

Broecker, W.S. 1992. The Glacial world according to wally. Eldigio Press.

Broecker, W., and S. Barker. 2007. A 190 ‰ drop in atmosphere's delta 14C during the "Mystery Interval" (17.5–14.5 kyear). *Earth and Planetary Science Letters* 256: 90–99.

Broecker, W., and G. Denton. 1990. Implications of global snowline lowering during glacial time to glacial theory. *Quaternary Science Reviews* 9: 305–341.

Broecker, W., S. Barker, E. Clark, I. Hajdas, G. Bonani, and L. Stott. 2004. Ventilation of the glacial deep Pacific Ocean. *Science* 306: 1169–1172.

Bryan, S.P., T.M. Marchitto, and S.J. Lehman. 2010. The release of C-14 depleted carbon from the deep ocean during the last deglaciation: evidence from the Arabian Sea. *Earth and Planetary Science Letters* 298: 244–254.

De Pol-Holz, R., L. Keigwin, J. Southon, D. Hebbeln, and M. Mohtadi. 2010. No signature of abyssal carbon in intermediate waters off Chile during deglaciation. *Nature Geoscience* 3: 192–195.

Eddy, J.A. 1976. Maunder minimum. *Science* 192: 1189–1202.

Fairbanks, R.G., R.A. Mortlock, T.C. Chiu, L. Cao, A. Kaplan, T.P. Guilderson, T.W. Fairbanks, A.L. Bloom, P.M. Grootes, and M.J. Nadeau. 2005. Radiocarbon calibration curve spanning 0 to 50,000 years BP based on paired Th-230/U-234/U-238 and C-14 dates on pristine corals. *Quaternary Science Reviews* 24: 1781–1796.

Finkel, R.C., and K. Nishiizumi. 1997. Beryllium 10 concentrations in the Greenland Ice Sheet Project 2: ice core from 3 to 40 ka. *Journal of Geophysical Research-Oceans* 102: 26699–26706.

Galbraith, E.D., S.L. Jaccard, T.F. Pedersen, D.M. Sigman, G.H. Haug, M. Cook, J.R. Southon, and R. Francois. 2007. Carbon dioxide release from the North Pacific abyss during the last deglaciation. *Nature* 449: 890–899.

Goldhagen, P., M. Reginatto, T. Kniss, J.W. Wilson, R.C. Singleterry, I.W. Jones, and W. Van Steveninck. 2002. Measurement of the energy spectrum of cosmic-ray induced neutrons aboard an ER-2 high-altitude airplane. *Nuclear Instruments and Methods in Physics Research Section a-Accelerators Spectrometers Detectors and Associated Equipment* 476: 42–51.

Hoffmann, D.L., J.W. Beck, D.A. Richards, P.L. Smart, J.S. Singarayer, T. Ketchmark, and C. J. Hawkesworth. 2010. Towards radiocarbon calibration beyond 28 Ka using speleothems from the Bahamas. *Earth and Planetary Science Letters* 289: 1–10.

Hogg, A.G., Q. Hua, P.G. Blackwell, M. Niu, C.E. Buck, T.P. Guilderson, T.J. Heaton, J.G. Palmer, P.J. Reimer, R.W. Reimer, C.S.M. Turney, and S.R.H. Zimmerman. 2013. SHCAL13 Southern Hemisphere Calibration, 0–50,000 years cal BP. *Radiocarbon* 55: 1889–1903.

Hua, Q., M. Barbetti, D. Fink, K.F. Kaiser, M. Friedrich, B. Kromer, V.A. Levchenko, U. Zoppi, A.M. Smith, and F. Bertuch. 2009. Atmospheric C-14 variations derived from tree Rings during the Early Younger Dryas. *Quaternary Science Reviews* 28: 2982–2990.

Hughen, K.A., J.R. Southon, S.J. Lehman, and J.T. Overpeck. 2000. Synchronous radiocarbon and climate shifts during the last deglaciation. *Science* 290: 1951–1954.

Hughen, K., S. Lehman, J. Southon, J. Overpeck, O. Marchal, C. Herring, and J. Turnbull. 2004a. C-14 activity and global carbon cycle changes over the past 50,000 years. *Science* 303: 202–207.

Hughen, K.A., J.R. Southon, C.J.H. Bertrand, B. Frantz, and P. Zermeno. 2004b. Cariaco basin calibration update: revisions to calendar and C-14 chronologies for core PL07-58PC. *Radiocarbon* 46: 1161–1187.

Hughen, K., J. Southon, S. Lehman, C. Bertrand, and J. Turnbull. 2006. Marine-derived C-14 calibration and activity record for the past 50,000 years updated from the Cariaco basin. *Quaternary Science Reviews* 25: 3216–3227.

Imbrie, J. 1982. Astronomical theory of the Pleistocene Ice ages—a brief historical review. *Icarus* 50: 408–422.

Kromer, B., M. Friedrich, K.A. Hughen, F. Kaiser, S. Remmele, M. Schaub, and S. Talamo. 2004. Late Glacial C-14 ages from a floating, 1382-ring pine chronology. *Radiocarbon* 46: 1203–1209.

Kromer, B., S.W. Manning, M. Friedrich, S. Talamo, and N. Trano. 2010. C-14 calibration in the 2nd and 1st millennia B.C. Eastern mediterranean radiocarbon comparison project (EMRCP). *Radiocarbon* 52: 875–886.

Laj, C., C. Kissel, A. Mazaud, E. Michel, R. Muscheler, and J. Beer. 2002. Geomagnetic field intensity, north Atlantic deep water circulation and atmospheric delta C-14 during the last 50 kyear. *Earth and Planetary Science Letters* 200: 177–190.

Magana, A., J. Southon, Nemec, Mojmir, L. Wacker, I. Hajdas, and H. Gaeggeler. 2010. Alternative methods for cellulose preparation for AMS measurement. Submitted to Nuclear Instruments and Methods in Physics Research B.

Marchitto, T.M., S.J. Lehman, J.D. Ortiz, J. Fluckiger, and A. van Geen. 2007. Marine radiocarbon evidence for the mechanism of deglacial atmospheric CO_2 rise. *Science* 316: 1456–1459.

Muscheler, R., J. Beer, G. Wagner, C. Laj, C. Kissel, G.M. Raisbeck, F. Yiou, and P.W. Kubik. 2004. Changes in the carbon cycle during the last deglaciation as indicated by the comparison of Be-10 and C-14 records. *Earth and Planetary Science Letters* 219: 325–340.

Muscheler, R., R. Beer, P.W. Kubik, and H.A. Synal. 2005. Geomagnetic field intensity during the last 60,000 years based on Be-10 and Cl-36 from the summit ice cores and C-14. *Quaternary Science Reviews* 24: 1849–1860.

Muscheler, R., B. Kromer, S. Bjorck, A. Svensson, M. Friedrich, K.F. Kaiser, and J. Southon. 2008. Tree rings and ice cores reveal C-14 calibration uncertainties during the Younger Dryas. *Nature Geoscience* 1: 263–267.

Petit, J.R., J. Jouzel, D. Raynaud, N.I. Barkov, J.M. Barnola, I. Basile, M. Bender, J. Chappellaz, M. Davis, G. Delaygue, M. Delmotte, V.M. Kotlyakov, M. Legrand, V.Y. Lipenkov, C. Lorius, L. Pepin, C. Ritz, E. Saltzman, and M. Stievenard. 1999. Climate and atmospheric history of the past 420,000 years from the Vostok ice core, Antarctica. *Nature* 399: 429–436.

Rahmstorf, S. 2002. Ocean circulation and climate during the past 120,000 years. *Nature* 419: 207–214.

Ramsey, C.B., R.A. Staff, C.L. Bryant, F. Brock, H. Kitagawa, J. van der Plicht, G. Schlolaut, M. H. Marshall, A. Brauer, H.F. Lamb, R.L. Payne, P.E. Tarasov, T. Haraguchi, K. Gotanda, H. Yonenobu, Y. Yokoyama, R. Tada, and T. Nakagawa. 2012. A complete terrestrial radiocarbon record for 11.2–52.8 kyear BP. *Science* 338: 370–374.

Reimer, P.J., E. Bard, A. Bayliss, J.W. Beck, P.G. Blackwell, C.B. Ramsey, C.E. Buck, H. Cheng, R.L. Edwards, M. Friedrich, P.M. Grootes, T.P. Guilderson, H. Haflidason, I. Hajdas, C. Hatte, T.J. Heaton, D.L. Hoffmann, A.G. Hogg, K.A. Hughen, K.F. Kaiser, B. Kromer, S.W.

Manning, M. Niu, R.W. Reimer, D.A. Richards, E.M. Scott, J.R. Southon, R.A. Staff, C.S.M. Turney, and J. van der Plicht. 2013. IntCal13 and Marine13 radiocarbon age calibration curves 0–50,000 years cal bp. *Radiocarbon* 55: 1869–1887.

Rodgers, K.B., S.E. Mikaloff-Fletcher, D. Bianchi, C. Beaulieu, E.D. Galbraith, A. Gnanadesikan, A.G. Hogg, D. Iudicone, B.R. Lintner, T. Naegler, P.J. Reimer, J.L. Sarmiento, and R.D. Slater. 2011. Interhemispheric gradient of atmospheric radiocarbon reveals natural variability of Southern Ocean winds. *Climate of the Past* 7: 1123–1138.

Shackleton, N.J., J.C. Duplessy, M. Arnold, P. Maurice, M.A. Hall, and J. Cartlidge. 1988. Radiocarbon age of last glacial Pacific deep water. *Nature* 335: 708–711.

Shindell, D.T., G.A. Schmidt, M.E. Mann, D. Rind, and A. Waple. 2001. Solar forcing of regional climate change during the Maunder minimum. *Science* 294: 2149–2152.

Sikes, E.L., C.R. Samson, T.P. Guilderson, and W.R. Howard. 2000. Old radiocarbon ages in the southwest Pacific Ocean during the last glacial period and deglaciation. *Nature* 405: 555–559.

Skinner, L.C., S. Fallon, C. Waelbroeck, E. Michel, and S. Barker. 2010. Ventilation of the deep Southern Ocean and deglacial CO_2 rise. *Science* 328: 1147–1151.

Southon, J., A.L. Noronha, H. Cheng, R.L. Edwards, and Y. Wang. 2012. A high resolution record of atmospheric C-14 based on Hulu Cave speleothem H82. *Quaternary Science Reviews* 33: 32–41.

Stott, L., J. Southon, A. Timmermann, and A. Koutavas. 2009. Radiocarbon age anomaly at intermediate water depth in the Pacific Ocean during the last deglaciation. Paleoceanography 24.

Suess, H.E., and T.W. Linick. 1990. The C-14 Record in bristlecone pine wood of the past 8000 years based on the dendrochronology of the late Ferguson, C.W. *Philosophical Transactions of the Royal Society a-Mathematical Physical and Engineering Sciences* 330: 403–412.

Toggweiler, J.R. 2009. Shifting westerlies. *Science* 323: 1434–1435.

Turney, C.S.M., L.K. Fifield, J.G. Palmer, A.G. Hogg, M.G.L. Baillie, R. Galbraith, J. Ogden, A. Lorrey, and S.G. Tims. 2007. Towards a radiocarbon calibration for oxygen isotope stage 3 using New Zealand Kauri (*Agathis australis*). *Radiocarbon* 49: 447–457.

Vaquero, J.M. 2007. Historical sunspot observations: a review. *Advances in Space Research* 40: 929–941.

Chapter 8
Accelerator Mass Spectrometry of Radiocarbon

Axel Steinhof

8.1 Introduction

Accelerator mass spectrometry (AMS) is a general technique for measuring isotopic ratios, combining mass spectrometry with an accelerator (Litherland 1980; Allen 1987; Synal and Wacker 2010). For the isotopic ratios of stable (not radioactive) isotopes, conventional mass spectrometry without the use of an accelerator usually has sufficient sensitivity. But in natural samples, the isotopic ratios of cosmogenic isotopes, such as ^{10}Be, ^{26}Al, ^{36}Cl, and ^{14}C, lie below 10^{-10} and are therefore beyond the detection limits of conventional mass spectrometry. In this field, ^{14}C is the isotope with the most AMS measurements because of its application to so many disciplines (Fink 2010).

Generally, the isotopic ratio of ^{14}C is given as fraction of the so-called modern ^{14}C isotopic ratio, denoted *F* (Fraction Modern). The corresponding ^{14}C/^{12}C ratio of this modern ^{14}C isotopic ratio can be deduced from the corresponding specific activity of 13.56 ± 0.07 dpm (disintegrations per minute) g^{-1} C (Karlén et al. 1964), referring to the year 1950 and the half-life of ^{14}C of 5730 ± 40 year (Godwin 1962) (Chap. 3). Given these values, an isotopic ratio ^{14}C/^{12}C of 1.18×10^{-12} is calculated for a sample with 1 *F*.

In the case of ^{14}C, counting radioactive decay is a suitable method to determine its concentration. In a decay of a ^{14}C atom, a low-energy β^- radiation (max. energy 156 keV) is emitted. The measurement problems related to this low-energy value are solved by a sophisticated detection technique. In a typical counting laboratory, background counts, which are low-energy decays not originating from the sample, are 1–2 dpm. As a result, even for modern samples, the necessary sample size is several grams of C, and the measurement times in the order of days to produce

A. Steinhof (✉)
Max Planck Institute for Biogeochemistry, Jena, Germany

E.A.G. Schuur et al. (eds.), *Radiocarbon and Climate Change*,
DOI 10.1007/978-3-319-25643-6_8

enough sample counts to sufficiently distinguish the measured ratio apart from background.

Measuring ^{14}C is a variant of the 'needle in a haystack' problem. In this analogy, a ^{14}C atom would be like a needle that beeps infrequently. Decay counting is equivalent to waiting in front of the haystack to measure beeps over time in order to determine how many needles are in the haystack. With AMS, the haystack is split into individual pieces and each one is checked to see whether it is either a stalk of hay or a needle. In a real AMS, a sample is ionized and the fraction of ^{14}C atoms is quantified. A current of (singly charged) C ions measuring 10 μA corresponds to 6.2×10^{13} ions s^{-1}. In the case of a modern sample, there are about 73 ^{14}C ions s^{-1} in that current. A measurement of 10,000 ^{14}C ions, corresponding to a precision of 1 %, would therefore require only 137 s. In contrast, one gram of modern C will generate 13.6 radioactive decays min^{-1}, so detecting 10,000 decays would take ~12 h assuming 100 % efficiency of counting. This highlights the second important advantage of AMS compared to decay counting: the required sample size. Instead of requiring several grams of C as in decay counting, fractions of a milligram are sufficient for a ^{14}C AMS measurement.

The uncertainties of the lifetime and of the specific activity lead to an uncertainty of 0.9 % in the measurement of the absolute ratio of $^{14}C/^{12}C$ in a modern sample. However, as with stable isotopes, relative isotope ratios among samples can be measured much more precisely than the absolute ratios. Therefore, the modern ^{14}C isotopic ratio is defined as 95 % of the radiocarbon standard SRM-4990B provided by the National Institute of Standards and Technology (NIST, MD, USA). There are indications from AMS measurements of the modern ^{14}C isotopic ratio that either (1) the specific activity; (2) the half-life; or (3) the value assigned to the standard are not consistent with each other by a few percent (Roberts and Southon 2007). But this does not matter for ^{14}C applications because all ^{14}C measurements are made relative to standards. And the calibration curves that relate the ^{14}C isotopic ratio to calendar years are also deduced from relative measurements of the same standards. Therefore, potential future changes in the definition of specific activity or of half-life would not influence any ^{14}C result reported as a ratio to the standard.

There are several major challenges in measuring isotopic ratios of 10^{-12} and below as in the case of natural $^{14}C/^{12}C$ abundance. An important part of an AMS facility is the detector for the ^{14}C atoms. Due to the low count rate of a few ^{14}C atoms s^{-1}, the ^{14}C atoms have to be detected as individual, single atoms. The resolution and count rate capability of the detector determines the requirements of the filtering features of the rest of the AMS facility. But other ions as well as the ion beams of ^{12}C and ^{13}C have to be kept away from the ^{14}C detector. The most critical contaminants are the isobaric interferences, which are ions of the same mass as ^{14}C. Particularly important contaminants include ^{14}N, which is very abundant, and molecular ions such as ^{13}CH and $^{12}CH_2$, because they cannot be suppressed by standard mass spectrometry. Depending on the resolution capability of the detector, rare processes that produce background ions also need to be considered. These processes include impacts of ions with residual gas in the beam line of the

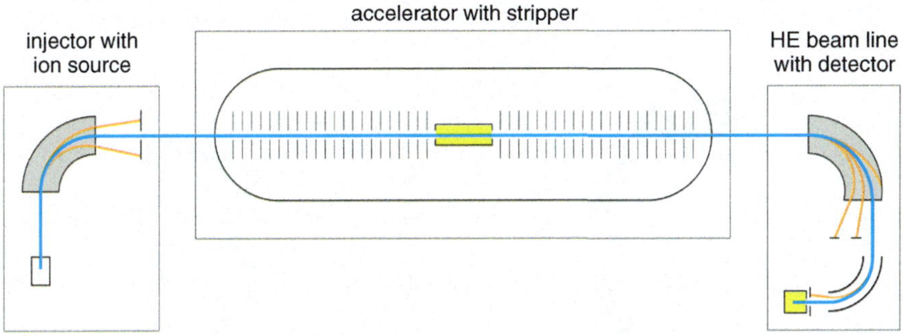

Fig. 8.1 Generalized schematic of the main parts of an AMS facility (not to scale). The *blue line* indicates the central ion trajectory. The *gray-shaded* areas represent magnetic sector fields. Because magnetic sector fields disperse the trajectories of ion beams with the wrong momentum (*orange lines*), these deviate from the central trajectory and are filtered out. The electrostatic analyzer (ESA), as indicated in the high-energy (HE) beam line, filters analogously the ion beams with the wrong energy (orange line). The stripper and the detector (*yellow*) are both gas filled

accelerator, which change the charge state or energy of the ions. An overview of how these challenges are met by ^{14}C AMS is presented in the following sections on the basis of the major components of a ^{14}C AMS facility. Sections 8.2–8.6 describe the basic principles underlying the different components used in ^{14}C AMS. The order of the descriptions follows the path of the ions (Fig. 8.1). Section 8.6 gives an overview of the various configurations for ^{14}C AMS.

8.2 Ion Source and Injector

In the ion source, the sample is split into atoms, or at least molecules, which are charged (ionized) by adding or withdrawing electrons. Different ion sources can emit either negative or positive ions (Hellborg 2005). The ion sources used for ^{14}C AMS all emit negative ions, because then ^{14}N is suppressed as it cannot form negative ions (Litherland 1980); most AMS sources are similar to the one developed by Middleton et al. (1994). The central part of the ion source is confined by the ionizer, its shroud, and the cover of the target or cathode (Fig. 8.2). In operation, a cesium (Cs) vapor is generated in this central part of the ion source with a pressure estimated to be smaller than 10^{-4} mbar. The ionizer is made of an appropriate metal such as tantalum, tungsten, or molybdenum that is characterized by a low work function, which is the energy required to remove an electron from inside the solid to a point outside the surface. It is heated to more than 1000 °C which thermally ionizes the Cs atoms to Cs^+ ions (Fig. 8.2). As a result of the electric potential (U_{tar}) applied between the ionizer and the sample target, these Cs^+ ions are focused onto the middle of the target where the sample material is positioned. Corresponding to the potential differences, the energy of the Cs^+ ions is

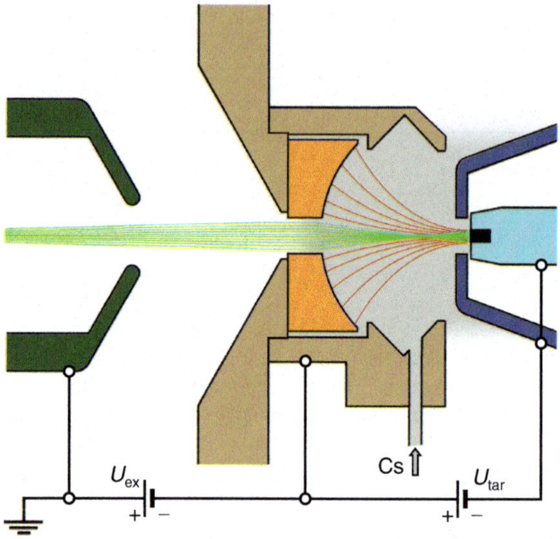

Fig. 8.2 Generalized schematic of a sputter ion source for negative ions (not to scale). In operation, Cs vapor is produced by heating a Cs reservoir or oven (not shown). The vapor is fed into the ion source through the tube (*gray arrow*). On the hot surfaces of the ionizer (*orange*), the Cs atoms are ionized to Cs^+ ions (*red lines*), which are focused by the target voltage (U_{tar}) onto the graphite (*black*) held within the metal target (*cyan*). The negative C ions (*light green*) produced by the Cs^+ ions as they sputter the graphite are accelerated first by U_{tar} and then by the extraction voltage (U_{ex})

several keV (thousand eV). An electron volt (eV) is the energy a single-charged ion or electron gains by being accelerated by a potential difference of 1 V. The Cs^+ ions hit the sample target with several keV of energy and eject atoms from its surface, a process known as sputtering. The sputtering process creates positive ions, neutral atoms, and negative ions. In the case of ^{14}C AMS, the material on the target is graphite made from the sample to be measured, which is then the source of the C ions extracted by the sputtering process. Due to the applied voltages, only the negative C ions are extracted and further accelerated away from the ion source. The energy E_{source} of the ions emitted from the ion source is

$$E_{source} = e \cdot q \cdot (U_{ex} + U_{tar}) \tag{8.1}$$

where the voltages U_{ex} and U_{tar} are explained in Fig. 8.2, q is the charge state and e the elementary charge ($e = 1.602 \times 10^{-19}$ C).

There are also ion sources that act upon CO_2 instead of graphite (Fahrni et al. 2013). In contrast to 'real' gas ion sources, sputtering processes ionize gas molecules absorbed on the surface of the target. In this case, the target has an insert made out of a material such as titanium or zirconium. The gas enters the target through holes in the target body and diffuses to the front of the target where sophisticated geometries increase the absorption probability. Using CO_2 directly has the advantage of avoiding the graphitization step in the sample preparation, but the disadvantage of producing a lower ion current and a higher background compared to using an ion source with graphite. In most laboratories, samples are still converted to graphite, whereas the direct ionization of CO_2 is used only for special samples such as those that are ultrasmall (see Sect. 8.7). The set-up of the ion source beside the target is similar, so that the most gas ion sources can also be operated with graphite and are therefore called hybrid ion sources (Ramsey and Hedges 1997).

The injector is the connection between the ion source and the accelerator. It also has the function of performing an initial mass separation of the ions using standard mass spectrometry technology. The typical ion energy in the injector is between 30 and 70 keV. The injector includes at least one magnetic sector field, lenses, steerers (deflectors), and sometimes also an electrostatic analyzer (ESA). The magnetic sector fields are also often called 'dipole magnets' in contrast to magnetic quadrupole lenses. The Lorentz force acts on a charged ion moving in a magnetic field. For an ion with charge q and momentum p in a homogeneous magnetic field with strength B oriented normally to the plane of motion, the radius of curvature of the trajectory (r) is proportional to:

$$r \propto \frac{p}{q \cdot B} \tag{8.2}$$

As the trajectories of ions with different momentum over charge ratios (p/q) have different radii of curvature, the ions experience different deflections. An appropriately designed magnetic sector field also performs angular focusing. This results from the fact that ions with a larger entrance angle have longer paths in the magnetic field and are therefore more deflected and vice versa (Fig. 8.3).

Equation 8.2 describes the principle of a classical mass spectrometer. The momentum (p) can be written as $p = \sqrt{2 \cdot A \cdot E}$ as a function of the mass (A) and the ion energy (E). All ions in the source are accelerated by the same potential and gain the same energy if they have the same charge, and as a result, the separation according to the momentum is a separation according to the different masses. As the acceptance of a mass spectrometer or an accelerator is limited, the ions with bending radii differing too much from the nominal value are stopped. To do the stopping in a controlled way, it is usually done with apertures (Fig. 8.3). The acceptance has to be chosen carefully during instrument design. If it is too large, the resolution is reduced, if it is too small, the transmission of the desired ion species is reduced or very sensitive to fluctuations of the energy, position, or angle of the ions.

Another analyzing ion optical element is an electrostatic analyser, usually consists of two curved condenser plates with a radial electrical field of strength E_{el} in between. Such a device deflects the ions on a circular path given by a radius r

$$r \propto \frac{E}{q \cdot E_{el}} \tag{8.3}$$

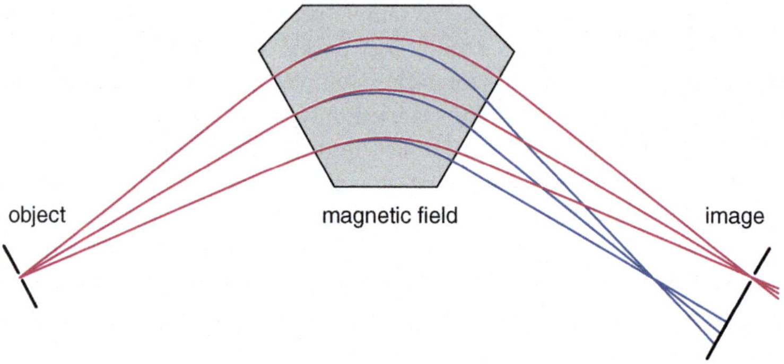

Fig. 8.3 Trajectories of a divergent ion beam in a homogeneous magnetic sector field. In this plane, the example shows a point-to-point focusing. The *blue* trajectories correspond to ions with a lower momentum than the *red* ones. The 'source' of the ions (object) can be either the aperture of an ion source or a slit in the beam line. The *blue* ions are stopped (filtered out) in this example, whereas the *red* ions pass through at the image

For a given charge state, the magnetic field selects according to the momentum over the charge and the electrical field according to the energy over the charge of the ions (E/q). If there are ions with double charge state produced by the ion source, they also gain twice the energy. Therefore, ions like $^{28}Si^{2+}$ have the same p/q and E/q ratio as $^{14}C^{+}$ and cannot be distinguished by a normal spectrometer based on magnetic sector fields or ESAs.

If the electric and magnetic fields are known, the ion trajectories can be calculated by solving the equations of motion (ray tracing). Ion trajectories can also be calculated by transfer matrices similar to light optics. The ion trajectories shown in Fig. 8.3 can be displayed as differences from the nominal value. In this case, the middle magenta trajectory would be the zero line, also called the central trajectory (Wollnik 1987; Liebl 2008). The magnet is represented by a simple lens with a focal length, and the parameter 'wavelength' in case of a light beam corresponds to three ion features: momentum p, charge state q, and energy E of the ion beam.

A typical mass spectrum of a negative ion source as used for ^{14}C AMS is shown in Fig. 8.4. It was obtained by scanning the magnetic sector field. The most important feature of this mass spectrum in the context of ^{14}C is that there is no peak resulting from $^{14}N^{-}$ or $^{28}Si^{2-}$ ions. Nitrogen does not form negative atomic ions due to its electron affinity (Litherland 1980), and in general negative ions with a double charge are not produced. These two ion species, if they could be formed, would not be separated from ^{14}C by electric or magnetic fields and would produce a background signal that would be very difficult to separate from the sample ^{14}C. The fact that neither is produced in the ion source is a critical feature of ^{14}C AMS. Nevertheless, even by using negative ions, a standard mass spectrometer is not sufficient to measure ^{14}C isotopic ratios. In particular, this is because the background signal resulting from $^{12}CH_2$ and ^{13}CH ions with the same mass as ^{14}C, and the overlapping 'tails' of the neighboring masses that are a consequence of the large

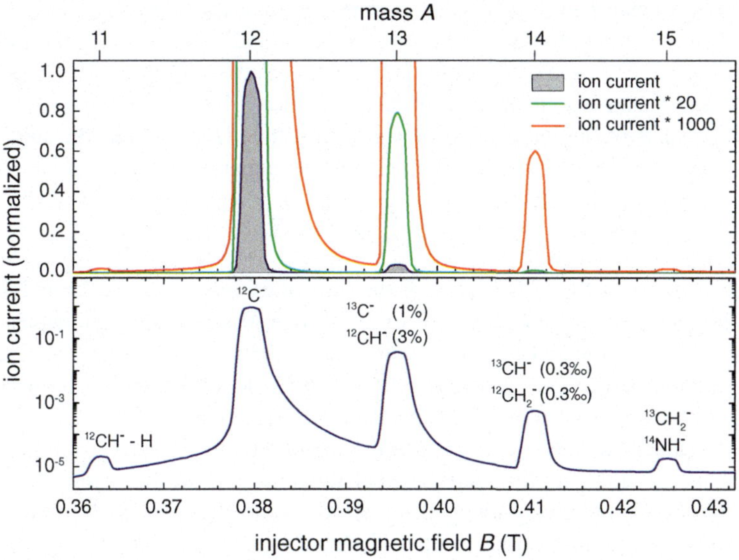

Fig. 8.4 Mass spectrum of an ion source for negative ions, with a linear ion current scale for the *top panel* and a logarithmic scale for the *lower panel*. The ion currents are displayed in three different amplifications (legend). The mass values given at the *top* of the figure are calculated from the observed magnetic fields assuming ions of charge state 1, but the ion identifications given in the *lower figure* are deduced by AMS methods. The relative intensities given in *brackets* in the lower panel are references to the $^{12}C^{-}$ current and are approximate values

abundance ratios. The tail of ^{12}C at the mass 14 position is still 10^{-5} of the ^{12}C peak intensity. This means that even for modern ^{14}C samples, there is 10 million times more ^{12}C than ^{14}C because of this tailing, which would obscure the detection of ^{14}C (Fig. 8.4). Thus, other features of the AMS are needed to resolve this issue.

8.3 Accelerator and Stripper

The basic concept of the accelerator used in ^{14}C AMS is electrostatic acceleration with a charge-exchange scheme to increase the final ion energy, called a tandem accelerator (Fig. 8.1). This concept is not necessarily state-of-the-art accelerator technology, but it has several advantages for ^{14}C AMS. Most importantly, the charge-exchange process removes interferences by breaking up molecules. The charge-exchange device, the so-called stripper, is not present in classical mass spectrometry. The vacuum in a typical ion beam is usually between 10^{-5} and 10^{-7} mbar to avoid interactions of the ions with other molecules. In contrast, the stripper within an AMS is either a thin foil or a volume of gas placed deliberately into the ion beam path. Within the stripper, the ions stochastically exchange electrons with the atoms from the foil or the gas. In a tandem accelerator, the stripper is placed on a positive electric potential called the terminal voltage (U_{TV}), which attracts the negative ions created in the ion source. The ion energy at the stripper (E_{strip}) consists of the ion energy after the ion source (E_{source}) plus the energy gain due to the attraction by the terminal voltage (the equation is based on $|q| = 1$):

$$E_{strip} = E_{source} + e \cdot U_{TV} \tag{8.4}$$

Due to the usually high energy of the ions, they lose more electrons than they gain and thus electrons are 'stripped' from the ions. With increasing thickness of the stripper foil or with higher gas pressure, an equilibrium is reached such that the charge state distribution of ions after the stripper depends only on the ion types, their energies, and the type of stripper atoms (Fig. 8.5).

The stripper has the effect of changing the charge of the ions from negative to positive ions, which are then further accelerated away from the terminal due to the repulsion of the positively charged ion from the positive terminal voltage. Aside from some energy loss in the stripper, the final energy E_{final} of an ion with a final charge state $+Q$ is therefore:

$$E_{final} = E_{source} + (Q+1) \cdot e \cdot U_{TV} \tag{8.5}$$

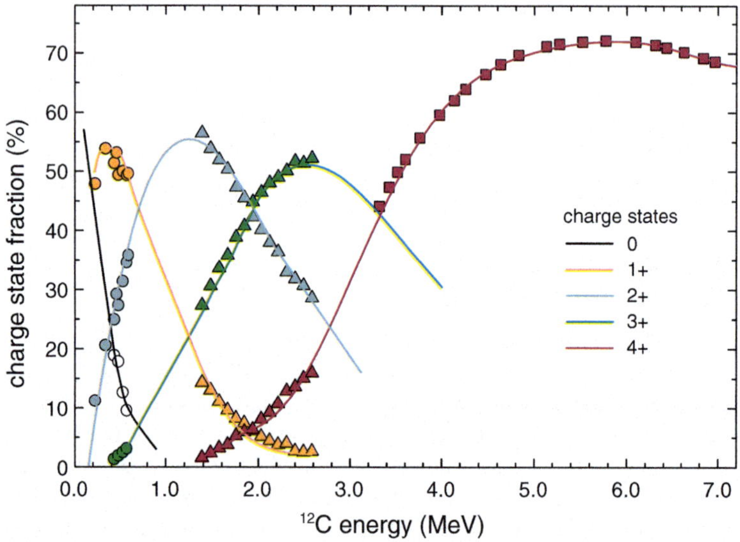

Fig. 8.5 Equilibrium charge state distributions of ^{12}C ions in argon as a function of the energy. The different charge states are indicated by the different colors of the *lines* and *symbols* (legend). Data from Jacob et al. (2000) (*circles*), Kiisk et al. (2002) (*triangles*), and Bonani et al. (1990) (*squares*). The energies of the values measured with ^{14}C Bonani et al. (1990) and ^{13}C Kiisk et al. (2002) were converted to the corresponding values for ^{12}C

Only one charge state can be analyzed by the high-energy beam line and the detector, and so ions in the nonselected charge states are lost. Therefore, an AMS facility should operate at an energy close to the maximum in stripping efficiencies. Based on the maximum terminal voltage that a given accelerator can reach, the best charge state, the optimum voltage, and hence the final energy of the ions can be calculated (Table 8.1). Higher final energy is better for discrimination of ^{14}C from background ions within the detector, and this depends nonlinearly on the charge state.

Table 8.1 Optimum energy for C ions in argon stripper gas according to the stripping efficiencies and the resulting final ion energy

Charge state after the stripper	Optimum terminal voltage (MV)	Final energy (MeV)
1+	0.37	0.74
2+	1.1	3.3
3+	2.5	10.0
4+	6.5	32.5

The energy out of the ion source and the energy loss in the stripper were ignored in these calculations

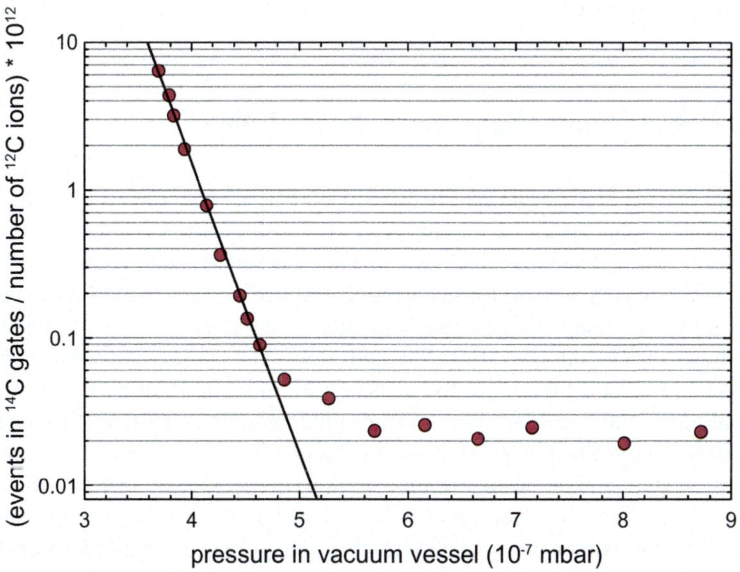

Fig. 8.6 Number of events identified as ^{14}C events as function of the stripper thickness, where gas pressure in the vacuum vessel is used as a proxy for the actual gas pressure in the stripper. The ion energy in this example was 228 keV and the stripper gas was N_2. At low stripper thickness, the majority of the events identified as ^{14}C result from $^{12}CH_2$ and ^{13}CH ions. The decrease in count rate with pressure, as indicated by the line fit, is due to the destruction of these molecules in the stripper. The remaining base level at high stripper thicknesses is in large part due to real ^{14}C ions. Figure modified from Synal et al. (2007)

Radiocarbon AMS facilities operating with charge states equal to 3+ or larger use the fact that the molecules CH_2 and CH are not stable in these charge states. They break apart as a result of 'Coulomb explosion' when the total ionic energy exceeds the molecular binding energy and thus do not contribute to the background in the detector. The thickness of the stripper is difficult to measure directly. Experimentally it is usually adjusted by optimizing the yield of the desired charge state. Due to the angular straggling in the stripper gas, the maximum of transmission occurs slightly below the equilibrium thickness (Niklaus et al. 1994; Kiisk et al. 2002).

Radiocarbon AMS with charge state smaller than 3+ is possible due to the molecular destruction of the CH_2 and CH molecules in multiple collisions with atoms of the stripper (Fig. 8.6). To achieve destruction of the molecules, the stripper itself has to be much thicker than what is needed for providing the required charge state. As a result, ions are also losing energy during this process and become scattered. Energy loss for individual ions does not occur as a single event, but instead, a small portion of energy is lost during every impact depending on the impact parameters. This leads to energy straggling where the energy distribution becomes broader within the ion beam. The same applies for the angular straggling due to scattering. The angular and the energy straggling are challenges for the AMS facilities using the 1+ charge state and relying on molecular destruction of the molecules in the stripper (see Sect. 8.7).

8.4 Detector and High-energy Beam Line

The high-energy (HE) beam line is the connection between the accelerator and the detector and, analogous to the injector, it has the task of filtering (Fig. 8.1). Therefore, a HE beam line includes one or more magnetic sector field, lenses, steerers, and usually also an ESA. Due to the higher ion energy after the accelerator and stripper, the magnetic and electric fields in these devices have to be stronger than those in the injector (Eqs. 8.2 and 8.3). The higher ion energy has the advantage of reducing the background as a result of reduced ion scattering from the residual gas in the vacuum system, because the respective cross sections of the ion beam decreases as the reciprocal of the ion energy. The background ions considered so far were shown in the mass spectra (Fig. 8.4) and ranged up to 10^{-5} with respect to the ^{12}C current. But because the $^{14}C/^{12}C$ isotopic ratio is lower by another 7 orders of magnitude or more, there are other background contaminations that need to be considered. For example, a well-understood example of contribution to the background results from a combination of charge-exchange and scattering of other C ions by residual gas atoms (Fig. 8.7). This example is shown for C ions, but analogous interactions can occur with other ions as well that can all reach the detector designed to measure ^{14}C (Steinhof 2014). There are also other background contributions, some of which are still not understood (Calcagnile and Quarta 2010) or are surprising at their first appearance (Southon et al. 2004).

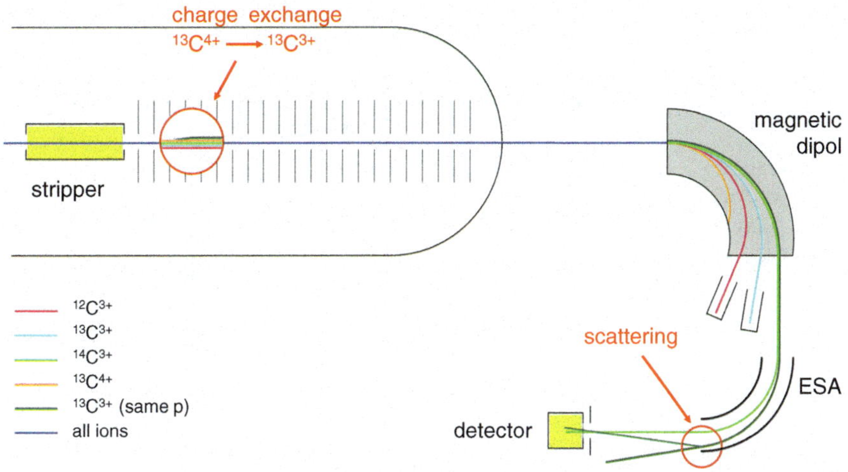

Fig. 8.7 Schematic demonstrates the background that results from a combination of charge exchange and scattering reactions, both with residual gas. The charge-exchange reaction from $^{13}C^{4+}$ to $^{13}C^{3+}$ occurs all along the path of the ion. The energy of the resulting $^{13}C^{3+}$ ions is maximal if the charge exchange occurs after the accelerating section (equal to the energy of the 4^+ ions from the accelerator) and minimal if it happens directly after the stripper (the ion has the nominal energy for 3^+). In between these limits, the whole continuum of energy is produced, including $^{13}C^{3+}$ ions with 14/13 times the nominal 3^+ energy (*dark green line*) resulting in a magnetic rigidity (momentum *p*) exactly the same as the one of $^{14}C^{3+}$ (*light green line*). These ^{13}C ions therefore pass the magnetic sector field, but are usually filtered out in the ESA (*dark green line*) because their energy is too high. However, if a second scattering reaction occurs in the ESA, causing either a change in direction or in charge state, some of these background contaminant ions can still reach the detector

The quality of the filtering of background ions in the HE beam line in combination with the stripper, the injector, and the ion source determines the required resolution and count rate capability of the detector. A very common type of detector is an ionization chamber (Knoll 2000), where the ion beam enters a volume filled with a suitable gas such as isobutene through an entrance window made out of a very thin foil (Fig. 8.8). As the ion beam passes through the foil and the gas it loses energy until it stops completely. A small fraction of the lost energy is transformed directly into the thermal energy of the gas, but the large fraction, the so-called electronic energy loss, produces electron/ion pairs in the gas. For C ion energies above several million electron volts (MeV), the electronic energy loss $(dE/dx)_e$ is given approximately by the simplified Bethe–Bloch equation (Eq. 8.6).

$$(dE/dx)_e \propto A \cdot Z^2/E \tag{8.6}$$

where *A* is the mass of the particle, *Z* its atomic number, and *E* its energy. In this regime, (dE/dx) varies as $1/E$, but at lower energies, the incoming ion is not fully

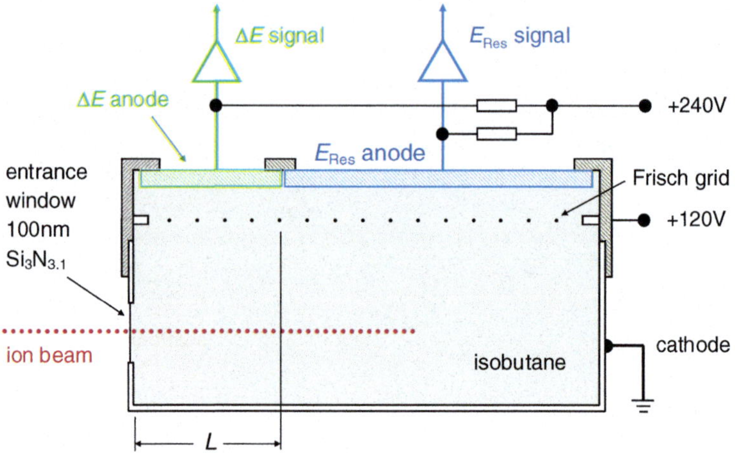

Fig. 8.8 Schematic of an ionization chamber detector with an anode split into two parts. Each anode collects only the electrons within the so-called sensitive volume directly below it. The sensitive volume of the ΔE-anode is indicated by the length L. A Frisch Grid is placed between the ionization volume and the anodes. The applied voltage in the figure refers to an internal gas pressure of 30 mbar of isobutane

stripped of electrons and ($\mathrm{d}E/\mathrm{d}x$) actually decreases with decreasing energy. More precise values are available in form of tables (Northcliffe and Schilling 1970) or computer programs (Ziegler et al. 2010).

Along the path of the ion beam, several million isobutane molecules are split into electron/ion pairs and separated from each other by electric fields (Fig. 8.8). Under the conditions applied in AMS detectors, the number of electron/ion pairs is proportional to the electronic energy loss. The electric field must not be overly strong, which would further ionize gas molecules, or be overly weak, which would allow recombination of electron/ion pairs. The electrons derived from the ion beam–gas interactions are collected on the anodes, and the signal is amplified electronically. Since the direction of the electric fields within the detector is normal to the anodes, each anode only collects the electrons of its respective gas volume directly under it. The detector signal for the first anode is proportional to the integral of the electronic energy loss over this volume

$$\Delta E \propto \int_0^L (\mathrm{d}E/\mathrm{d}x)_e \mathrm{d}x \tag{8.7}$$

where L is the length of the ΔE-anode in the beam direction. This signal ΔE is called the energy-loss signal. The second anode is of sufficient length (for the applied isobutane pressure) so that all ions stop under it. Therefore its signal is the residual energy (E_{Res}) and given by

$$E_{\text{Res}} \propto \int_{L}^{\infty} (\mathrm{d}E/\mathrm{d}x)_e \mathrm{d}x \tag{8.8}$$

The simultaneous measurement of energy loss and residual energy of an ion (ΔE, E_{Res}) by the detector allows the determination of the nuclear charge of this ion (Z). The optimum separation of two nuclides (ΔE, E_{Res}) in a detector is given if the length L is equal to the crossing point of the respective curves. An optimized separation of ^{14}C from ^{14}N is then based on the length of the ΔE-anode L (Fig. 8.9).

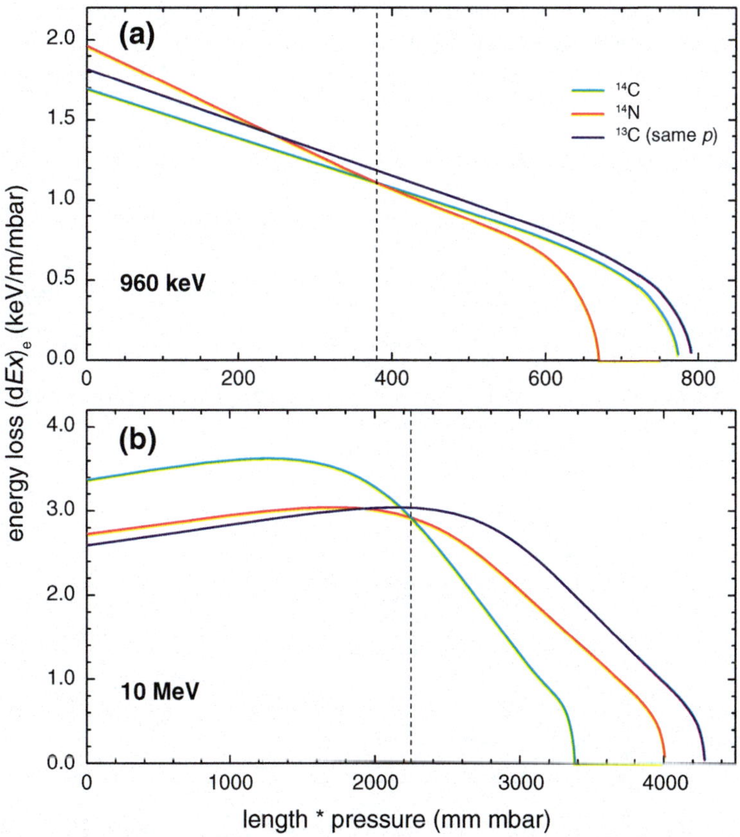

Fig. 8.9 Electronic energy loss of different ions within an AMS detector calculated with energy loss tables from Ziegler et al. (2010). The energy of the ^{14}C and ^{14}N ions are 960 keV (*top*) and 10 MeV (*bottom*) referring to AMS machines with differing configurations. The x-axis is the penetration depth in the gas of the detector as a function of gas pressure and detector length. The *vertical dashed lines* show the length of the ΔE-anode L (in mm * mbar), optimized for the separation of ^{14}C (*green*) and ^{14}N (*red*) ions. The ^{13}C ion (*blue*) energy is 14/13 times that of ^{14}C and so has the same momentum as the other ions. For this example, a silicon nitride foil entrance window with a thickness of 50 nm (*top*) and 100 nm (*bottom*) was taken into account

Based on the electronic energy-loss curves, the values for the ΔE and E_{Res} signals can be estimated for different ions (Table 8.2). The sum of both signals scales with the ion energy, but the differences between ions are much smaller than either signal. Whether these small differences in the signals are sufficient for separation depends on the resolution of the detector. The resolution depends on (1) the noise of the electronics for data acquisition, mainly the preamplifier; (2) the energy straggling of the ions in the entrance window of the detector; and (3) the ionization process of the detector gas (Fig. 8.10). The total energy resolution (R) is given by the root of the sum of the squares of the single contributions. The contribution of the ionization process results from the stochastic nature of the production of the electron/ion pairs in the detector gas. The mean number of charge pairs N produced is simply the energy loss in the detector divided by the energy required per charge pair (21 eV for isobutane), and the fluctuations about this mean value are given by the square root of N times the so-called Fano factor (0.2 for isobutane). This

Table 8.2 Energy loss (ΔE) and residual energy (E_{Res}) for ^{14}C ions with an ion energy of 960 and 10 MeV

	960 keV		10 MeV	
Ion	ΔE	E_{Res}	ΔE	E_{Res}
^{14}C	534	300	6610	3203
Difference $^{14}C–^{13}C$	37	44	−179	955
Difference $^{14}C–^{14}N$	50	−72	1238	−1271

Calculations are based on electronic energy-loss curves and the optimal length of the ΔE-electrode for best separation of ^{14}C from contaminant ions at each energy. For the ^{14}N and ^{13}C ions, the differences are reported relative to the ^{14}C signals

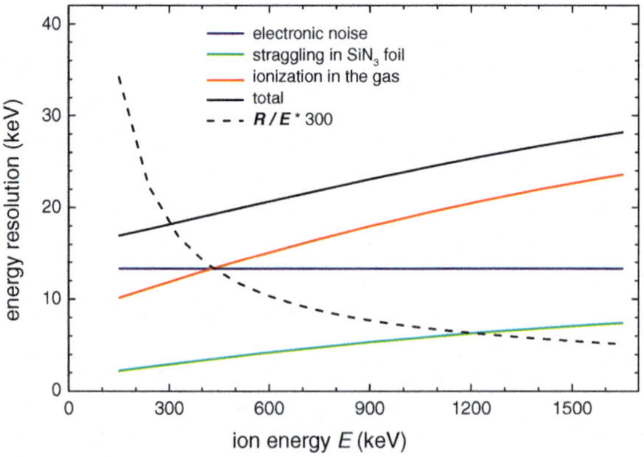

Fig. 8.10 Different contributions to the total energy resolution (R) of an ionization chamber given as FWHM. The curves are based on data measured by Müller (2009) with ^{13}C ions and isobutane as detector gas. The *dashed line* shows the resolution divided by the ion energy multiplied by a factor of 300

contribution is therefore not a technological, but a physical, limitation. The resolution divided by the ion energy (*R*/*E*) decreases with increasing ion energy, and different ions (Table 8.2) can be distinguished even at an ion energy of 960 keV (Fig. 8.10). The requirements on the filtering of the beam lines at an ion energy of 10 MeV are much smaller due to this effect of ion energy on *R*/*E*, and good separation among different ions is achieved (Fig. 8.11).

Other detector technologies such as solid-state PIN (positive, intrinsic, negative) diodes or surface barrier detectors (Knoll 2000) are used in some AMS configurations. Analogous to the gas ionization chamber, the ions create electron/hole pairs in the silicon of the diode as they collide. These semiconductor detectors theoretically should give a higher resolution based on a smaller energy required to produce a charge pair (3.6 eV instead of 21 eV per charge pair, similar Fano factor). But in contrast to this expectation, gas ionization chambers have a better resolution for ions with energies below a few MeV (Döbeli et al. 2004). Furthermore, solid-state detectors cannot be made thin enough for the determination of ΔE (for ion energies up to several MeV) and can therefore only be used to measure the residual or the total energy. They also suffer from radiation damage after long periods of operation and need to be replaced more frequently (Müller et al. 2011), but are typically easier to operate because no gas handling system is needed.

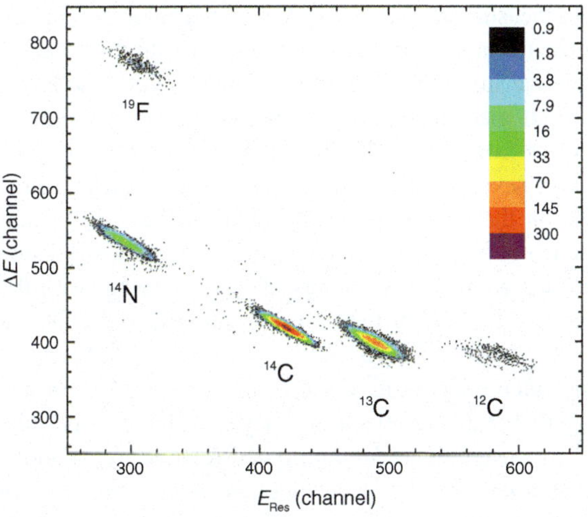

Fig. 8.11 The energy loss (ΔE), and residual energy (E_{Res}) spectrum measured on a low-ratio ^{14}C secondary standard (IAEA-C8), chosen specifically to reduce the dominance of the ^{14}C peak. The number of events per pixel is shown with *color* on a logarithmic scale (legend). The processes by which the non-^{14}C ions reach the detector and the identification of these ions are discussed in Steinhof (2014)

8.5 From Radiocarbon Counts to Radiocarbon Isotopic Ratios

The discussion so far has focused on the detection of the number of ^{14}C atoms (N_{14}) that are mixed with background ions. To calculate the isotopic ratio of ^{14}C relative to ^{12}C or ^{13}C, the respective number of either ^{12}C or ^{13}C atoms is also needed. The choice of reference isotope differs across laboratories, but because the results are given in *F*, this choice does not influence the result (Chap. 3). In the following equation, ^{13}C is shown as the reference. Like ^{12}C, the mass-separated ^{13}C beam has a negligible background and the number of ^{13}C ions (N_{13}) is large enough for a current measurement. In contrast to ^{14}C ions, the intensities are so high, that ^{13}C ions cannot be counted individually. Therefore N_{13} is calculated from the ion current integral:

$$N_{13} = \frac{\int_0^{tm} I_{13}(t)\mathrm{d}t}{q \cdot 1.602 \times 10^{-19}\,\mathrm{C}} \tag{8.9}$$

where $I_{13}(t)$ is the ^{13}C ion current as function of the time t and becomes integrated over the measurement time tm in which ^{14}C is counted, q is the charge state of the ^{13}C ions in the current measurement.

The ion currents are measured in Faraday cups. These are isolated cups connected to current measurement devices. When ions enter the cup and hit a cup wall, they stop, and their charges are collected and measured as electrical current. The resulting signal is the sum of the flux of incoming charges and any leakage currents. The most important mechanism for generating leakage currents is the ejection of secondary electrons from the back wall of the Faraday cup by the incoming ions. Each ion can eject several electrons, but the fraction of electrons actually leaving the cup depends on the geometry and is usually reduced sufficiently by placing a negatively biased suppression electrode at the cup entrance. This negative potential barrier also prevents parasitic low-energy electrons 'flying' with the ion beam from entering the Faraday cup. With increasing requirements on the measurement precision, other effects such as the reduction of the potential barrier due to the space charge of intense incoming positive ion beams or the effect of positive secondary ions ejected together with the electrons may also have to be taken into account. However, as ^{14}C measurements are computed relative to a standard, any errors in the current measurement that are a constant percentage of the true current cancel out when the ratio of sample to standard is calculated.

Correcting for mass-dependent fractionation requires that the ratio of ^{13}C to ^{12}C atoms must also be measured as a ratio of the respective currents (Eq. 8.9). Fractionation occurs in all parts of the AMS facility especially in the ion source and the stripper, and also during the sample preparation. This methodological fractionation is on top of physical and biological fractionations that have occurred in the environment, for example, the enzymatic fractionation of photosynthesis. For most applications of ^{14}C, it is important to correct for the effect of all of these fractionations, both naturally occurring processes as well as those induced by the measurement process. For this correction, it is assumed that the fractionation of ^{14}C with respect to ^{13}C is equal to that of ^{13}C with respect to ^{12}C (Southon 2011). The correction is performed by a normalization to $\delta^{13}C = -25$ ‰ with respect to the VPDB standard (Vienna Pee Dee Belemnite standard for $\delta^{13}C$, see Chap. 3). For this correction, the origin of the fractionation does not matter, but fractionation by some processes such as grazing of an ion beam on the wall of the vacuum chamber cannot be corrected and must be avoided.

The ^{14}C isotopic ratio of an unknown sample (R_{unk}) is determined with respect to the nominal ratio of a standard (R_{std}) (Eq. 8.10).

$$R_{unk} = \frac{\frac{1-25\,‰}{1+\delta^{13}C_{unk}} \cdot \frac{N_{14,unk}}{N_{13,unk}}}{\frac{1-25\,‰}{1+\delta^{13}C_{std}} \cdot \frac{N_{14,std}}{N_{13,std}}} \cdot R_{std} = \frac{R_{13,std}}{R_{13,unk}} \cdot \frac{N_{13,std}}{N_{13,unk}} \cdot \frac{N_{14,unk}}{N_{14,std}} \cdot R_{std} \tag{8.10}$$

For the right side of Eq. 8.10, the $\delta^{13}C$ values were replaced by the respective current ratios $R_{13} = I_{13}/I_{12}$. In the ^{14}C community worldwide, two 'primary' standards are used: the standard SRM-4990B, called OX-I, and the standard SRM-4990C, called OX-II or NOX (for **N**ew **OX**alic acid), both distributed by NIST. As noted in Chap. 3, the nominal value $F = 1/0.95$ for OX-I refers traditionally to a normalization to $\delta^{13}C = -19$ ‰. The application of Eq. 8.10 therefore requires a renormalization to $\delta^{13}C = -25$ ‰, giving a value of $N_{14,std}$ for OX-I of 1. 0398 F. For NOX $N_{14,std} = 1.3406\ F$. A set of secondary standards or reference materials ranging from 0.0 to 1.50 F is offered by the IAEA (International Atomic Energy Agency, Vienna, Austria) (Rozanski et al. 1992). These standards are usually used for quality control purposes.

A critical point in ^{14}C AMS is the drift in measurement over time. The underlying problem is that the standards are measured at different times than the unknown samples (Steinhof 2013). The time between measuring standards and samples is reduced by measuring each sample and standard multiple times, and by measuring a number of different standards within the sample set to control for drift. This scheme helps cancel out errors due to fractionation fluctuations over time. A typical AMS run will mix samples with primary and secondary standards and blanks. In particular, enough primary standards (OX-I or NOX) are interspersed with samples at regular intervals (e.g., fourth to sixth measurement) to check for drift and averaged for normalization.

Due to ^{14}C AMS being a relative measurement that compares samples with standards, precise measurements require stable conditions. The stability of good commercial power supplies is sufficient for most voltages and currents driving the AMS system with the exception of the terminal voltage. Therefore the terminal voltage is feedback-stabilized by monitoring the position (and hence the energy) of a stable isotope beam after the sector magnet in the HE beamline—the so-called slit stabilization (White et al. 1981; Steinhof et al. 2010). One source of instability is the ion source itself. Oversupply of Cs can lead to sparks and arcing, and during operation, the graphite target becomes cratered by the sputtering of the Cs^+ ions, which may alter the

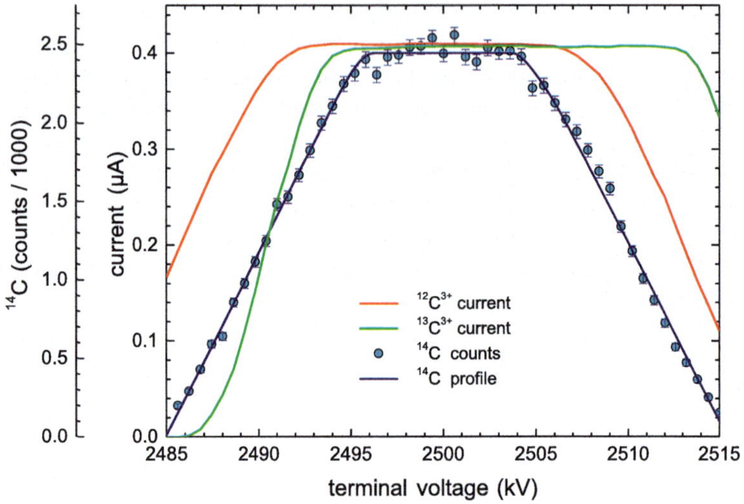

Fig. 8.12 Scan of the terminal voltage for all three isotopes of C. This scan gives a (relative) transmission profile, provided that the current output of the ion source is constant. The ^{14}C count error bars show ± 1 standard deviation, and the ^{12}C current is similar in value to the ^{13}C due to the mechanical chopper that decreases the ^{12}C current

shape of the outgoing beam. Stability also has to be guaranteed over several target changes, and the ion current is not necessarily constant over the whole measurement. Ion current variation is no problem itself, but a high ion current increases the emittance of the ion beam due to the repulsive forces between the charges of the ions (space charge). The emittance is the volume of the ion beam in the phase space built up by the two transverse beam dimensions (i.e., beam height and width) and the corresponding velocity components (Wollnik 1987). Because of these instabilities, the transmission profile, which records the relative transmissions versus the variations of ion optical parameters, should be insensitive against small variations around the nominal value. In the ideal case, a flattop transmission profile is observed (Fig. 8.12) indicating that the emittance of the ion beam is smaller than the acceptance of the AMS facility. During the fine adjustment of the AMS facility, all ion optical parameters are scanned and set to the respective optimum values in the common center of the flattop transmission profiles of the three C isotopes.

8.6 Various Configurations for Radiocarbon Accelerator Mass Spectrometry

Radiocarbon AMS instruments can be divided in two groups with different configurations that depend on the destruction of the isobaric molecules. In one group, the AMS instruments use a charge state greater than or equal to 3+ after the stripper, in the second group, they use a thick stripper to destroy the molecules that could interfere with ^{14}C detection (Sect. 8.3). Radiocarbon AMS facilities using a charge state greater than or equal to 3+ after the stripper have a terminal voltage of 2.5 MV or larger to achieve a good stripping efficiency. This also ensures that the ion energy at the detector is sufficient for good background discrimination. There are AMS facilities capable of applying a terminal voltage high enough to measure C ions even in the 4+ charge state, (Bonani et al. 1990; Southon et al. 1990; Fink et al. 2004) but a majority use the 3+ charge state (Kutschera et al. 1997; Gottdang et al. 2001).

Only a few machines use the 2+ charge state along with a thick stripper (Chamizo et al. 2008; Hong et al. 2010) Using the 2+ charge state requires more energy than the 1+ charge state and introduces a $^{7}Li^{+}$ background not present with the 1+ charge state. These $^{7}Li^{+}$ ions are not suppressed by the HE beam line because they have the same momentum to charge and energy-to-charge ratios as the $^{14}C^{2+}$ ions. The discrimination in the detector is difficult because the two $^{7}Li^{+}$ ion fragments arrive simultaneously. The final ion energy of 3 MeV of these instruments makes this background separation possible. The choice by an AMS facility to use the 2+ charge state is often motivated by the capability to measure other isotopes such as ^{10}Be, ^{27}Al, and ^{129}I.

The first machine performing ^{14}C AMS using the 1+ charge state was called TANDY (Synal et al. 2000). This machine also destroys the isobaric molecules using a thick stripper, and the terminal voltage of 500 kV is produced by a Pelletron accelerator charging system in a SF_6 filled tank, similar to large tandem accelerators (Goslar et al. 2004; Southon et al. 2004; Kobayashi et al. 2007). A further development was the Mini Carbon Dating System (MICADAS) (Synal et al. 2007) which avoids using a Pelletron charging facility with its moving parts and SF_6. It operates in a vacuum-insulated tank at a terminal voltage of 200 kV produced by an external electronic solid-state power supply, and also uses the destruction of molecules by the stripper. The main difficulty of this configuration, aside from identification of ^{14}C against the background, was the angular straggling of the ion beam in the thick stripper. The angular straggling is 85 times worse in case of the MICADAS machine compared to 2.5 MV machines. This estimate is based on the ratios of the stripper thicknesses and the straggling, which is proportional to the reciprocal ion energies. However, this technical problem has been solved and many MICADAS facilities are operating successfully worldwide. The low terminal voltage of MICADAS leads to a higher background level than at the AMS facilities with higher terminal voltage. A correlation of the background with the ion current resulting from the ^{13}C fragments of ^{13}CH molecules was discovered, which allows a further background correction. When this is applied, MICADAS is capable of measuring a background level between 0.003 and 0.01 F for combusted and graphitized samples (Synal et al. 2007), comparable with results from higher energy systems that do not require this correction.

The sample preparation including chemical cleaning, combustion, and graphitization of the CO_2 can leave behind, or even introduce, residual ^{14}C contamination (see Chap. 9). As (nearly) all samples have to pass through sample preparation, it is sufficient for an AMS facility to have the intrinsic machine background significantly smaller than the preparation background. In this case a further reduction of the intrinsic machine background improves the overall measurement precision marginally.

There are also attempts to measure ^{14}C on 250 kV single-stage accelerators (Skog 2007; Freeman et al. 2010; Prasad et al. 2013). At these facilities, the stripper, the HE beam line, and the detector are placed on an isolated high-voltage platform or deck, and the second stage of acceleration is omitted. The final ion energy is 45 keV from the ion source plus 245 keV from the high-voltage deck for a total of 290 keV. In this configuration, there are some unresolved interferences and a nonzero detector 'dark-count' rate, but a single-stage system has reached a machine background below 0.0025 *F* (Skog 2007). Like MICADAS, such an accelerator avoids using a Pelletron charging facility and SF_6. Furthermore, it has the advantage that the polarity of the single-stage can be changed and it can therefore also operate with positive ions for other applications.

The terminal voltage or charge state is not the only criteria upon which to group AMS facilities. High precision ^{14}C measurements require that all three C isotopes are sent through the accelerator and measured close to each other. Two methods were developed to send more than one isotope through the accelerator: simultaneous injection and sequential injection, each having strengths and weaknesses. In the case of simultaneous injection, at least two magnets (four are more common) are used in the injector. The first magnet or magnet pair splits the ion beams according to mass, and the second set recombines the ion beams of mass 12, 13, and 14 before injecting them into the accelerator. Other ions such as oxygen are not recombined because they are stopped on slits and walls of the vacuum chamber. The intensity of the ^{12}C ion beam is reduced by a factor of 100 by a mechanical chopper that blocks part of the beam between the two magnets where the beams of the different isotopes are spatially separated (Fig. 8.13). The chopper has three advantages: (1) the ^{12}C current is reduced so as to not influence the stability of the accelerator; (2) the intensity of the ^{12}C and ^{13}C ion beam at the HE side is of similar intensity and their ratio can be measured more precisely; and (3) the background resulting from the ^{12}C beam is reduced by the chopper.

In the sequential injection technique, the three C isotopes are injected one after the other into the accelerator (Fig. 8.14). The beams of ^{12}C and ^{13}C are sent in short microsecond pulses to avoid high ion currents in the accelerator. These time intervals are so short that the accelerator 'detects' only the smaller mean ion current (Suter et al. 1984). The repetition rate of the switching between different isotopes is several times per second so that the majority of time is left for injecting the rare ^{14}C. Such a fast switching facility requires an electrically isolated vacuum chamber of the injector magnet. The switching is achieved by applying voltages known as bouncing voltages to the chamber, which modifies the ion energy and thereby causes the injection magnet to select ions of a different mass. Because data acquisition by the ^{14}C detector ceases during the injection of ^{12}C and ^{13}C ions, the background is not increased.

The various configurations are difficult to compare. First, there are many important parameters including the ion current, the transmission, the stability of the

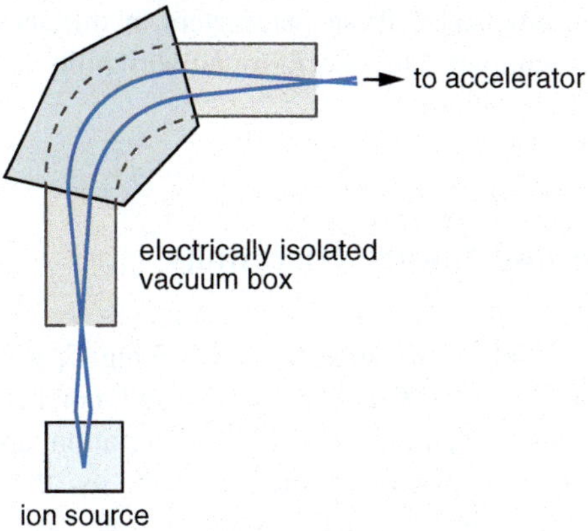

Fig. 8.13 Schematic of a 'bouncer' injector for a ^{14}C AMS system. A voltage applied to the insulated vacuum box is varied (bounced) to select different masses for injection into the accelerator. Switching times are rapid (tens of microseconds), and the system is cycled through mass 12, 13, and 14 many times per second to provide quasi-simultaneous measurements of the three isotopes

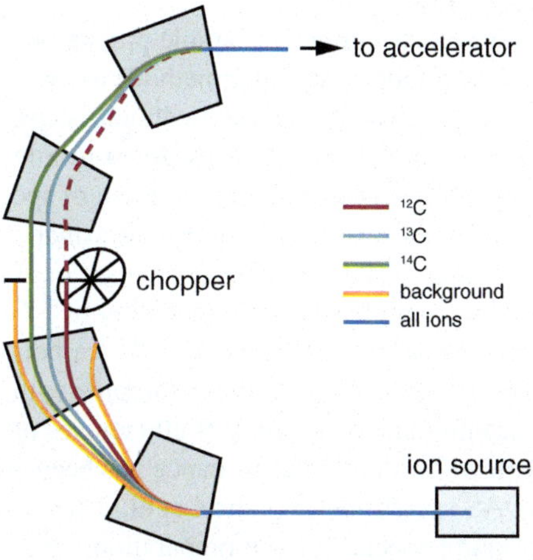

Fig. 8.14 Schematic of a recombinator used for simultaneous injection of the three C isotopes into the accelerator. Unwanted beams are stopped at the midpoint of the system where the beams have been spatially separated by magnets. A rapidly rotating chopper wheel with a small aperture reduces the time-averaged intensity of the transmitted ^{12}C beam by a factor of 100 to avoid overloading the accelerator

transmission and the variability of the background. Even though the background is subtracted during the data analysis and so its value is not important, its variability adds to the uncertainty. Second, not all of these parameters are fully controlled, but partly influence each other and vary at the same facility from day to day. Third, different applications emphasize these parameters in different ways, for example while dating very old samples a low background variability is more important than the stability of the transmission.

8.7 Conclusion and Future Directions

A typical ^{14}C sample with a size greater than 0.5 mg C, a precision for modern samples not better than 0.003 *F* and a background not below 0.004 *F* can be measured at many AMS facilities worldwide. Variation among laboratories is quantified during intercomparison projects, such as the Fifth International Radiocarbon Inter-comparison (FIRI) (Scott et al. 2010) with the Sixth Inter-comparison now underway. In past intercomparisons, roughly 10 % of the results were identified as outliers, which are around twice as frequent as would be expected from simply random fluctuations. The majority of outliers come from only 14 % of the laboratories (Scott 2006). Taking into account further that part of the outliers in the Fourth Inter-comparison (VIRI) were caused by the sample preparation it can be estimated that more than 90 % of the laboratories give, within the uncertainties, equivalent AMS results.

The expertise of various laboratories in sample preparation can be very different. Some laboratories have developed particular methods to measure samples from soil, bone, air, etc. each with its own specific pretreatment steps (Chap. 9). Other laboratories specialize on very small sample sizes down to a few μg C, which opens new applications to ^{14}C AMS such as the measurement of specific C compound that have been isolated from a bulk mixture. For these ultrasmall samples, the efficiency of the measurement becomes very important. In the case of 100 % efficiency for the AMS process, the required sample size for a precision of 1 % (10,000 counted ^{14}C atoms) for a modern sample would be only 10^{16} carbon atoms, or 0.2 μg C. Precisions of 1 % can be reached for graphite samples with ~5 μg sample mass (Smith et al. 2010), meaning that measurement efficiencies of 4 % are possible (see also Santos et al. 2010), while similar performance has been achieved with a gas ion source on the MICADAS spectrometer (Fahrni et al. 2013). Samples that are a few tens of μg C or less require special sample preparation, great care during the AMS measurement, and also specialized data analysis to account for a background that becomes mass dependent (Donahue et al. 1990; Brown and Southon 1997; Santos et al. 2010). However, the measurement of very small samples is becoming routine in more and more AMS laboratories, as researchers explore additional science applications calling for this capability.

Along with attempts to reduce sample size, there are also attempts to increase precision. In a proof of principle study, Guilderson et al. (2006) reached a precision for modern samples of 1 ‰ at the CAMS spectrometer at Livermore, CA (Southon et al. 1990). With the lower-energy MICADAS system, a precision of 2 ‰ was reached (Wacker et al. 2010). The 1 ‰ level of uncertainty is now a challenge that remains as a benchmark for many AMS facilities. Other directions for AMS include measurement of samples with ^{14}C isotopic ratios much higher than natural samples. These samples are typically produced as a result of tracer experiments for medicine where ^{14}C-enriched compounds are given to a test subject. These samples typically need less precision, so increasing instrument throughput (number of samples per time) is often a primary goal.

Before the year 2000, the golden rule of AMS was that charge states after the stripper greater than or equal to 3+ were needed to destroy isobaric molecules. This meant that a terminal voltage of 2.5 MV was required, which was considered relatively small at the time. After that time, a machine configuration utilizing a thick stripper was developed and much smaller, lower-energy systems became available. Regardless of system size, the main feature of AMS is the use of very sensitive mass spectrometry, while combining other methods to filter background contaminants. Up to now, this method has always included an accelerator, but new work has focused on leaving out the accelerator, keeping only the stripper as a filter (Synal et al. 2013). There have also been attempts to measure ^{14}C concentrations with lasers, such as with intracavity optogalvanic spectroscopy (Eilers et al. 2013) or by saturated-absorption cavity ring-down (Galli et al. 2013), which bypasses the need for mass spectrometry. These techniques to reach AMS sensitivity without an accelerator are still at the forefront of technological development and are occurring in parallel with refinements to current ^{14}C AMS systems that will likely continue to lead to better or less expensive machines in the future. Together these technological developments will open new avenues for science discovery by making new applications possible for ^{14}C measurements.

References

Allen, K.W. 1987. *Ultra high sensitivity mass spectrometry with accelerators*. London, England: Royal Society Carlton House Terrace.

Bonani, G., P. Eberhardt, H.J. Hofmann, T.R. Niklaus, M. Suter, H.A. Synal, and W. Wolfli. 1990. Efficiency improvements with a new stripper design. *Nuclear Instruments and Methods in Physics Research Section B-Beam Interactions With Materials and Atoms* 52: 338–344.

Brown, T.A., and J.R. Southon. 1997. Corrections for contamination background in AMS C-14 measurements. *Nuclear Instruments and Methods in Physics Research Section B-Beam Interactions with Materials and Atoms* 123: 208–213.

Calcagnile, L., and G. Quarta. 2010. E/Q and ME/Q(2) contributions to machine background in sequential injection radiocarbon AMS. *Nuclear Instruments and Methods in Physics Research Section B-Beam Interactions with Materials and Atoms* 268: 830–833.

Chamizo, E., J.M. Lopez-Gutierrez, A. Ruiz-Gomez, F.J. Santos, M. Garcia-Leon, C. Maden, and V. Alfimov. 2008. Status of the compact 1 MV AMS facility at the Centro Nacional de Aceleradores (Spain). *Nuclear Instruments and Methods in Physics Research Section B-Beam Interactions with Materials and Atoms* 266: 2217–2220.

Döbeli, M., C. Kottler, M. Stocker, S. Weinmann, H.A. Synal, M. Grajcar, and M. Suter. 2004. Gas ionization chambers with silicon nitride windows for the detection and identification of low energy ions. *Nuclear Instruments and Methods in Physics Research Section B-Beam Interactions with Materials and Atoms* 219: 415–419.

Donahue, D.J., T.W. Linick, and A.J.T. Jull. 1990. Isotope ratio and background corrections for accelerator mass spectrometry radiocarbon measurements. *Radiocarbon* 32: 135–142.
Eilers, G., A. Persson, C. Gustavsson, L. Ryderfors, E. Mukhtar, G. Possnert, and M. Salehpour. 2013. The radiocarbon intracavity optogalvanic spectroscopy setup at uppsala. *Radiocarbon* 55: 237–250.
Fahrni, S.M., L. Wacker, H.A. Synal, and S. Szidat. 2013. Improving a gas ion source for C-14 AMS. *Nuclear Instruments and Methods in Physics Research Section B-Beam Interactions with Materials and Atoms* 294: 320–327.
Fink, D., M. Hotchkis, Q. Hua, G. Jacobsen, A. Smith, U. Zoppi, D. Child, C. Mifsud, H. van der Gaast, A. Williams, and M. Williams. 2004. The antares AMS facility at ANSTO. *Nuclear Instruments and Methods in Physics Research Section B-Beam Interactions With Materials and Atoms* 223: 109–115.
Fink, D. 2010. AMS-11 in Rome, 2008: Past achievements, current and future trends. *Nuclear Instruments and Methods in Physics Research Section B-Beam Interactions with Materials and Atoms* 268: 1334–1342.
Freeman, S., G.T. Cook, A.B. Dougans, P. Naysmith, K.M. Wilcken, and S. Xu. 2010. Improved SSAMS performance. *Nuclear Instruments and Methods in Physics Research Section B-Beam Interactions with Materials and Atoms* 268: 715–717.
Galli, I., S. Bartalini, P. Cancio, P. De Natale, D. Mazzotti, G. Giusfredi, M.E. Fedi, and P.A. Mando. 2013. Optical detection of radiocarbon dioxide: first results and AMS intercomparison. *Radiocarbon* 55: 213–223.
Godwin, H. 1962. Half-life of radiocarbon. *Nature* 195: 984.
Goslar, T., J. Czernik, and E. Goslar. 2004. Low-energy C-14 AMS in Poznan radiocarbon laboratory, Poland. *Nuclear Instruments and Methods in Physics Research Section B-Beam Interactions With Materials and Atoms* 223: 5–11.
Gottdang, A., M. Klein, and D. Mous. 2001. Accelerator mass spectrometry at high voltage engineering Europa (HVEE). *Radiocarbon* 43: 149–156.
Guilderson, T.P., E.B. Roark, P.D. Quay, S.R.F. Page, and C. Moy. 2006. Seawater radiocarbon evolution in the gulf of Alaska: 2002 observations. *Radiocarbon* 48: 1–15.
Hellborg, R. 2005. *Electrostatic accelerators*. Berlin: Springer.
Hong, W., J.H. Park, K.S. Sung, H.J. Woo, J.K. Kim, H.W. Choi, and G.D. Kim. 2010. A New 1MV AMS Facility at Kigam. *Radiocarbon* 52: 243–251.
Jacob, S., M. Suter, and H. Synal. 2000. Ion beam interaction with stripper gas—key for AMS at sub MeV. *Nuclear Instruments and Methods in Physics Research Section B-Beam Interactions With Materials and Atoms* 172: 235–241.
Karlén, I., et al. 1964. Absolute determination of the activity of two C14 dating standards. *Arkiv for Geofysik* 4: 465–471.
Kiisk, M., B. Erlandsson, M. Faarinen, R. Hellborg, K. Hakansson, P. Persson, G. Skog, and K. Stenstrom. 2002. The charge state distribution of a carbon beam measured at the lund pelletron accelerator with the newly installed terminal pumping system in use. *Nuclear Instruments and Methods in Physics Research Section a-Accelerators Spectrometers Detectors and Associated Equipment* 481: 1–8.
Knoll, G. 2000. *Radiation Detection and Measurement*. New York: John Wiley and Sons Inc.
Kobayashi, K., E. Niu, S. Itoh, H. Yamagata, Z. Lomtatidze, I. Jorjoliani, K. Nakamura, and H. Fujine. 2007. The compact C-14 AMS facility of Paleo Labo Co., Ltd, Japan. *Nuclear Instruments and Methods in Physics Research Section B-Beam Interactions With Materials and Atoms* 259: 31–35.
Kutschera, W., P. Collon, H. Friedmann, R. Golser, P. Hille, A. Priller, W. Rom, P. Steier, S. Tagesen, A. Wallner, E. Wild, and G. Winkler. 1997. VERA: A new AMS facility in Vienna. *Nuclear Instruments and Methods in Physics Research Section B-Beam Interactions with Materials and Atoms* 123: 47–50.
Liebl, H. 2008. *Applied charged particle optics*. Berlin Heidelberg: Springer.
Litherland, A.E. 1980. Ultrasensitive mass spectrometry with accelerators. *Annual Review of Nuclear and Particle Science* 30: 437–473.

Middleton, R., D. Juenemann, and J. Klein. 1994. Isotopic fractionation of negative-ions produced by CS sputtering in a high-intensity source. *Nuclear Instruments and Methods in Physics Research Section B-Beam Interactions With Materials and Atoms* 93: 39–51.

Müller, A. 2009. Entwicklung von universellen AMS Anlagen bei tiefen Energien.

Müller, A., A. Cassimi, M. Dobeli, M. Mallepell, I. Monnet, M. Simon, M. Suter, and H. Synal. 2011. A new mini gas ionization chamber for IBA applications. *Nuclear Instruments and Methods in Physics Research Section B-Beam Interactions With Materials and Atoms* 269: 3037–3040.

Niklaus, T.R., G. Bonani, Z. Guo, M. Suter, and H.A. Synal. 1994. Optimizing tandem accelerator stripping efficiency by simulation of charge changing processes. *Nuclear Instruments and Methods in Physics Research Section B-Beam Interactions With Materials and Atoms* 92: 115–121.

Northcliffe, L.C., and R. Schilling. 1970. Range and stopping-power tables for heavy ions. *Atomic Data and Nuclear Data Tables* 7: 233–463.

Prasad, G.V.R., J.E. Noakes, A. Cherkinsky, R. Culp, and D. Dvoracek. 2013. The new 250 kv single stage AMS system at Cais, University of Georgia: performance comparison with a 500 kv compact tandem machine. *Radiocarbon* 55: 319–324.

Ramsey, C.B., and R.E.M. Hedges. 1997. Hybrid Ion Sources: Radiocarbon Measurements from Microgram to Milligram. *Nuclear Instruments & Methods in Physics Research Section B-Beam Interactions with Materials and Atoms* 123: 539–545.

Roberts, M.L., and J.R. Southon. 2007. A Preliminary Determination of the Absolute C-14/C-12 Ratio of OX-I. *Radiocarbon* 49: 441–445.

Rozanski, K., W. Stichler, R. Gonfiantini, E.M. Scott, R.P. Beukens, B. Kromer, and J. Vanderplicht. 1992. The IAEA C-14 intercomparison exercise 1990. *Radiocarbon* 34: 506–519.

Santos, G.M., J.R. Southon, N.J. Drenzek, L.A. Ziolkowski, E. Druffel, X. Xu, D. Zhang, S. Trumbore, T.I. Eglinton, and K.A. Hughen. 2010. Blank assessment for ultra-small radiocarbon samples: chemical extraction and separation versus AMS. *Radiocarbon* 52: 1322–1335.

Scott, E.M. 2006. Assuring measurement quality: the international ^{14}C laboratory inter-comparison program. *Pages News* 14: 7–9.

Scott, E.M., et al. 2010. The fifth international radiocarbon intercomparison (VIRI): an assessment of laboratory performance in stage 3. *Radiocarbon* 52: 859–865.

Skog, G. 2007. The single stage AMS machine at Lund University: status report. *Nuclear Instruments and Methods in Physics Research Section B-Beam Interactions with Materials and Atoms* 259: 1–6.

Smith, A.M., Q. Hua, A. Williams, V. Levchenko, and B. Yang. 2010. Developments in micro-sample C-14 AMS at the ANTARES AMS facility. *Nuclear Instruments and Methods in Physics Research Section B-Beam Interactions with Materials and Atoms* 268: 919–923.

Southon, J. 2011. Are the fractionation corrections correct: are the isotopic shifts for C-14/C-12 ratios in physical processes and chemical reactions really twice those for C-13/C-12? *Radiocarbon* 53: 691–704.

Southon, J.R., M.W. Caffcc, J.C. Davis, T.L. Moore, I.D. Proctor, B. Schumacher, and J.S. Vogel. 1990. The new LLNL AMS spectrometer. *Nuclear Instruments and Methods in Physics Research Section B-Beam Interactions with Materials and Atoms* 52: 301–305.

Southon, J., G. Santos, K. Druffel-Rodriguez, E. Druffel, S. Trumbore, X. Xu, S. Griffin, S. Ali, and M. Mazon. 2004. The keck carbon cycle AMS laboratory, University of California, Irvine: initial operation and a background surprise. *Radiocarbon* 46: 41–49.

Steinhof, A. 2013. Data analysis at the JENA C-14 laboratory. *Radiocarbon* 55: 282–293.

Steinhof, A. 2014. Analysis of the background of the Jena ^{14}C-AMS facility. *Nuclear Instruments & Methods in Physics Research, Section B: Beam Interactions with Materials and Atoms* 331: 238–242.

Steinhof, A., I. Hejja, and T. Wagner. 2010. Improvements of the Jena AMS system. *Nuclear Instruments and Methods in Physics Research Section B-Beam Interactions with Materials and Atoms* 268: 902–905.

Suter, M., R. Balzer, G. Bonani, and W. Wolfli. 1984. A fast beam pulsing system for isotope ratio measurements. *Nuclear Instruments and Methods in Physics Research Section B-Beam Interactions With Materials and Atoms* 5: 242–246.

Synal, H., and L. Wacker. 2010. AMS measurement technique after 30 years: possibilities and limitations of low energy systems. *Nuclear Instruments and Methods in Physics Research Section B-Beam Interactions With Materials and Atoms* 268: 701–707.

Synal, H., S. Jacob, and M. Suter. 2000. The PSI/ETH small radiocarbon dating system. *Nuclear Instruments and Methods in Physics Research Section B-Beam Interactions With Materials and Atoms* 172: 1–7.

Synal, H., M. Stocker, and M. Suter. 2007. MICADAS: a new compact radiocarbon AMS system. *Nuclear Instruments and Methods in Physics Research Section B-Beam Interactions With Materials and Atoms* 259: 7–13.

Synal, H.A., T. Schulze-Koenig, M. Seiler, M. Suter, and L. Wacker. 2013. Mass spectrometric detection of radiocarbon for dating applications. *Nuclear Instruments and Methods in Physics Research Section B-Beam Interactions with Materials and Atoms* 294: 349–352.

Wacker, L., G. Bonani, M. Friedrich, I. Hajdas, B. Kromer, M. Nemec, M. Ruff, M. Suter, H.A. Synal, and C. Vockenhuber. 2010. MICADAS: routine and high precision radiocarbon dating. *Radiocarbon* 52: 252–262.

White, N.R., et al. 1981. The radiocarbon facility at the Research Lab for Archaeology in Oxford—a review. Symposium on Accelerator Mass Spectrometry, Argonne National Laboratory, Argonne, Illinois, 1981, Argonne, Argonne National Laboratory (ANL)—Physics Division.

Wollnik, H. 1987. *Optics of charged particles.*

Ziegler, J.F., M.D. Ziegler, and J.P. Biersack. 2010. SRIM—the stopping and range of ions in matter (2010). *Nuclear Instruments and Methods in Physics Research Section B-Beam Interactions with Materials and Atoms* 268: 1818–1823.

Chapter 9
Preparation for Radiocarbon Analysis

S.E. Trumbore, X. Xu, G.M. Santos, C.I. Czimczik, S.R. Beaupré, M.A. Pack, F.M. Hopkins, A. Stills, M. Lupascu and L. Ziolkowski

9.1 Introduction

This chapter presents a brief overview of the steps required to prepare a sample for radiocarbon (^{14}C) measurement by accelerator mass spectrometry (AMS). These include the following: (1) collection of an appropriate sample that can answer the question being asked; (2) pretreatment of samples to isolate the most representative fraction of the bulk carbon (C) or to separate total C into different components; (3) conversion of C in the sample to CO_2 and/or graphite for measurement by AMS; and (4) assessing errors, especially those associated with ^{14}C contamination that occur during processing.

Careful sampling and adequate pretreatment (physical and/or chemical) are very important stages of the ^{14}C measurement process and are critical for interpreting the data. As we have seen in Chap. 3, ^{14}C measurements of homogeneous pools are interpreted differently from those of heterogeneous (mixed) C pools. The same principles apply to ^{14}C measurements of individual samples. To properly interpret results, one must understand if the sample represents a closed, homogenous system or an

S.E. Trumbore (✉) · X. Xu · G.M. Santos · C.I. Czimczik · M.A. Pack · F.M. Hopkins · A. Stills · M. Lupascu
Department of Earth System Science, University of California, Irvine, Irvine, CA, USA

S.R. Beaupré
Department of Geology and Geophysics, Woods Hole Oceanographic Institution, Woods Hole, MA, USA

L. Ziolkowski
Department of Earth and Ocean Sciences, University of South Carolina, Columbia, SC, USA

S.E. Trumbore
Max Planck Institute for Biogeochemistry, Jena, Germany

S.R. Beaupré
School of Marine and Atmospheric Sciences, Stony Brook University, Stony Brook, MA, USA

E.A.G. Schuur et al. (eds.), *Radiocarbon and Climate Change*,
DOI 10.1007/978-3-319-25643-6_9

average over a heterogeneous mixture. The pretreatment examples described here are meant to either remove contaminating, extraneous C from what is assumed to be a homogeneous, closed system sample, or to separate a heterogeneous sample into component C fractions with different physical, chemical, or biological properties. Procedures vary considerably in complexity depending on the nature of the samples and the question being asked.

One major advantage of AMS over decay counting is the ability to measure very small amounts of C (μg-mg C). While small size capabilities increase the types of samples that can be easily measured for ^{14}C, they also present new analytical challenges. For example, a milligram size subsample of bulk sediment or soil may not be representative of the bulk sample if it is not well homogenized. Another challenge is the increase in the relative contribution of extraneous C introduced during many laboratory preparative steps. The following sections present general information on sample processing and examples of some of the more commonly used procedures to illustrate these issues. More specific discussion of methods, including new developments in pretreatment procedures, can be found in the proceedings of the International Radiocarbon and AMS Conferences, and in the very rapidly growing literature. These updates should be reviewed in addition to the material presented here when starting to plan new ^{14}C measurements.

9.2 General Procedures for Sample Preparation

In this book, we assume the user is someone involved (or likely to become involved) in the preparation of samples that are sent to and analyzed by an AMS laboratory. There are a number of steps that a sample generally goes through during this process (Fig. 9.1). Of these, defining the scope of the problem, as well as sample collection and pretreatment, is mostly performed by the user, while the conversion of the sample into a form that can be put into the ion source is often done at the AMS laboratory. The user is responsible for ensuring that appropriate tests of their collection and pretreatment procedures, which could introduce extraneous C (i.e., C not derived from the sample itself), are included with the samples submitted to an AMS laboratory. The AMS laboratory is responsible for tests of their graphitization and measurement procedures that could introduce additional extraneous C. That being said, it is not uncommon that users and AMS facilities join their expertise to attain better results.

9.2.1 Sample Selection

The most important criterion for ^{14}C analysis is selecting a sample that will actually provide the best possible answer to the question being asked. For the purposes of ^{14}C dating to determine the chronology of a paleoclimatic or paleoenvironmental proxy, it is best to date C in the material measured for the proxy itself—e.g., pollen, foraminifera

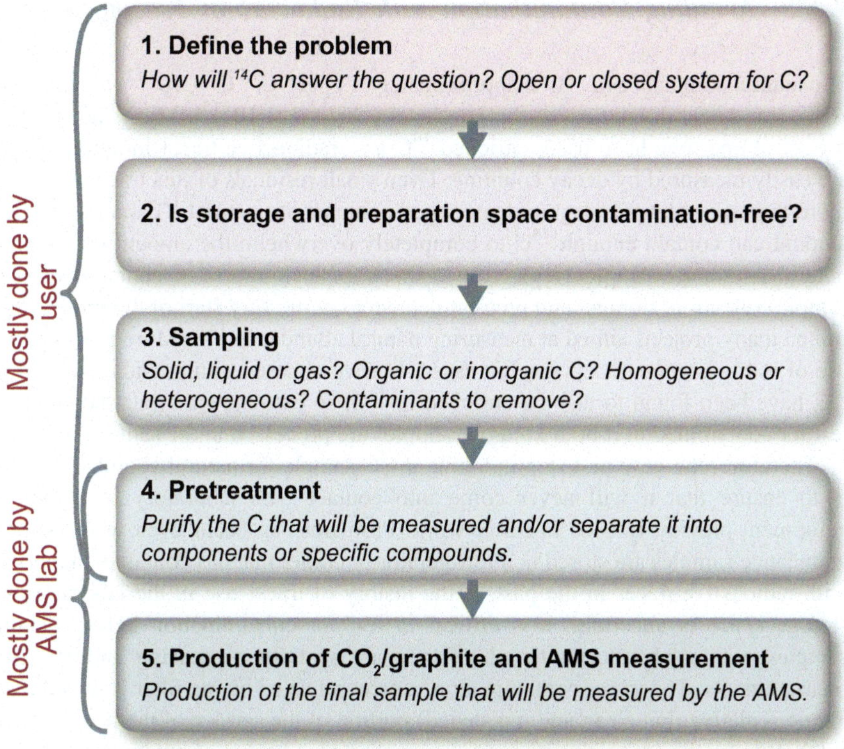

Fig. 9.1 Schematic showing steps common to all ^{14}C measurements

shells, or biomarker molecules. This is preferable to dating the bulk carbonate or organic matter of the sample matrix that contains these fossils because the matrix may consist of a mix of materials with vastly different ages. When trying to understand the age structure of C in a heterogeneous sample such as sedimentary or soil organic matter, it may be best to fractionate the bulk sample physically or chemically into components that differ in stabilization mechanism. The limitations of the ^{14}C method must also be taken into account during this stage of sample selection. For example, samples originating in the last 300 years have additional uncertainties solely because of the limited precision of ^{14}C dating during this time period. Finally, and most importantly, any methods used to isolate and pretreat the sample must be assessed.

Sampling techniques will obviously vary widely according to the question, and to the sample phase (solid, dissolved, gaseous) and type (sediment, air, organic, inorganic). Three important considerations are: (1) samples should be of sufficient size to yield enough C (normally at least tens of micrograms to several milligrams C) after treatments have been applied to allow for ^{14}C analysis; (2) whether the material being sampled is an open or a closed system for C; and (3) the degree of heterogeneity of C within the sample. When materials are expected to be heterogeneous, there should be suitable replication to assess intra-sample spatial variability.

9.2.2 Avoiding Contamination with Radiocarbon Tracer

The natural abundance of ^{14}C is approximately 1 ppt ($^{14}C/^{12}C \approx 10^{-12}$) and modern AMS can detect differences at the ppq level ($^{14}C/^{12}C \approx 10^{-15}$). Many laboratories use, or at one time have used, enriched ^{14}C as a radioactive tracer in amounts that are easily measured by decay counting. Even small residuals of this tracer C can be more than six orders of magnitude enriched compared to natural ^{14}C levels. A trace amount can contain enough ^{14}C to completely overwhelm the amount of ^{14}C in a natural abundance sample. Neglecting to check for such tracer-^{14}C contamination in sample collection, storage, and pretreatment areas at the very start of the process has ruined many projects aimed at measuring natural abundance ^{14}C. More recently, the use of stable isotope tracers has increased. In several cases, compounds enriched in ^{13}C have been found to also be enriched in ^{14}C, so extreme care must be taken to test for contamination if both kinds of samples are processed in the same laboratory.

Therefore, the very first step in taking a ^{14}C sample for natural abundance work is to ensure that it will never come into contact with laboratory or sampling equipment previously used in tracer work. Problems can occur even when natural abundance samples are stored in a refrigerator or cooler that has contained materials with radioactive tracer in the past. If the history of tracer use in the laboratory or storage space is uncertain, it is critical to test for contamination before risking samples or the laboratories to which they are sent for processing and/or AMS measurements. To test for contamination, a precombusted quartz filter is wiped over a representative (but clean and moistened) portion of the surface of the item or area. If labeled volatile organic compounds are suspected, a beaker of precombusted sand can be left out in the laboratory to adsorb some of this material. The dried filter or sand is then combusted together with ^{14}C-free C (such as coal), and measured by AMS. If there is no tracer, the sample will have close to normal background levels (note that care should be taken to avoid adding modern C through dirt to the filters when swiping). By this method, any contaminating ^{14}C, even at low C quantities, can be detected against the background C of the carrier that is known to be free of ^{14}C. Most facilities provide some type of swipe kit for starters. Contacting facilities for help would be an important first step, before collecting samples.

9.2.3 Pretreatment and Conversion to Carbon Dioxide

The goal of sample pretreatment is to isolate and purify the pool of C for which the ^{14}C measurement is desired. Such procedures can vary in complexity based on the type of sample, and the question being asked. We will illustrate some common methods in the case histories that follow, but common to all procedures is the inclusion of tests designed to quantify potential artifacts associated with sample pretreatment that can affect the ^{14}C measurement, its uncertainty, and its ultimate interpretation (Sect. 9.3).

Once the desired C is isolated, it is converted to carbon dioxide (CO_2), which is then purified in the amount appropriate for AMS measurement. The method of conversion varies according to the state of the C in the sample—organic, inorganic, or gaseous.

Carbon dioxide from organic samples is obtained by combustion of the clean sample material at high temperature (>900 °C). This combustion is performed in some laboratories using a commercial combustion analyzer that uses high-temperature flash combustion with oxygen (O_2) (e.g., McIntyre et al. 2009). In other laboratories, or for low %C samples, a sealed quartz tube combustion method is used. The sample is transferred to a quartz tube (prebaked at 900 °C for at least three hours) along with pre-baked cupric oxide (CuO), and silver (Ag) wire. The oxygen from the CuO provides the primary O_2 source for the oxidation. The Ag wire surface helps to remove sulfur as well as chlorine that may be present in the raw sample material and may interfere with C oxidation and/or graphitization that follows. The quartz tube with the sample and chemicals is evacuated and flame-sealed with small torch (Sofer 1980; Boutton et al. 1983; Santos 2008). The sample is then transferred to a muffle furnace and combusted at 900 °C for 2–3 h to completely oxidize all organic C to CO_2.

Inorganic C (dissolved or solid) is converted to CO_2 by acid hydrolysis. Usually, the carbonate sample of appropriate size (i.e., to yield ~0.5–1.0 mg C) is placed in a vial with a leakproof septum. The vial is evacuated, and 80 % orthophosphoric acid, used because of its low vapor pressure, is added in amounts sufficient to completely dissolve the sample and release all dissolved CO_2.

Once CO_2 is produced from a sample, it may be necessary to separate it from other gases such as nitrogen (N_2), O_2, and water (H_2O) vapor. In commercial combustion analyzers, this is accomplished using chemical traps. For sealed-tube combustion and other applications, such as purifying CO_2 from air samples, a vacuum line is used for cryogenic separation of gases (Fig. 9.2). Water and other molecules of sufficiently low vapor pressure are removed from the gas stream by passing the gas mixture through a trap that is submerged in a dry ice/alcohol slush bath (−78 °C), while the CO_2 is collected downstream in a trap submerged in a liquid nitrogen (LN_2) bath (−196 °C). Other gases, such as O_2 and N_2, will not freeze at low pressure and can be pumped away. Note that when CO_2 from air samples or combustion of samples in an O_2 stream is combined with cryogenic purification, care must be taken that the total pressure remains low enough so that O_2 does not liquefy in the LN_2 trap—this can lead to explosions. The CO_2 yield from the sample is typically measured as the pressure of pure CO_2 in a known, measured volume within the vacuum line. If desired, the sample can be split and an aliquot removed for analysis of the stable isotopes ($^{13}C/^{12}C$ ratio of CO_2) by isotope ratio mass spectrometry at this point; combustion analyzers often have a split of the gas stream sent to a stable isotope ratio mass spectrometer for online ^{13}C measurement.

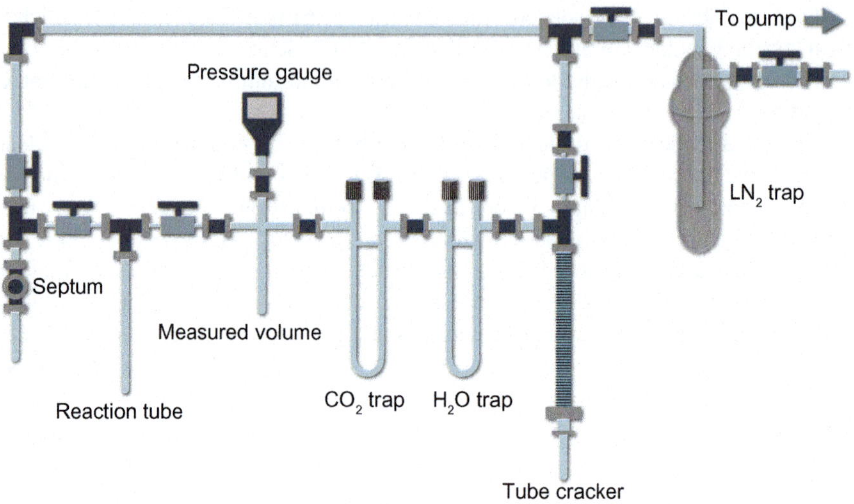

Fig. 9.2 Typical components of a vacuum line for purification of CO_2 from gas mixtures. Samples—for example, a quartz combustion tube—are introduced through the 'tube cracker.' Air in this part of the vacuum line is pumped away; then, the sample tube is broken, releasing the enclosed gases. Water is trapped in a dry ice/alcohol slush trap, and CO_2 in a trap submerged in liquid nitrogen. Other gases (N_2 and O_2) are pumped away. The frozen CO_2 is warmed and then trapped again with LN_2 in the measuring volume. It is allowed to expand and the pressure is measured in the known volume to determine the yield of CO_2. An appropriate quantity of the purified CO_2 is then transferred to a reaction vessel for reduction to graphite. Figure modified from Xu et al. (2007)

9.2.4 Graphitization Methods

An increasing number of AMS systems introduce purified CO_2 directly into ion source (e.g., Wacker et al. 2013; McIntyre et al. 2013). However, at the time of writing this book, most AMS laboratories still employ filamentous graphite coating a metal catalyst as a target (Chap. 8). The metal provides not only a binding agent to hold the graphite securely in the target holder after pressing, but also can efficiently transport heat away from the surface where sputtering occurs. A number of graphitization methods have been developed to meet this need since the development of AMS, though all rely on reduction of CO_2 in the presence of a catalyst. The two most commonly used reducing reagents to convert CO_2 to graphite are hydrogen (H_2) and zinc (Zn), and the catalysts are iron (Fe) and cobalt (Co) powders. The reaction requires a temperature >500 °C at the catalyst and can take place either in sealed reaction vessels attached to a vacuum line or offline in sealed tubes.

The hydrogen reduction method (Lowe 1984; Vogel et al. 1984) is the most widely used method for graphite production from CO_2. The net reaction is:

$$CO_2 + H_2 \rightarrow H_2O + C\ (\text{graphite}) \tag{9.1}$$

A typical reactor system for H_2 reduction is shown in Fig. 9.3. A number of studies have optimized the conditions of this reaction, such as the powder mesh size, manufacturer, and the mass ratio of C to catalyst (Santos et al. 2007a), the reaction temperature, size and type of reaction tube (quartz vs. Pyrex), the method of H_2O removal (cryogenic versus chemical sorption), and the reaction kinetics and chemistry (McNichol et al. 1992; Santos et al. 2004). Most laboratories have optimized their procedures for the performance of the produced graphite in their ion source/AMS system, so any researchers wishing to produce their own graphite must first work with their specific AMS laboratory to test and optimize any procedure. Graphite should not be stored in laboratory air as it will sorb gases such as H_2O and CO_2 which will change its characteristics; it should be stored in dry, CO_2-free conditions, ideally not for more than a few weeks before it is measured.

Zinc reduction has been developed as an alternative to H_2 reduction for AMS ^{14}C measurements (Jull et al. 1986; Slota et al. 1987). Vogel (1992) adapted the sealed-tube Zn method for biomedical ^{14}C-labeled samples where Zn and titanium hydride (TiH_2), which releases H_2 at ~440 °C temperature, are used as reducing agents and Fe or Co as the catalyst. In these systems (see Fig. 9.4 for an example), Zn not only is the reducing agent for CO_2, but also it cycles H_2O, the side product in the graphitization process, back to H_2 so that it helps to push the reaction to completion. The net reaction is thus:

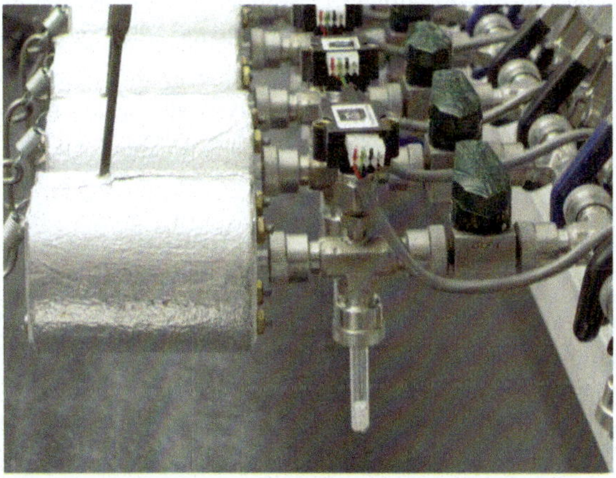

Fig. 9.3 Graphitization reactors used at KCCAMS/UCI laboratory [this design was adapted from the reactors used in CAMS/LLNL as described in Santos et al. (2004)]. The reactor has connections to a tube holding the catalyst (covered by the oven, to the *left*), a tube holding magnesium perchlorate for trapping evolved water (*pointing down*), a transducer to monitor pressure (*top*), and a valve to introduce CO_2 and H_2 and then isolate the reactor (*right*). *Source* G. Santos

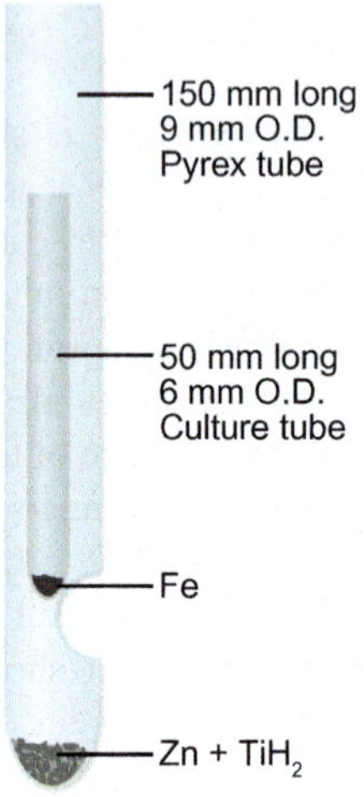

Fig. 9.4 Typical reactor for the sealed-tube Zn reduction method of graphite production. The open end of the Pyrex tube is attached to the vacuum line. After an appropriate amount of purified CO_2 is frozen into it, the tube is flame-sealed and put into an oven (550 °C) for graphitization. Samples can be stored indefinitely in the sealed tubes, either as CO_2 or reduced graphite. Figure modified from Xu et al. (2007)

$$CO_2 + Zn \rightarrow ZnO + C\,(\text{graphite}) \tag{9.2}$$

This method was further modified to introduce a small amount of H_2 to speed up the reduction reaction (Jacobsen et al. 1997; Hua et al. 2001) and adapted for biomedical AMS (Ognibene et al. 2003) as well as natural abundance samples (Xu et al. 2007). Graphite can be stored indefinitely in the sealed tubes.

9.3 Evaluating Measurement Uncertainties

The fundamental limitations to answering any scientific question are measurement error and uncertainty, reflecting errors introduced in addition to those associated with the measurement itself. Errors can influence both *precision*—the degree to which the ^{14}C value of a sample is repeatable, and *accuracy*—the degree to which the measured value reflects the actual amount of ^{14}C in the sample. Errors can be *random*, including spatial variability from where the samples were collected, or they can be *systematic*, including the introduction of extraneous C during processing. Evaluating these factors is essential to developing high-precision methods,

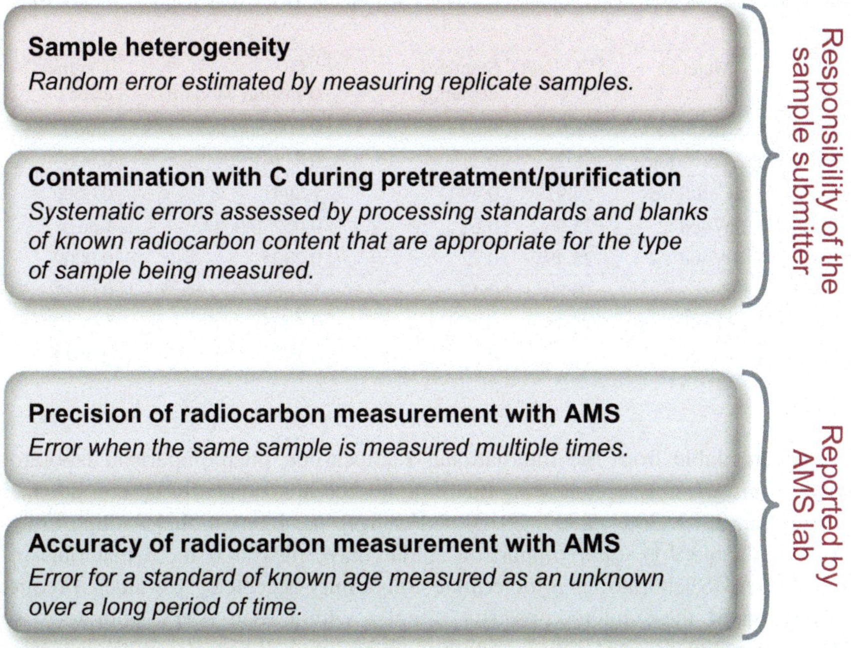

Fig. 9.5 Processes for determining the error in a ^{14}C measurement. Ultimately, the final reported error depends on all of these factors

performing quality control on established protocols, and accurately reporting the ^{14}C content of the original sample, as it existed at the time of sample collection.

Each step of AMS ^{14}C analyses—sampling, pretreatment, and analysis, needs to be evaluated for introduction of uncertainty into the final measurement (Fig. 9.5). Sources of error, common to all samples, include the errors associated with the AMS laboratory processing of samples. Typically these include any pretreatment done at the AMS laboratory, conversion to CO_2/purification of CO_2/graphitization of CO_2 and AMS measurement. These steps are continuously assessed in ^{14}C laboratories and reported as an error in the analysis report sent to the user. AMS laboratories evaluate accuracy using a number of 'secondary' standards of known ^{14}C content, and testing how well they reproduce the relationship between these and the primary ^{14}C standards, oxalic acid I or oxalic acid II. These secondary standards comprise a range of ^{14}C contents and types of material (organic, inorganic C) or are made for specific purposes and types of samples (e.g., CO_2 in air, available from the National Oceanic and Atmospheric Administration, NOAA, USA). In addition, ^{14}C-free materials (referred to as 'blanks') from materials such as coal and calcite that are known to be isolated from atmospheric inputs for $\gg$100,000 years are measured alongside the unknown samples. Secondary standards and blanks are

Table 9.1 Examples of standard materials available from the IAEA (International Atomic Energy Agency)

ID	Material	$\delta^{13}C$ (‰)	Standard deviation	^{14}C (Fraction Modern)	1 sigma error
IAEA-C1	Marble	2.42	0.33	0.0000	0.0002
IAEA-C2	Travertine	−8.25	0.31	0.4114	0.0003
IAEA-C3	Cellulose	−24.91	0.49	1.2941	0.0006
IAEA-C4	Wood	−23.96	0.62	0.002–0.0044	Range
IAEA-C5	Wood	−25.49	0.72	0.2305	0.0002
IAEA-C6	Sucrose	−10.80	0.47	1.506	0.0011
IAEA-C7	Oxalic acid	−14.48	0.20	0.4953	0.0012
IAEA-C8	Oxalic acid	−18.31	0.20	0.1503	0.0017

readily available from the International Radiocarbon Intercomparison laboratory evaluations, the most recent being the Sixth Intercomparison (SIRI) (Scott 2003a, b; Scott et al. 2003, 2004, 2007, 2010) and/or from the International Atomic Energy Agency (Table 9.1). Intercomparison studies have measured these materials in a range of AMS laboratories to determine community-accepted ^{14}C values. Frequent evaluation of these intercomparison samples within each AMS laboratory is a requirement for assessing the variability arising from different sample laboratory techniques, equipment, and intrinsic sample heterogeneity, as well as for correcting sample measurements for background contamination.

For many applications, the error in the reported AMS analysis is minor compared to sampling and pretreatment/purification sources of error (Fig. 9.5). In ecological samples, often the largest source of uncertainty is spatial or temporal heterogeneity of the samples that are collected, so sampling design must include appropriate replication to assess those errors. An example would be when comparing the ^{14}C content of soil organic matter in an area where the same soil is subject to different land uses. To confidently ascribe differences in C cycling to land use, the range of spatial variation within a given land use must be known. Other kinds of samples, such as sediment cores, may be more difficult to replicate spatially, but sediment accumulation is assumed to record temporal variability. These temporal variations must be reproducible in different cores taken from similar locations to be trusted.

In addition to variations associated with heterogeneity in the field, samples that involve considerable laboratory preprocessing, such as compound-specific analyses, must be tested for extraneous C (contamination) introduced by solvents or chromatography, which can overwhelm other sources of error. Thus even when a researcher desires to measure only a single sample, additional tests (and therefore samples) are always required in order to accurately assess the overall uncertainty of the ^{14}C measurement. These include tests for (1) introduction of extraneous C during processing; and (2) processing of known ^{14}C materials to ensure that corrections applied to correct for inclusion of these contaminants are sufficient and appropriate.

9.3.1 Introduction of Extraneous Carbon During Processing

Every process associated with a ^{14}C measurement, from precleaning collection bottles to atom counting on the AMS instrument, can introduce extraneous C into the sample. The amount and ^{14}C content of this extraneous C (also commonly referred to as the *background*) is unique to each sample preparation method. Failure to assess its magnitude and correct for it leads to the erroneous interpretation of ^{14}C measurements.

To illustrate this point, assume that the mass of background C (m_{bg}) with ^{14}C content (F_{bg}, where F is fraction modern), mixes with the mass of unadulterated sample C of mass m_s and fraction modern F_s. The total measured mass (m_{mix}) and fraction modern of the mixture (F_{mix}) can be predicted by the following equations for conservation of mass:

$$m_{mix} = m_{bg} + m_s \tag{9.3}$$

$$F_{mix} = \left(F_{bg}m_{bg} + F_s m_s\right)/m_{mix} \tag{9.4}$$

The influence of contamination on ^{14}C values is more clearly understood by substituting Eq. 9.3 into Eq. 9.4 and rearranging

$$F_{mix} - F_s = \left(F_{bg} - F_s\right) \times \left(m_{bg}/m_{mix}\right) \tag{9.5}$$

In general, Eq. 9.5 demonstrates that the magnitude of error ($F_{mix} - F_s$) increases with both the proportion of contamination in the mass of the mixture (m_{bg}/m_{mix}) and the difference between isotope ratios of the sample and background ($F_{bg} - F_s$) (Fig. 9.6). Neglecting the fortuitous case where sample and background are isotopically identical ($F_{bg} - F_s = 0$), this equation suggests that minimizing the proportion of contamination (m_{bg}/m_{mix}) is one of the most promising means for improving the accuracy of ^{14}C measurements (Table 9.2).

While background mass can be reduced in principle, it cannot be eliminated in practice. Therefore, all ^{14}C measurements must be corrected for background contamination. As the size of samples AMS can measure has decreased, now to a few micrograms, extraneous C material can make up the majority of the sample and thus the ability to correct for background dominates the uncertainty associated with the sample measurement. For example, while normally we make the implicit assumption that $m_{bg} \ll m_s$, and therefore, the mass we measure on a balance ($m_{s+bg} = m_{mix}$) is $\sim m_s$, this is not necessarily the case for very small sample (see Fig. 9.6).

Background corrections are further complicated by the modern practice of submitting samples to dedicated AMS facilities for ^{14}C quantification. AMS facilities correct all samples for backgrounds associated with their internal processes (e.g., combustion/acidification, graphitization, AMS and in many cases common chemical or physical fractionation pretreatment processes (Vogel et al. 1987; Donahue et al.

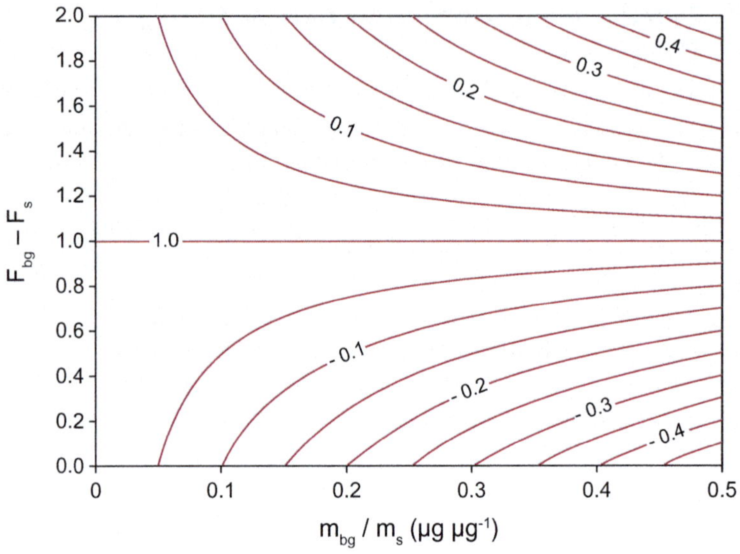

Fig. 9.6 Contour plot of ^{14}C errors ($F_{mix} - F_s$, contour lines with intervals of 0.05 Fraction Modern) expected due to various proportions of contamination (m_{bg}/m_s, mg/mg) and isotopic disparity among samples and contaminants ($F_{bg} - F_s$, expressed in Fraction Modern), as per Eq. 9.5

Table 9.2 Estimating the accuracy of ^{14}C measurement. Abbreviations are defined with equations 9.3 and 9.4

m_s (µg)	m_{bg} (µg)	m_{mix} µg	F_s	F_{bg}	F_{mix}	$F_{mix}–F_s$	Error 100* ($F_{mix}–F_s$)/F_s
990	10	1000	0.100	0.100	0.100	0	0
990	10	1000	0.100	1.100	0.110	–0.010	−10
90	10	100	0.100	1.100	0.200	–0.100	−100
90	10	100	0.100	0.200	0.110	0.010	−10
990	10	1000	1.100	0.100	1.089	0.011	1.0
90	10	100	1.100	0.100	1.000	0.100	9.1

1990; Brown and Southon 1997; Santos et al. 2007b). However, the sample may have been subjected to additional chemical or physical fractionation pretreatment processes by the user prior to submission. The corrections applied by the AMS laboratory cannot correct for background introduced during this additional pre-treatment. Therefore, the person submitting the sample must also subject a standard material to the same procedure to test for additional potential background contamination. If one of the commercially available standards (e.g., Table 9.1) is not appropriate, a new standard material must be produced and tested. Consequently, blank corrections are a two-step endeavor with shared responsibility among AMS facilities and their clients (Fig. 9.5). It is imperative that the user submitting a sample to an AMS facility discusses potential issues in advance.

9.3.2 *Correcting for Background Carbon*

Correcting ^{14}C measurements for background contamination requires estimation of both the mass and ^{14}C content of the extraneous C (F_{bg}; Eq. 9.5). There are two general approaches to assess F_{bg}: direct measurement and dilution. When the background mass is sufficiently large (more than ~10 μg C), it can be directly collected and quantified by processing materials analogous to samples but lacking in C. For example, Beaupre et al. (2007) collected CO_2 produced in a dissolved organic carbon (DOC) combustion system by processing CO_2-free distilled water exactly as if it were a sea water sample. In such cases, m_{bg} and F_{bg} are determined directly and Eqs. 9.3 and 9.5 can be used to calculate m_s and F_s from the overall C yield (m_{mix}) and measured ^{14}C content (F_{mix}). More typically, however, the background is too small for direct measurement (<10 μg C). In this case, its ^{14}C content and mass are estimated together by dilution with a known mass of representative standards that differ in ^{14}C content from the sample—usually materials known to be ^{14}C-free or modern (see Fig. 9.7). In such cases, m_{bg} and F_{bg} are solved for in Eq. 9.5, but other terms (m_s and F_s) must be known. Usually, these are obtained by ensuring that pure, unprocessed materials of sufficiently large quantity are analyzed so that $m_{bg} \ll m_s$ and $F_s \sim F_{mix}$.

A similar approach uses linear regression to circumvent determination of the absolute background mass or isotope ratio. This process is analogous to constructing a Keeling plot (Keeling 1958) by substituting Eq. 9.1 into Eq. 9.4 and rearranging:

$$F_{s+bg} = \left(F_{bg} - F_s\right)m_{bg}\frac{1}{m_{s+bg}} + F_s \tag{9.6}$$

If the background mass and isotope ratio are constant, then measured isotopes ratios $F_{mix} = F_{s+bg}$ and corresponding inverse masses $\left(1/m_{mix} = 1/m_{s+bg}\right)$ should be linearly related with a slope = $(F_{bg} - F_s)m_{bg}$, and an intercept = F_s (Fig. 9.8). Plotting measurements obtained from variously sized, identically processed, aliquots of the same sample should yield a background corrected isotope ratio F_s as the intercept (Hayes 2002). The reliance on inverse masses suggests that error and uncertainty in determining the corrected isotope ratio will be reduced when analyzing a range of masses that includes large as well as small samples.

The ^{14}C signature of the background (F_{bg}) is not always predictable. Normally, C materials assumed to have zero ^{14}C (e.g., 'blanks,' such as calcite or coal) are processed to detect ^{14}C-enriched contamination ($F_s = 0$, so $F_{mix} = F_{bg} > 0$; Eq. 9.5). However, the ^{14}C content of contaminating C is not necessarily modern,

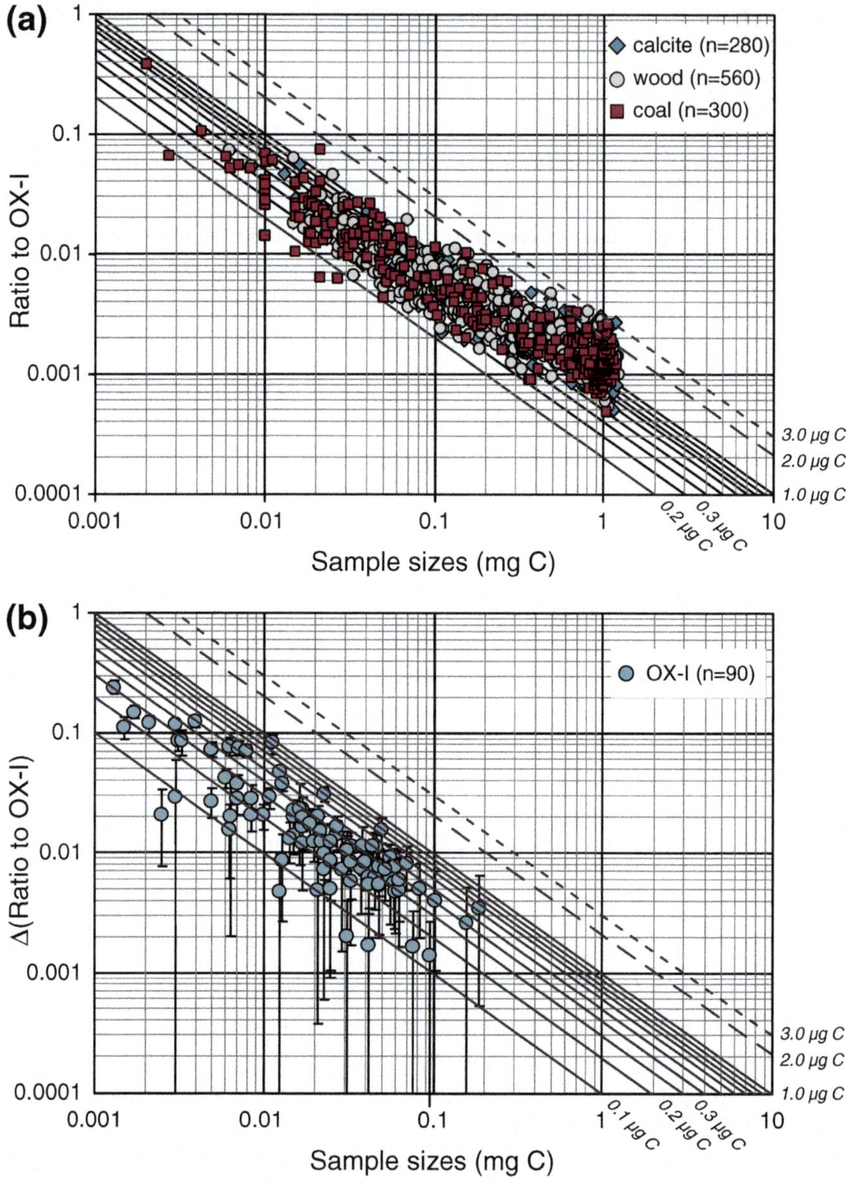

Fig. 9.7 Example using analysis of different amounts of standard materials of known-^{14}C (F_S) to assess the amounts of 'modern' and 'dead' contamination. The *top plot* shows results obtained from combustion of (^{14}C-free) coal, wood, and acidified calcite samples of different sizes to assess the amount of modern C contamination. The plot on the *bottom* shows the results from combustion and measurement of small and ultrasmall OX-I samples (0.0015 to 0.2 mg C). In each case, these have been normalized using 1-mg C OX-I standards. On the y-axis, (capital greek Delta) is the difference of this ratio from 1. The *solid lines* represent the effects of fixed amounts in microgram C of modern (*top*) or ^{14}C-free C contamination (*bottom*) introduced during sample processing. *Error bars* represent propagated errors. Figure modified from Santos et al. (2010b)

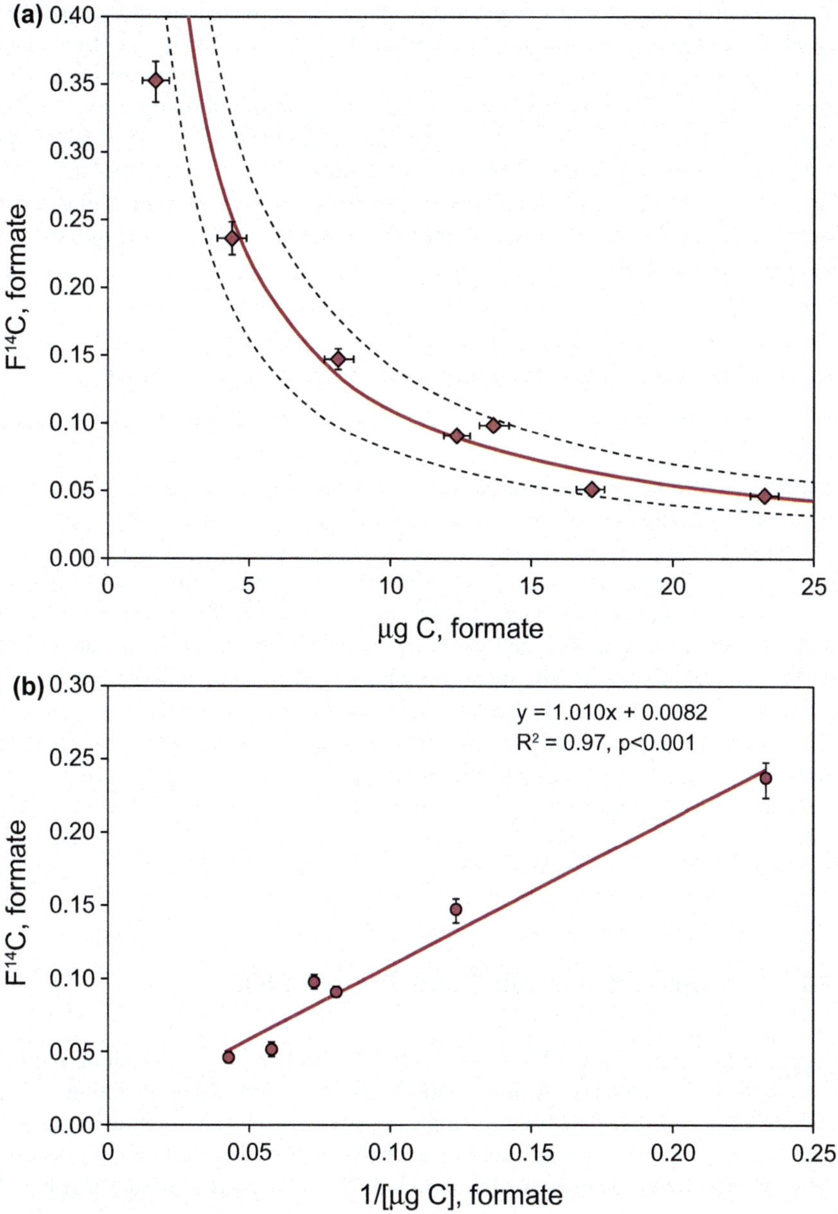

Fig. 9.8 Example of Keeling plot approach to estimating *F* for extraneous C. Figure modified from Lang et al. (2013)

or even bomb-C. For example, contamination by petroleum-derived organic solvents will add ^{14}C-free C, while contamination by $^{14}CO_2$ in laboratory air can be a mixture of background air and CO_2 derived from evaporating dry ice. Therefore, complementary analyses of modern standards ($F \sim 1$), which are most sensitive to fossil C contamination, should be combined with analysis of ^{14}C-free blanks ($F = 0$) to model the background as a combination of modern and fossil C components (Santos et al. 2007b). Similarly, the standard dilution approaches to estimating m_{bg} and F_{bg} should ideally be applied using more than one standard with different isotope signatures (Hwang and Druffel 2005; Santos et al. 2010b; Fig. 9.7). The background should be assessed regularly so that its variability can be included in the overall analysis of the uncertainty for the ^{14}C measurement of unknown samples.

9.3.3 *Processing of Materials with Known Radiocarbon Content*

One way to test the adequacy of background corrections, and to assess the need for a thorough investigation of sources of extraneous C, is to include samples of known ^{14}C content in all processing prior to submission to an AMS laboratory (Table 9.1). In some cases, this requires the production of a new standard material—e.g., soil, CO_2 in air, etc. Laboratory intercomparisons have been performed using materials such as bone, peat, charcoal, and wood of known ^{14}C contents. However, soil and sediment standards are usually made and maintained by individual laboratories. To be a useful standard material, there should be sufficient amounts for use over many years (i.e., to make hundreds of analyses), and the material needs to be homogeneous so that measured differences reflect processing errors rather than heterogeneity of the starting material. Once more, readers are advised to consult an AMS laboratory for advice on how to obtain or produce standardized materials to ensure that their results are useful and credible.

9.4 Examples of Sampling and Pretreatment

Sampling and pretreatment both have the goal of collecting and isolating the C of interest for ^{14}C analysis. A fundamental consideration when planning sample preparation is whether the C in the sample is homogeneous or a mixture of materials with different ages or cycling times. The answer will influence how ^{14}C measurements will be interpreted, as well as the kind of tests needed to ensure that the ^{14}C result is not affected by introduction of extraneous C. As noted throughout this book, ^{14}C can be used in several different ways: (1) to determine the age of isolation of C from exchange with the atmospheric CO_2; (2) as an indicator of components in a mixture; and (3) to estimate the residence time of a C atom in a (homogeneous or heterogeneous) reservoir that is open to exchange with the atmosphere. We have used these definitions to provide representative examples of a few methods in current use. The first example (Sect. 9.4.1) involves preparation of homogeneously aged C for dating (i.e., assumed to be a closed C system) that must be separated from potential contamination of different-aged, extraneous C. The second class of procedures (Sect. 9.4.2) involves homogenous C compounds, such as dissolved inorganic carbon (DIC), CO_2, or CH_4, with isotopic signatures that reflect open system dynamics (i.e., a range of source and sink processes). The final example (Sect. 9.4.3) involves heterogeneous C compounds, such as organic matter, with a spectrum of ^{14}C signatures that reflect differences in C cycling. The examples we give are by no means complete descriptions of laboratory protocols. The reader should consult the recent literature for details of available methods before embarking on a new project.

9.4.1 Homogeneous Carbon Used for Age Determination (Closed System)

9.4.1.1 Acid–Base–Acid Treatment

The 'acid–base–acid' (ABA) treatment (Olsson 1986) has been used extensively to remove contaminants that may have entered what should be closed C systems—e.g., archeological samples or paleoclimate proxy materials. Over the long time scale of burial, such micro- and macrofossils (e.g., seeds or charcoal) can adsorb organic matter from the surrounding soil or sediment. Extraction with sequential acid and base solutions is effective at removing adsorbed mobile C compounds that are often younger in age than the macrofossil. While often effective, the ABA method is an operationally defined treatment, i.e., it is not easy to say exactly what it does chemically, and therefore, AMS laboratories frequently report different procedures to reach the same outcome (Santos and Ormsby 2013).

Briefly, the sample is initially washed with an acid solution (e.g., 1.0 N HCl) to dissolve any geological carbonate accumulated from dust or soil. The second step is an alkaline (e.g., 1.0 N NaOH) wash that dissolves any alkali-soluble soil C (humic substances). Often the dissolution of these humic substances will change the solution color from clear to brown or black. The alkaline washing step is repeated until the solution becomes clear (i.e., all humics are removed). The third step is another acid wash (1.0 N HCl) to remove any atmospheric C that may have been absorbed during the alkaline wash. Lastly, the sample pH is neutralized with distilled water, which also removes the chloride introduced during the ABA process. Sodium chloride (salt) residues can corrode combustion tubes, and it may interfere with the graphitization reaction.

The efficiency of ABA treatment is hard to assess. For example, if the ^{14}C age of charcoal treated with ABA is older than untreated charcoal, it is clear that there was younger C added to the sample. However, it is not certain that 100 % of this material was removed by the ABA procedure. Ultimately in such cases, the age obtained may be interpreted as a minimum, since the investigator cannot be confident that all younger C was removed.

9.4.1.2 Calcium Carbonate Materials

The $CaCO_3$ shells formed by single-celled foraminifera are commonly used proxies to record environmental conditions in the sea water where they live. These shells normally are too small to measure ^{14}C on only a single shell—so they are picked individually from sediments and sorted by species (it is especially important to separate species that dwell in different depths in the water column or are affected by bioturbation of sediment). Calcareous foraminifer tests are often covered with secondary authigenic carbonates that must be removed prior to analysis. Such growths are often visible from microscopic examination. To remove these authigenic carbonates, the foraminifer shells can be cleaned with 0.1 N HCl for no longer than 10 s with mild agitation in an ultrasonic bath. This treatment can also be used for cleaning other biogenic calcareous samples, e.g., corals and bivalve shells. For such samples the strength of the acidification applied for cleaning should be adjusted accordingly to the degree of alteration. Samples are then thoroughly rinsed in deionized water, dried and sealed in a vial that can be evacuated, and acidified with 80 % orthophosphoric acid to release CO_2. For dating older calcareous macrofossils, care must be taken to check for diagenetic alteration of the carbonate. This is particularly important for old fossil coral samples, where the aragonite gets altered to calcite distorting their ^{14}C ages.

9.4.1.3 Cellulose Extraction

Wood samples contain a number of compounds, including primary structural elements such as cellulose and lignin, but also secondary lignin, resins, starches, lipids that are added later or can mix radially within tree stems. Cellulose is a polysaccharide chain that does not exchange C atoms with the atmosphere or within the tree after its formation and is assumed to form within a single year during the growing season. Hence, it is the most often used component for dating wood and is assumed to provide a faithful record of atmospheric $^{14}CO_2$ variations in the past for construction of calibration curves (Chaps. 4 and 7). For cellulose sampled in postbomb trees, it has been shown that extractable sugars and starches contained within wood can postdate the cellulose matrix where they are stored (Richardson et al. 2012). While it is normally assumed that cellulose is constructed from recent photosynthetic products, in some cases older storage reserves (up to several years old) can be mobilized and used to construct cellulose in early wood (Keel et al. 2006) or roots (Vargas et al. 2009). However, these are not important issues for pre-1950 wood—the ^{14}C method can only make such fine temporal distinctions in the bomb period.

Pretreatments to isolate cellulose from wood and other plant materials are operationally defined—they are designed to extract other components but may still not produce a pure cellulose product (Gaudinski et al. 2005). One common method, called the Jayme–Wise cellulose extraction, developed by Leavitt and Danzer (1993) and commonly used for stable isotope measurements has been widely adopted for ^{14}C measurements, for example, for processing wood samples to produce the ^{14}C calibration curves.

The Jayme–Wise method initially extracts wood samples using polar solvents to remove lipids (e.g., toluene–ethanol azeotrope followed by ethanol). The samples are then bleached, using sodium hypochlorite to remove lignins (see Gaudinski et al. 2005) for a detailed description and references to various laboratory methods). Southon and Magana (2010) recently showed that in many cases and for ^{14}C analyses (but not for stable isotopes), the Jayme–Wise procedure can be replaced with ABA treatment followed by bleaching and offer a simplified laboratory procedure that does not require Soxhlet solvent extractions. The yield of cellulose-C can also be an indication of the state of preservation of the wood. Hollow or highly decomposed wood samples that give a low yield indicate that the remaining cellulose may not give an accurate age for the wood formation. Regardless of which method is used, standard wood materials (^{14}C-free and modern as well as close in age to those being analyzed) should be processed in parallel with samples to test for introduction of extraneous C during the procedure.

9.4.1.4 Bone

Although more often associated with archeology than with C cycle studies, bone samples are a common material used for ^{14}C measurement. The material normally measured in bone is collagen, the principle protein in mammalian bone. Collagen forms the matrix on which mineral components in the bone precipitate, and is also involved in formation of tooth enamel and antlers. The ^{14}C of intact collagen can thus provide the age of bone formation. However, bone can also absorb organic contaminants from the soil in which it is preserved, and collagen can be entirely leached from bones when they are submerged in water. Hence, one of the most important steps is selecting a bone sample that is hard (i.e., non-porous) and intact (not broken into small pieces) and has not been in an environment where collagen is easily leached.

The problems of obtaining accurate ^{14}C ages on fossil bone material are well known (see examples given in DeNiro 1985; Stafford et al. 1987, 1990; Brown et al. 1988; Ajie et al. 1990, 1992; Hedges and Vanklinken 1992; Burky et al. 1998; Taylor et al. 2001; George et al. 2005; Higham et al. 2006). These studies document significant variability in the degree to which endogenous C-containing fractions in bone are retained or lost and replaced with various amounts of exogenous organic materials. Subfossil bones in general retain 1–5 % of the collagen content of modern bone and exhibit a non-collagen amino acid profile. For fossil bone with residual collagen <1 % of that of modern bone, significant anomalies in the ^{14}C values are typically encountered, with the magnitude of the age offsets tending to increase with decreasing in situ collagen concentration. Some laboratories measure the N content (and/or C/N ratio) of bone (collagen is made up of amino acids and thus should have relatively high N content compared to contaminating material) as a prescreening procedure to decide whether it is worth trying to measure ^{14}C in the sample (Brock et al. 2012). In the cases where bone samples are severely degraded, no reliable results can be achieved.

For bones containing sufficient quantities of collagen, it is generally agreed that standard pretreatment methods can effectively isolate and purify the residual collagen or collagen-derived organic materials (Brown et al. 1988; Hedges and Vanklinken 1992; Ramsey et al. 2004; Higham et al. 2006). A typical treatment is to first mechanically clean and decalcify the bone (Longin 1971). The crude collagen residue is then hydrolyzed to gelatin, which is then ultrafiltered to remove small contaminating molecules (Brown et al. 1988). The ^{14}C determination is thus performed on the larger molecular fraction most likely to represent intact proteins formed in the bone. Extraction methods are tested for contamination using bones of known age and therefore known ^{14}C content.

9.4.2 Homogeneous Compounds in Open Systems

In this section, we discuss most of the inorganic forms of C, which exist in a single chemical form and thus can be measured once extracted from their matrix (water, air, or soil). Although the C is in a single form, its ^{14}C content can reflect a range of processes, from transport to mixing. Special attention is required to sample in a manner that will identify the sources that are mixing to produce the measured ^{14}C signature. We will focus on DIC, and the gases CO_2 and CH_4 in air.

9.4.2.1 Dissolved Inorganic Carbon

Dissolved inorganic carbon is defined as the sum of CO_2 and the carbonate species that it forms upon hydration when dissolved in water: carbonic acid (H_2CO_3), bicarbonate ion (HCO_3^-), and carbonate ion (CO_3^{2-}). Decreasing the pH of a water sample to ≤2 will convert the majority of these species into CO_2 that is easily separated from water by bubbling with an inert gas (e.g., Fig. 9.9) or sampling of the headspace gas (Gao et al. 2014).

The amount of water needed for an AMS sample depends on the DIC concentration in the sample. Concentrations of DIC in ocean waters vary within narrow limits of ~2050–2460 μM (Key et al. 2004), but show a wide range in terrestrial waters (~50–330,000 μM). In marine sediment pore waters, DIC values near the sediment–water interface are similar to ocean water, but can reach 55,000 μM deeper in the sediment (e.g., Reeburgh 1980; Borowski et al. 1997). Hence, the volume of water to be sampled can vary from a few milliliters for pore waters to more than a liter for some fresh waters. For samples with higher pH, CO_2 contamination can occur through dissolution of atmospheric CO_2, so it is important not to expose these samples to air. A C-free preservative (e.g., mercuric chloride or cupric chloride) is added to inhibit microbial processes that could alter the chemical and isotopic composition of the samples during storage. Freezing samples at −20 °C

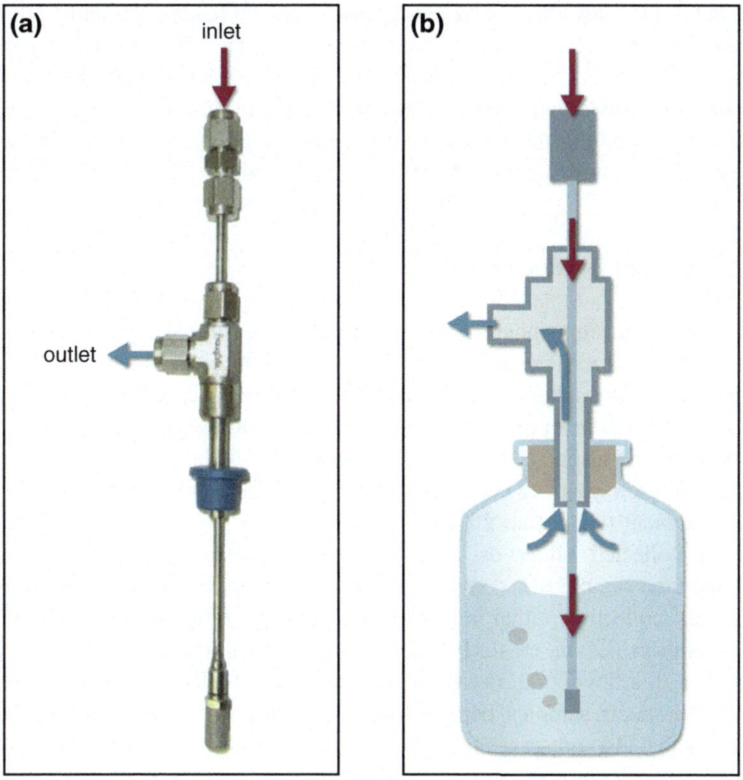

Fig. 9.9 Example of a sparging system for extraction of CO_2 from dissolved inorganic C. Acid is added through the butyl rubber stopper of a 120-ml septum vial (for sea water samples). N_2 is used to sparge the CO_2, which is trapped cryogenically on a vacuum line similar to the one shown in Fig. 9.2. Figure modified from Pack et al. (2011)

is an alternative to chemical poisons, but requires a large headspace ($\sim$9 %, required for crystalline expansion of water during freezing) and thus is not ideal unless a CO_2-free headspace can be provided. Finally, sample bottles are sealed with gas tight stoppers or caps. Please note that some types of stopper/septa could release trace amounts of dead C when in contact with hot acid vapor during sample acidification or during long-term sample storage (Gao et al. 2014).

For very low-DIC fresh waters, where large volumes of water are required, the DIC species can be precipitated in the field as insoluble carbonate salts by using a base and a salt solution containing the necessary cations [e.g., using $BaCl_2$ or $SrCl_2$ to create $BaCO_3$ or $SrCO_3$; see Drimmie et al. (1991)]. After the carbonate has precipitated, most of the water can be removed and the smaller sample transported to the laboratory for acidification. The disadvantage of this method is the potential for the basic samples to absorb atmospheric CO_2—samples must be airtight when stored and transported, and methods to test for introduction of atmospheric $^{14}CO_2$ need to be carefully devised.

Standard materials for dissolved inorganic C are not readily available. For testing methods, it is therefore recommended to produce DIC by dissolving carbonate materials of known ^{14}C content in water that has been acidified and presparged with N_2 (to remove dissolved CO_2). Alternatively, if the amount of extraneous CO_2 introduced in handling the sample is sufficiently large, backgrounds can be determined by re-extracting samples that have been previously acidified and bubbled.

9.4.2.2 Carbon Dioxide in Air

Analyzing ^{14}C in atmospheric CO_2 is important because it provides the starting point of the ^{14}C clock (see also Chap. 4). The establishment of several continuous, high-precision atmospheric $^{14}CO_2$ records by Levin and Kromer (1997, 2004), Levin et al. (2010) from the Northern Hemisphere and Manning et al. (1990) from the Southern Hemisphere, have made it possible to date modern C-containing materials to a few years or better for the bomb period, namely from the end of the 1950s to present (Hua et al. 2013).

Air is collected either as a single sample (a snapshot in time) or as a time-integrated sample that continuously collects air over a defined period of time. Continuous collection either separates CO_2 by bubbling through a (CO_2-free) NaOH solution (Levin et al. 1980), or by slowly filling an evacuated canister equipped with a flow-controller (such as a capillary restrictor) to maintain constant airflow. Whole-air samples benefit from H_2O removal using traps such as Mg$(ClO_4)_2$, which also preserves the ^{18}O signature of CO_2. Some drying agents, such as anhydrous calcium sulfate (commonly sold as 'Drierite'), may absorb CO_2 and should be avoided (Elia et al. 1986). Another method is to use molecular sieves, such as zeolite, to actively pump through (Gaudinski et al. 2000) or passively (adsorption by diffusion, Garnett and Hartley 2010) trap atmospheric CO_2. Molecular sieve methods tend to introduce trace amounts of old C when the material is not cleaned thoroughly prior to use and the method should be tested carefully for memory between samples. In addition, molecular sieves can fractionate ^{13}C and alter the ^{18}O signature of trapped CO_2. Tree rings (Hua and Barbetti 2004), and more recently, annual plants, such as corn leaf (Hsueh et al. 2007) and annual grass (Wang and Pataki 2010) have also been used to obtain time-integrated samples of atmospheric CO_2, but these are biased toward daytime and/or growing season conditions. Perennial plants run the risk of sampling stored C as well as fresh products of photosynthesis (see Chap. 7), though this is mostly important during the bomb period (since 1950).

Extraction of CO_2 from air samples will depend on the sampling method used. For gas samples collected by canister, the common way is to cryogenically extract CO_2 using a vacuum line like that in Fig. 9.2. Carbon dioxide trapped in NaOH solution is released by acidification. Samples trapped on molecular sieve must be heated to release CO_2 (Gaudinski et al. 2000). For testing of storage and extraction procedures, standard materials are available. The CO_2 in commercially available mixes of known CO_2 concentration is often ^{14}C-free, while NOAA has produced modern CO_2 in air standards to ensure intercalibration among the laboratories measuring $^{14}CO_2$ in air. From 2007 to 2010, three rounds of an international intercomparison of measurements of $\Delta^{14}CO_2$ in these whole-air standards by groups using AMS had been completed and the results were reported in Miller et al. (2013).

9.4.2.3 Methane in Air and Water

A number of authors have reported methods for ^{14}C measurement in CH_4, including Eisma et al. (1994), Kessler and Reeburgh (2005), Klouda et al. (1986), Lowe et al. (1991), Nakagawa et al. (2002), Petrenko et al. (2008a, b), Quay et al. (1991) and Wahlen et al. (1989). These methods involve several common steps: (1) removal of water vapor, CO_2, and carbon monoxide (CO) or other volatile hydrocarbons (such as ethane or propane); (2) separation/concentration of sufficient quantities of CH_4; and (3) oxidation of the CH_4 to CO_2.

As with DIC, the best CH_4 sampling method depends on CH_4 concentration. Low mixing ratios of CH_4 (about 1.7 ppm) in the atmosphere mean that the volume of clean air sampled (20–70 L) must be much larger than for $^{14}CO_2$ (380 ppm; 2–5 L). Usually, an oil-free pump or an air compressor is used to compress ambient air into a stainless steel cylinder after removing H_2O and CO_2. Air samples with very high CH_4 concentrations, including bubbles from sediments or wetlands require much smaller sample volumes (few 10 s of milliliters; Zimov et al. 1997; Pack et al. 2014). In contrast, the collection of gas for isotopic analysis requires melting several tons of ice to recover enough CH_4 for a ^{14}C measurement. Recently, Petrenko et al. (2008a) obtained ^{14}C measurements of CH_4 released from only 30–35 L of air using melt extraction. The extraction of CH_4 from sea water requires only milliliters of water for high CH_4 pore waters but more than 100 L for low CH_4 sea water.

9.4.3 Heterogeneously Aged Carbon

Most organic matter—whether in soils, sediments, in dissolved or particulate form, is a heterogeneous mixture of many different compounds, often with different origins and histories. For example, the bulk ^{14}C content of marine sediment organic matter has been demonstrated to consist of a mixture of relatively young components derived from marine phytoplankton mixed with older components derived from terrestrial sources (Eglinton et al. 1997). Even for the same compound, the ^{14}C content of the different constituent C atoms may be very different, as C can be recycled through microbes that ingest different-aged C substrates. Studies of ^{14}C in organic matter therefore attempt to quantify the age structure of the C present in various kinds of organic matter and to test hypotheses about the persistence or reactivity of organic matter in terrestrial or marine environments.

Table 9.3 Methods of preparing or separating organic matter prior to ^{14}C analysis

Basis of method	Soil	Sediment/particulate	Dissolved	Char
Sampling	Depth, horizon	Depth (sediment), filtration or sediment trap	Filtration, ultrafiltration, concentration on XAD resin	Aerosol (trap on filter), ice (filter melt)
Physical properties	Density, size	Size	Molecular weight, size	Size
Chemical properties	Inorganic/organic solubility in acid + base, oxidizing agents	Inorganic/organic compound classes, compound specific, pyrolysis	Inorganic/organic successive oxidation, sorption to surfaces	Oxidation methods, compound specific (BPCA)
Biological properties	Incubations (microbial availability), compound-specific biomarkers	Pore water DOC/DIC and modeling to infer production, consumption, compound-specific biomarkers	Incubations	Incubations inferred from experiments with char addition

Methods commonly used to fractionate organic matter prior to ^{14}C analysis are summarized in Table 9.3. These methods include separations based on physical properties (separation by size or density), chemical properties (e.g., water- or acid-soluble), biological properties (using microbes to access the most 'labile' organic matter), and separation of specific compounds (e.g., by chromatography of extracts). The idea is that these fractions that differ in properties may also differ in their intrinsic rates of C cycling. However, it is important to emphasize that none of the techniques in Table 9.3 isolates soil organic matter that is purely homogeneous in its isotopic and elemental composition (Swanston et al. 2005; Sollins et al. 2006; Heim and Schmidt 2007). Further, all methods are prone to C loss via oxidation, hydrolysis, or leaching, and to artifact formation, since some treatments can redistribute C among fractions.

9.4.3.1 Soil Organic Matter

Soils generally contain distinct horizontal layers, or horizons. One of the most important of these is the organic layer that is at the surface of many soils, where aboveground plant litter accumulates and decomposes. This layer is characterized by a low mineral content and is often a mix of large-sized pieces of readily recognizable material and much smaller-sized material that is not identifiable. Thus, size separation in the surface organic layer is one way to separate materials with differing dynamics. In very thick organic layers such as those found in poorly drained peatlands, macrofossils can be used to help quantify rates of accumulation and decomposition (Trumbore and Harden 1997; Czimczik and Trumbore 2007).

For many scientists, the 'real' soil organic matter begins in the mineral soil, below the organic layer. One important decision in sampling mineral soils is whether to sample by specific depth increments (e.g., 0–5 cm, 5–10 cm, etc.) or to follow the natural horizon boundaries while letting the depth interval vary. In addition, a decision needs to be made whether to sample the whole depth interval (i.e., take a depth integrating sample) or to take a sample only from a specific point. For designing a repeat soil survey to quantify changes in soil C stocks and ^{14}C over time, Schrumpf et al. (2011) used integrated samples (also used for bulk density determination) and specific depth intervals, as the depths of horizon boundaries vary laterally, and bulk density can vary with season. Although depth-integrated sampling combined with simultaneous determination of bulk density is best for inventories and understanding the bulk C cycle in soils, recent work has shown that there is very large spatial variation in ^{14}C at even smaller spatial scales: For example, younger C is associated with the walls of earthworm channels (Don et al. 2008), or on the outside of clod structures. More work is required to understand the spatial controls of organic matter dynamics in soil, which will require development of alternative sampling strategies.

Once the soil sample is collected, a range of methods have been used to characterize the organic matter it contains. Size separation involves sieving and decisions about the degree to which roots need to be picked out, and whether aggregates should be separately analyzed (e.g., Six et al. 2000). Density separation involves flotation in liquids of varying density to separate organic matter fractions based on their degree of association with mineral particles (Trumbore and Zheng 1996; Monnier et al. 1962; Christensen 1992; Golchin et al. 1995). In addition, due to variations in the specific gravity of different minerals, density has been used to separate mineral species in clay (Spycher and Young 1979; Jaynes and Bigham 1986) and silt (Shang and Tiessen 1998). Non-crystalline (amorphous) minerals, such as allophane, imogolite, and ferrihydrite, have lower density (1.6–1.9 g cm^{-3}), while minerals like goethite and hematite have much higher density (>3 g cm^{-3}), and the ^{14}C content of organic matter associated with these mineral phases can differ (Basile-Doelsch et al. 2007). As with size separation methods, the preremoval of materials like fine roots in sample collection and preparation will have a large effect on the amount and ^{14}C content of the lowest density fraction (Castanha et al. 2008).

It is clear from many studies that the ^{14}C signatures of both low- and high-density fractions integrate contributions from both faster and slower cycling pools of organic matter, i.e., they are themselves made up of heterogeneously aged material. A number of other methods have been applied to continue to break these fractions down into more homogeneously aged components. These include incubations (using microbes to separate labile from refractory C; e.g., Hopkins et al. 2012), sequential pyrolysis (e.g., Plante et al. 2013), chemical oxidation (e.g., Jagadamma et al. 2010), or sequential hydrolysis (e.g., Trumbore and Zheng 1996), with the goal of isolating increasingly refractory residues. Perhaps because chemical recalcitrance and age are not necessarily correlated (Schmidt et al. 2011), no single method has been shown to be effective in all soil environments to date.

This is a very active area of continuing research, and more discussion can be found in Chap. 6 on terrestrial C cycling.

Production of standard materials to test the methods used in soils is difficult, and mostly, researchers have relied on replication of the same sample in terms of yield and ^{14}C of the resulting ^{14}C signature of different density fractions prepared in parallel. This will test precision, rather than determine overall accuracy. Most soil C is close to modern in its ^{14}C signature, so contamination with modern background material is usually considered too small to affect the results. The spatial heterogeneity of soils is quite large, thus errors associated with this variability are generally assumed to be large compared to other potential analytical errors.

9.4.3.2 Dissolved Organic Carbon

DOC is operationally defined as the C in that fraction of organic matter in water that passes through a filter with pore size diameter of 0.2–1.0 μm. This size-based definition provides a practical means (i.e., filtration) of isolating DOC from complex mixtures. Freshwaters contain anywhere from 8 to ~274,000 μM DOC, while ocean waters typically range from 35 to 100 μM DOC and sediment pore waters can reach values of 800 μM in deeper sediment layers. Reviews for ^{14}C-DOC applications and measurement techniques in marine and fresh waters can be found in McNichol and Aluwihare (2007), Raymond and Bauer (2001b), and Chap. 5.

During collection, water samples are filtered to remove particles (the filter size is normally defined as 0.45 μm but can vary between >0.2–1 μm in practice) before transfer to precombusted sample bottles. If the filter will be discarded, other types of non-paper filters may be used as long as they are properly cleaned. Filtered sea water can be directly frozen for storage and maintained fidelity for at least 12 years (Beaupre and Druffel 2009).

Researchers also use alternative collection methods that isolate fractions of DOC, though inevitably this isolates only a portion of the total DOC pool. In the past, humic fractions of DOC have been isolated onto cross-linked polystyrene (XAD) resins (e.g., Thurman and Malcolm 1981; Hedges et al. 1986; Druffel et al. 1992; Hendry and Wassenaar 2005; McNichol and Aluwihare 2007). More recently, cross-filtration and ultrafiltration methods used on large water volumes have been used to isolate high molecular weight fractions of DOC (e.g., Santschi et al. 1995; Crum et al. 1996; Guo et al. 1996; Benner et al. 1997; Aluwihare et al. 2002; McNichol and Aluwihare 2007; Walker et al. 2011). This ultrafiltered DOC (consisting of components >1 kDa) comprises up to 50–60 % of freshwater and estuarine DOC, but usually makes up no more than 20–30 % of marine DOC (Raymond and Bauer 2001b).

The first step in analyzing DOC is to acidify the water to a pH of 2 so that all dissolved inorganic C is converted to CO_2. The water is then sparged with a CO_2-free gas to completely remove inorganic C. This procedure also removes volatile organic compounds.

The next step is to convert DOC to CO_2; there are some special considerations because DOC is in a dilute liquid phase. A number of different methods have been developed with the goal of quantitatively oxidizing all of this diverse pool. These methods include the following: (1) ultraviolet (UV) oxidation (Armstrong et al. 1966; Raymond and Bauer 2001a; Williams et al. 1969; Beaupre et al. 2007); (2) sealed-tube high-temperature combustion of dried sample (STC) (e.g., Fry et al. 1993; Peterson et al. 1994; Schiff et al. 1997); (3) high-temperature catalytic oxidation (HTCO) (Bauer et al. 1992; Druffel et al. 1992); (4) wet chemical oxidation (Menzel and Vaccaro 1964; Sharp 1973); and (5) combined UV and wet chemical oxidation (Bauer et al. 1991; Bouillon et al. 2006). Factors such as sample size, DOC concentration, salt content, quantitative recovery, and system blanks all should be considered before choosing an appropriate method. Ultraviolet oxidation and STC methods are most commonly used for ^{14}C-AMS analysis of marine and freshwater samples, respectively, because these methods can quantitatively oxidize DOC from liters of sample and provide enough C for ^{14}C-AMS analysis while maintaining reasonable blanks and precision (Peterson et al. 1994; McNichol and Aluwihare 2007). In some cases, background introduced during processing significantly influences these measurements, so it is important to make the appropriate tests to evaluate extraneous C corrections for each method (Hedges et al. 1993; Peterson et al. 1994).

9.4.3.3 Compound-Specific Methods

Methods to separate organic matter range from procedures intended to break it down into broadly defined chemical constituents (e.g., bulk lipids, water soluble components, acid- or base-hydrolyzable components), or highly specific ones (e.g., C_{25} alkene, or bacterial DNA). Because organic matter is normally such a heterogeneous mixture, one specific compound isolated from it will reflect the dynamics of a rather small fraction of the total organic matter. Nonetheless, the compounds isolated may represent the behavior of a class of materials (e.g., aquatic versus terrestrial derived organic matter preserved in marine sediments; Eglinton et al. 1997), or allow tracking of a low-level ^{14}C tracer through a complex biochemical pathway, as is frequently done in biomedical AMS applications (Turteltaub and Vogel 2000).

Several analytical steps are needed to perform compound-specific analysis: (1) extraction of a subset of the organic compounds from the bulk using properties such as solubility in polar or nonpolar solvents; (2) treatment to narrow the range of available compounds (e.g., using ion exchange resin columns); (3) separation of individual compounds, usually by gas or high-pressure liquid chromatography; (4) collection of the peak or peaks of interest on elution from the chromatography column, with repeated injections of sample until a sufficient amount for ^{14}C analysis is obtained. In step (3), sometimes the compounds must be derivatized (e.g., a component added to make the compound volatile for gas chromatography). In such cases, a correction needs to be made for the isotopic signature of the added C atoms (Rethemeyer et al. 2013). Details of analysis will vary with the specific question, type of matrix material and compound to be isolated. As this is an area that is evolving rapidly, the reader should consult the literature for details on procedures for specific compounds. Increasingly, the use of CO_2 sources in AMS instruments allows online oxidation of small samples for ^{14}C measurement.

Compound-specific samples are often small and require multiple separations (multiple injections on a gas- or liquid chromatography column) to obtain enough of the compound for a ^{14}C measurement. As previously noted, extraneous C added during processing can have a range of ^{14}C values, so it is important to process several standards made from the pure compound of interest with known ^{14}C value (previously measured on the pure compound itself). These must span at least the ^{14}C-free (to test for addition of modern C) and modern (to test for addition of old C). Known ^{14}C standard additions can be used to test all steps of the separation and purification procedures [see Santos et al. (2007b) for more detail or Santos et al. (2010b)].

9.4.3.4 Organic Compounds in Biosilica

Another active research area is the sources of C within biosilica structures, such as diatoms, deep-sea sponges, and phytoliths (biosilica of higher plants). The first two have been more frequently studied (Richmond and Sussman 2003 and references therein), while less study has been made of terrestrial plants. In both cases, it is assumed that all C embedded in the biogenic silica has the same source as that for the living organism that precipitates the silica. For phytoliths (minute silica structures precipitated inside the plant), that source has commonly been assumed to be atmospheric CO_2. Thus, the ^{14}C signature of C trapped in phytoliths should reflect those of the atmosphere at the time the plant was growing. Claims were recently made that the C occluded in phytoliths play a significant role in atmospheric C sequestration (Song et al. 2014 and references therein). Anomalously old phytolith C age results have challenged the idea that C in phytoliths has purely photosynthetic origins (Santos et al. 2010a). Because plants also take up small amounts of C from soils during nutrient uptake, the unexpected old ages of some of the phytolith C suggest that soil C could become trapped in phytoliths as well (Santos et al. 2012). If this is the case, opportunities for sequestering C via phytoliths are probably overestimated.

Because the C content in biosilica extracts is very small ($\ll$1 %C by dry weight), inadequate extraction methods can leave organic matter residues of different ^{14}C ages (Corbineau et al. 2013), and the clean biosilica is prone to atmospheric CO_2 surface adsorption. Thus, to obtain reliable ^{14}C results from such exotic materials, it is extremely important that researchers work in tandem with the AMS laboratories to test hypotheses, methodologies as well as overall accuracy and precision.

9.5 Isotopically Enriched Samples

Experiments with ^{14}C-enriched compounds (e.g., ^{14}C-labeled CO_2) can be very useful in C cycling and other studies. When using these radioactive compounds, it is important that all regulations concerning radioactive material are followed and precautions are taken to avoid the contamination of natural abundance measurements and AMS laboratories. In the USA, the Nuclear Regulatory Commission (NRC) oversees the use of radioactive material using Title 10 of the Federal Code of Regulations. Thirty-five US states have an agreement with the NRC (these states are called 'Agreement States') and regulate the use of most radioactive material within their borders using the same standards as the NRC. The Agreement States list regulations for radioactive material use in their state code of regulations. For example, California lists regulations for radioactive material in Title 17 of the California Code of Regulations. The use of radioactive materials in US ocean waters is also regulated by the NRC.

The following basic precautions need to be taken when using ^{14}C-enriched compounds: isolate the work in a designated isotope facility (e.g., a van on a ship), assign designated equipment for use with labeled material and keep the equipment in the designated isotope facility, bring labeled samples to AMS facilities as pressed graphite in clean packaging, and run frequent blanks and standards.

An additional issue for samples where the ^{13}C or ^{14}C content is manipulated independently is the recognition that mass-dependent fractionation corrections used in calculating ^{14}C data are no longer appropriate (Torn and Southon 2001).

9.6 Conclusions and Future Directions

As the applications for AMS measurements of ^{14}C increase, so do the complexity and specificity of sample pretreatment methods. Nonetheless, the major challenges in ^{14}C preparation methods remain the same: (1) matching the C to be analyzed with the question being asked; and (2) ensuring that the total uncertainty associated with the ^{14}C measurement is determined. In addition, there are new choices available to a researcher trying to use ^{14}C to understand the Earth's C cycle. These choices range from using compound specific methods to understanding how a tiny fraction of organic matter behaves to using operationally-defined measures to separate constituent fractions that may behave quite differently from the average bulk material. The few examples provided here are meant to make the reader aware of both of these issues. As procedures become more and more complex, one must appreciate the complexity of testing new methods, so the results accurately reflect the true ^{14}C signature in a way that it can be appropriately interpreted.

The future will undoubtedly see many new analytical approaches as sample sizes required for AMS decrease even further. Methods for ^{14}C dating will benefit increasingly from the ability of compound-specific methods to isolate C compounds that have not been diagenetically altered or contaminated. A few very difficult measurements remain undone—for example, the ^{14}C signature of dissolved CH_4 in deep-ocean water—and will require dedicated souls who can collect and process the large amounts of sample needed while convincingly demonstrating that the result does not just reflect contamination. Finally, as the bomb ^{14}C tracer becomes diluted enough by the background ^{14}C and fossil-derived C that the atmosphere $^{14}CO_2$ drops below 1950 ^{14}C levels, we will rely more on low-level ^{14}C labeling to understand short-term C flows. During the coming decade, it is increasingly important to maintain sample archives to keep the ability to document the flow of bomb ^{14}C through the Earth system as new methods become available. The number of ^{14}C samples analyzed is usually fewer than the number desired, since it is the cost of the measurement, rather than imagination, that often imposes limits.

References

Ajie, H.O., I.R. Kaplan, P.J. Slota, and R.E. Taylor. 1990. AMS radiocarbon dating of bone osteocalcin. *Nuclear Instruments & Methods in Physics Research Section B-Beam Interactions with Materials and Atoms* 52: 433–437.

Ajie, H.O., I.R. Kaplan, P.V. Hauschka, D. Kirner, P.J. Slota, and R.E. Taylor. 1992. Radio dating of bone osteocalcin—isolating and characterizing a non-collagen protein. *Radiocarbon* 34: 296–305.

Aluwihare, L.I., D.J. Repeta, and R.F. Chen. 2002. Chemical composition and cycling of dissolved organic matter in the Mid-Atlantic Bight. *Deep-Sea Research Part Ii-Topical Studies in Oceanography* 49: 4421–4437.

Armstrong, F.A.J., P.M. Williams, and J.D.H. Strickland. 1966. Photo-oxidation of organic matter in sea water by ultraviolet radiation. *Nature* 211: 481–483.

Basile-Doelsch, I., R. Amundson, W.E.E. Stone, D. Borschneck, J.Y. Bottero, S. Moustier, F. Masin, and F. Colin. 2007. Mineral control of carbon pools in a volcanic soil horizon. *Geoderma* 137: 477–489.

Bauer, J.E., R.I. Haddad, and D.J. Desmarais. 1991. Method for determining stable isotope ratios of dissolved organic-carbon in interstitial and other natural marine waters. *Marine Chemistry* 33: 335–351.

Bauer, J.E., P.M. Williams, and E.R.M. Druffel. 1992. Recovery of submilligram quantities of carbon dioxide from gas streams by molecular sieve for subsequent determination of isotopic (^{13}C and ^{14}C) natural abundances. *Analytical Chemistry* 64: 824–827.

Beaupre, S.R., and E.R.M. Druffel. 2009. Constraining the propagation of bomb-radiocarbon through the dissolved organic carbon (DOC) pool in the northeast Pacific Ocean. *Deep-Sea Research Part I Oceanographic Research Papers* 56: 1717–1726.

Beaupre, S.R., E.R.M. Druffel, and S. Griffin. 2007. A low-blank photochemical extraction system for concentration and isotopic analyses of marine dissolved organic carbon. *Limnology and Oceanography-Methods* 5: 174–184.

Benner, R., B. Biddanda, B. Black, and M. McCarthy. 1997. Abundance, size distribution, and stable carbon and nitrogen isotopic compositions of marine organic matter isolated by tangential-flow ultrafiltration. *Marine Chemistry* 57: 243–263.

Borowski, W.S., C.K. Paull, and W. Ussler. 1997. Carbon cycling within the upper methanogenic zone of continental rise sediments: An example from the methane-rich sediments overlying the Blake Ridge gas hydrate deposits. *Marine Chemistry* 57: 299–311.

Bouillon, S., M. Korntheuer, W. Baeyens, and F. Dehairs. 2006. A new automated setup for stable isotope analysis of dissolved organic carbon. *Limnology and Oceanography-Methods* 4: 216–226.

Boutton, T.W., W.W. Wong, D.L. Hachey, L.S. Lee, M.P. Cabrera, and P.D. Klein. 1983. Comparison of quartz and Pyrex tubes for combustion of organic-samples for stable carbon isotope analysis. *Analytical Chemistry* 55: 1832–1833.

Brock, F., R. Wood, T.F. Higham, P. Ditchfield, A. Bayliss, and C.B. Ramsey. 2012. Reliability of nitrogen content (% N) and carbon: nitrogen atomic ratios (C: N) as indicators of collagen preservation suitable for radiocarbon dating. *Radiocarbon* 54: 879–886.

Brown, T.A., and J.R. Southon. 1997. Corrections for contamination background in AMS C-14 measurements. *Nuclear Instruments and Methods in Physics Research, Section B: Beam Interactions With Materials and Atoms* 123: 208–213.

Brown, T.A., D.E. Nelson, J.S. Vogel, and J.R. Southon. 1988. Improved collagen extraction by modified Longin method. *Radiocarbon* 30: 171–177.

Burky, R.R., D.L. Kirner, R.E. Taylor, P.E. Hare, and J.R. Southon. 1998. C-14 dating of bone using gamma-carboxyglutamic acid and alpha-carboxyglycine (aminomalonate). *Radiocarbon* 40: 11–20.

Castanha, C., S. Trumbore, and R. Amundson. 2008. Methods of separating soil carbon pools affect the chemistry and turnover time of isolated fractions. *Radiocarbon* 50: 83–97.

Christensen, B. T. 1992. Physical fractionation of soil and organic matter in primary particle size and density separates. In *Advances in soil science*, 1–90. Berlin: Springer.

Corbineau, R., P. Reyerson, A. Alexandre, and G.M. Santos. 2013. Towards producing pure phytoliths that are suitable for carbon isotopic analysis. *Review of Palaeobotany and Palynology* 197: 179–185.

Crum, R.H., E.M. Murphy, and C.K. Keller. 1996. A non-adsorptive method for the isolation and fractionation of natural dissolved organic carbon. *Water Research* 30: 1304–1311.

Czimczik, C.I., and S.E. Trumbore. 2007. Short-term controls on the age of microbial carbon sources in boreal forest soils. *Journal of Geophysical Research-Biogeosciences* 112: 8.

DeNiro, M.J. 1985. Postmortem preservation and alteration of invivo bone-collagen isotope ratios in relation to paleodietary reconstruction. *Nature* 317: 806–809.

Don, A., B. Steinberg, I. Schöning, K. Pritsch, M. Joschko, G. Gleixner, and E.-D. Schulze. 2008. Organic carbon sequestration in earthworm burrows. *Soil Biology & Biochemistry* 40: 1803–1812.

Donahue, D.J., T.W. Linick, and A.J.T. Jull. 1990. Isotope-ratio and background corrections for accelerator mass spectrometry radiocarbon measurements. *Radiocarbon* 32: 135–142.

Drimmie, R.J., R. Aravena, L.I. Wassenaar, P. Fritz, M.J. Hendry, and G. Hut. 1991. Radiocarbon and stable isotopes in water and dissolved constituents, Milk River aquifer, Alberta, Canada. *Applied Geochemistry* 6: 381–392.

Druffel, E.R.M., P.M. Williams, J.E. Bauer, and J.R. Ertel. 1992. Cycling of dissolved and particulate organic matter in the open ocean. *Journal of Geophysical Research* 97: 15639–15659.

Eglinton, T.I., B.C. Benitez-Nelson, A. Pearson, A.P. McNichol. 1997. Variability in radiocarbon ages of individual organic compounds from marine sediments. *Science* 277(5327): 796.

Eisma, R., K. Vanderborg, A.F.M. Dejong, W.M. Kieskamp, and A.C. Veltkamp. 1994. Measurements of the C-14 content of atmospheric methane in the Netherlands to determine the regional emissions of (CH_4)-C-14. *Nuclear Instruments and Methods in Physics Research Section B: Beam Interactions with Materials and Atoms* 92: 410–412.

Elia, M., T. McDonald, and A. Crisp. 1986. Errors in measurements of CO_2 with the use of drying agents. *Clinica Chimica Acta* 158: 237–244.

Fry, B., S. Saupe, M. Hullar, and B.J. Peterson. 1993. Platinum-catalyzed combustion of DOC in sealed tubes for stable isotopic analysis. *Marine Chemistry* 41: 187–193.

Gao, P., X. Xu, L. Zhou, M.A. Pack, S. Griffin, G.M. Santos, J.R. Southon, and K. Liu. 2014. Rapid sample preparation of dissolved inorganic carbon in natural waters using a headspace-extraction approach for radiocarbon analysis by accelerator mass spectrometry. *Limnology and Oceanography-Methods* 12: 174–190.

Garnett, M.H., and I.P. Hartley. 2010. A passive sampling method for radiocarbon analysis of atmospheric CO_2 using molecular sieve. *Atmospheric Environment* 44: 877–883.

Gaudinski, J.B., S.E. Trumbore, E.A. Davidson, and S.H. Zheng. 2000. Soil carbon cycling in a temperate forest: radiocarbon-based estimates of residence times, sequestration rates and partitioning of fluxes. *Biogeochemistry* 51: 33–69.

Gaudinski, J.B., T.E. Dawson, S. Quideau, E.A.G. Schuur, J.S. Roden, S.E. Trumbore, D.R. Sandquist, S.W. Oh, and R.E. Wasylishen. 2005. Comparative analysis of cellulose preparation techniques for use with C-13, C-14, and O-18 isotopic measurements. *Analytical Chemistry* 77: 7212–7224.

George, D., J. Southon, and R. Taylor. 2005. Resolving an anomolous radiocarbon determination on mastodon bone from Monte Verde, Chile. *American antiquity*:766–772.

Golchin, A., J.M. Oades, J. Skjemstad, and P. Clarke. 1995. Structural and dynamic properties of soil organic-matter as reflected by 13C natural-abundance, pyrolysis mass-spectrometry and solid-state 13C NMR-spectroscopy in density fractions of an oxisol under forest and pasture. *Soil Research* 33: 59–76.

Guo, L.D., P.H. Santschi, L.A. Cifuentes, S.E. Trumbore, and J. Southon. 1996. Cycling of high-molecular-weight dissolved organic matter in the middle Atlantic bight as revealed by carbon isotopic (C-13 and C-14) signatures. *Limnology and Oceanography* 41: 1242–1252.

Hayes, J. M. 2002. Practice and principles of isotopic measurements in organic geochemistry. http://www.nosams.whoi.edu/docs/IsoNotesAug02.pdf. 2nd revision, Aug 2002. Accessed Mar 2006, 1–25.

Hedges, R.E.M., and G.J. Vanklinken. 1992. A review of current approaches in the pretreatment of bone for radiocarbon dating by AMS. *Radiocarbon* 34: 279–291.

Hedges, J.I., J.R. Ertel, P.D. Quay, P.M. Grootes, J.E. Richey, A.H. Devol, G.W. Farwell, F.W. Schmidt, and E. Salati. 1986. Organic C-14 in the Amazon River system. *Science* 231: 1129–1131.

Hedges, J.I., B.A. Bergamaschi, and R. Benner. 1993. Comparative analyses of DOC and DON in natural waters. *Marine Chemistry* 41: 121–134.

Heim, A., and M.W.I. Schmidt. 2007. Lignin turnover in arable soil and grassland analysed with two different labelling approaches. *European Journal of Soil Science* 58: 599–608.

Hendry, M. J. and L. I. Wassenaar. 2005. Origin and migration of dissolved organic carbon fractions in a clay-rich aquitard: C-14 and delta C-13 evidence. *Water Resources Research* 41.

Higham, T.F.G., R.M. Jacobi, and C.B. Ramsey. 2006. AMS radiocarbon dating of ancient bone using ultrafiltration. *Radiocarbon* 48: 179–195.

Hopkins, F.M., M.S. Torn, and S.E. Trumbore. 2012. Warming accelerates decomposition of decades-old carbon in forest soils. *Proceedings of the National Academy of Sciences of the United States of America* 109: E1753–E1761.

Hsueh, D. Y., N. Y. Krakauer, J. T. Randerson, X. M. Xu, S. E. Trumbore, and J. R. Southon. 2007. Regional patterns of radiocarbon and fossil fuel-derived CO_2 in surface air across North America. *Geophysical Research Letters* 34.

Hua, Q., and M. Barbetti. 2004. Review of tropospheric bomb ^{14}C data for carbon cycle modeling and age calibration purposes. *Radiocarbon* 46: 1273–1298.

Hua, Q., G.E. Jacobsen, U. Zoppi, E.M. Lawson, A.A. Williams, A.M. Smith, and M.J. McGann. 2001. Progress in radiocarbon target preparation at the ANTARES AMS Centre. *Radiocarbon* 43: 275–282.

Hua, Q., M. Barbetti, and A.Z. Rakowski. 2013. Atmospheric Radiocarbon for the Period 1950-2010. *Radiocarbon* 55: 2059–2072.

Hwang, J., and E.R.M. Druffel. 2005. Blank correction for ^{14}C measurements in organic compound classes of oceanic particulate matter. *Radiocarbon* 47: 75–87.

Jacobsen, G., Q. Hua, J. Tarshishi, D. Fink, M. Hotchkis, E. Lawson, A. Smith, and C. Tuniz. 1997. AMS radiocarbon analysis of micocrosamples In: Handbook of the Sixth Australian Archaeometry Conference, 36, Sydney, Australia.

Jagadamma, S., R. Lal, D.A.N. Ussiri, S.E. Trumbore, and S. Mestelan. 2010. Evaluation of structural chemistry and isotopic signatures of refractory soil organic carbon fraction isolated by wet oxidation methods. *Biogeochemistry* 98: 29–44.

Jaynes, W., and J. Bigham. 1986. Concentration of iron oxides from soil clays by density gradient centrifugation. *Soil Science Society of America Journal* 50: 1633–1639.

Jull, A.J.T., D.J. Donahue, A.L. Hatheway, T.W. Linick, and L.J. Toolin. 1986. Production of graphite targets by deposition from CO/H_2 for precision accelerator C-14 measurements. *Radiocarbon* 28: 191–197.

Keel, S.G., R.T. Siegwolf, and C. Körner. 2006. Canopy CO_2 enrichment permits tracing the fate of recently assimilated carbon in a mature deciduous forest. *New Phytologist* 172: 319–329.

Keeling, C.D. 1958. The concentration and isotopic abundances of atmospheric carbon dioxide in rural areas. *Geochimica et Cosmochimica Acta* 13: 322–334.

Kessler, J.D., and W.S. Reeburgh. 2005. Preparation of natural methane samples for stable isotope and radiocarbon analysis. *Limnology and Oceanography-Methods* 3: 408–418.

Key, R. M., A. Kozyr, C. L. Sabine, K. Lee, R. Wanninkhof, J. L. Bullister, R. A. Feely, F. J. Millero, C. Mordy, and T. H. Peng. 2004. A global ocean carbon climatology: Results from Global Data Analysis Project (GLODAP). *Global Biogeochemical Cycles* 18. doi:10.1029/2004GB002247.

Klouda, G.A., L.A. Currie, D.J. Donahue, A.J.T. Jull, and M.H. Naylor. 1986. Urban atmospheric (CO)-C-14 and (CH_4)-C-14 measurements by accelerator mass spectrometry. *Radiocarbon* 28: 625–633.

Lang, S.Q., G.L. Früh-Green, S.M. Bernasconi, and L. Wacker. 2013. Isotopic (δ13C, Δ14C) analysis of organic acids in marine samples using wet chemical oxidation. *Limnology and Oceanography-Methods* 11: 161–175.

Leavitt, S.W., and S.R. Danzer. 1993. Method for batch processing small wood samples to holocellulose for stable-carbon isotope analysis. *Analytical Chemistry* 65: 87–89.

Levin, I. and B. Kromer. 1997. Twenty years of atmospheric $^{14}CO_2$ observations at Schauinsland Station, Germany. *Radiocarbon* 39.

Levin, I., and B. Kromer. 2004. The tropospheric $^{14}CO_2$ level in mid-latitudes of the northern hemisphere (1959–2003). *Radiocarbon* 46: 1261–1272.

Levin, I., K.O. Munnich, and W. Weiss. 1980. The effect of anthropogenic CO_2 and C-14 sources on the distribution of C-14 in the atmosphere. *Radiocarbon* 22: 379–391.

Levin, I., T. Naegler, B. Kromer, M. Diehl, R.J. Francey, A.J. Gomez-Pelaez, L.P. Steele, D. Wagenbach, R. Weller, and D.E. Worthy. 2010. Observations and modelling of the global distribution and long-term trend of atmospheric $14CO_2$. *Tellus B* 62: 26–46.

Longin, R. 1971. New method of collagen extraction for radiocarbon dating.

Lowe, D.C. 1984. Preparation of graphite targets for radiocarbon dating by tandem accelarator mass spectrometer (TAMS). *The International Journal of Applied Radiation and Isotopes* 35: 349–352.

Lowe, D.C., C.A. Brenninkmeijer, S.C. Tyler, and E.J. Dlugkencky. 1991. Determination of the isotopic composition of atmospheric methane and its application in the Antarctic. *Journal of Geophysical Research* 96: 15455–15467.

Manning, M.R., D.C. Lowe, W.H. Melhuish, R.J. Sparks, G. Wallace, C.A.M. Brenninkmeijer, and R.C. McGill. 1990. The use of radiocarbon measurements in atmospheric studies. *Radiocarbon* 32: 37–58.

McIntyre, C.P., S.P. Sylva, and M.L. Roberts. 2009. Gas chromatograph-combustion system for C-14-accelerator mass spectrometry. *Analytical Chemistry* 81: 6422–6428.

McIntyre, C.P., A.P. McNichol, M.L. Roberts, J.S. Seewald, K.F. von Reden, and W.J. Jenkins. 2013. Improved precision of ^{14}C measurements for CH_4 and CO_2 using GC and continuous-flow AMS achieved by summation of repeated injections. *Radiocarbon* 55: 677–685.

McNichol, A.P., and L.I. Aluwihare. 2007. The power of radiocarbon in biogeochemical studies of the marine carbon cycle: Insights from studies of dissolved and particulate organic carbon (DOC and POC). *Chemical Reviews* 107: 443–466.

McNichol, A., A. Gagnon, G. Jones, and E. Osborne. 1992. Illumination of a black box: Analysis of gas composition during graphite target preparation. *Radiocarbon* 34: 321–329.

Menzel, D.W., and R.F. Vaccaro. 1964. The measurement of dissolved organic and particulate carbon in seawater. *Limnology and Oceanography* 9: 138–142.

Miller, J., S. Lehman, C. Wolak, J. Turnbull, G. Dunn, H. Graven, R. Keeling, H.A.J. Meijer, A.T. Aerts-Bijma, S.W.L. Palstra, A.W. Smith, C. Allison, J. Southon, X. Xu, T. Nakazawa, S. Aoki, T. Nakamura, T. Guilderson, B. LaFranchi, H. Mukai, Y. Terao, M. Uchida, and M. Kondo. 2013. Initial Results of an Inter-comparison of AMS-based Atmospheric $^{14}CO_2$ measurements. *Radiocarbon* 55: 1475–1483.

Monnier, G., L. Turc, and C. Jeanson-Luusinang. 1962. Une méthode de fractionnement densimétrique par centrifugation des matières organiques du sol, 55–63 *in* Annales Agronomiques.

Nakagawa, F., N. Yoshida, Y. Nojiri, and V.N. Makarov. 2002. Production of methane from alasses in eastern Siberia: Implications from its C-14 and stable isotopic compositions. *Global Biogeochemical Cycles* 16.

Ognibene, T.J., G. Bench, J.S. Vogel, G.F. Peaslee, and S. Murov. 2003. A high-throughput method for the conversion of CO2 obtained from biochemical samples to graphite in septa-sealed vials for quantification of C-14 via accelerator mass spectrometry. *Analytical Chemistry* 75: 2192–2196.

Olsson, I.U. 1986. Radiometric dating. Handbook of Holocene palaeoecology and palaeohydrobiology:273–312.

Pack, M., M. Heinta, W. Reeburgh, S. Trumbore, D. Valentine, X. Xu, and E. Druffel. 2011. A method for measuring methane oxidation rates using low-levels of 14C-labeled methane and accelerator mass specrometry. *Limnology and Oceanography: Methods* 9: 245–260.

Pack, M.A., X. Xu, M. Lupascu, J.D. Kessler, and C.I. Czimczik. 2014. A rapid method for preparing low-volume CH_4 and CO_2 gas samples for ^{14}C-AMS analysis. *Organic Geochemistry* 78: 89–98.

Peterson, B., B. Fry, M. Hullar, S. Saupe, and R. Wright. 1994. The distribution and stable carbon isotopic composition of dissolved organic-carbon in estuaries. *Estuaries* 17: 111–121.

Petrenko, V.V., J.P. Severinghaus, E.J. Brook, J. Muhle, M. Headly, C.M. Harth, H. Schaefer, N. Reeh, R.F. Weiss, D. Lowe, and A.M. Smith. 2008a. A novel method for obtaining very large ancient air samples from ablating glacial ice for analyses of methane radiocarbon. *Journal of Glaciology* 54: 233–244.

Petrenko, V.V., A.M. Smith, G. Brailsford, K. Riedel, Q. Hua, D. Lowe, J.P. Severinghaus, V. Levchenko, T. Bromley, R. Moss, J. Muehle, and E.J. Brook. 2008b. A new method for analyzing (14)C of methane in ancient air extracted from glacial ice. *Radiocarbon* 50: 53–73.

Plante, A.F., S.R. Beaupré, M.L. Roberts, and T. Baisden. 2013. Distribution of radiocarbon ages in soil organic matter by thermal fractionation. *Radiocarbon* 55: 1077–1083.

Quay, P.D., S.L. King, J. Stutsman, D.O. Wilbur, L.P. Steele, I. Fung, R.H. Gammon, T.A. Brown, G.W. Farwell, P.M. Grootes, and F.H. Schmidt. 1991. Carbon isotopic composition of atmospheric CH_4: Fossil and biomass burning source and strengths. *Global Biogeochemical Cycles* 5: 25–47.

Ramsey, C.B., T. Higham, A. Bowles, and R. Hedges. 2004. Improvements to the pretreatment of bone at Oxford. *Radiocarbon* 46: 155–163.

Raymond, P.A., and J.E. Bauer. 2001a. Riverine export of aged terrestrial organic matter to the North Atlantic Ocean. *Nature* 409: 497–500.

Raymond, P.A., and J.E. Bauer. 2001b. Use of C-14 and C-13 natural abundances for evaluating riverine, estuarine, and coastal DOC and POC sources and cycling: a review and synthesis. *Organic Geochemistry* 32: 469–485.

Reeburgh, W.S. 1980. Anaerobic methane oxidation: rate depth distributions in Skan Bay sediments. *Earth and Planetary Science Letters* 47: 345–352.

Rethemeyer, J., R.-H. Fülöp, S. Höfle, L. Wacker, S. Heinze, I. Hajdas, U. Patt, S. König, B. Stapper, and A. Dewald. 2013. Status report on sample preparation facilities for ^{14}C analysis at

the new CologneAMS center. *Nuclear Instruments & Methods in Physics Research, Section B: Beam Interactions with Materials and Atoms* 294: 168–172.
Richardson, A.D., M.S. Carbone, T.F. Keenan, C.I. Czimczik, D.Y. Hollinger, P. Murakami, P.G. Schaberg, and X.Xu. 2012. Seasonal dynamics and age of stemwood nonstructural carbohydrates in temperate forest trees.
Richmond, K.E., and M. Sussman. 2003. Got silicon? The non-essential beneficial plant nutrient. *Current Opinion in Plant Biology* 6: 268–272.
Santos, G.M. 2008. KCCAMS facility: combustion protocol.
Santos, G.M., and K. Ormsby. 2013. Behavioral variability in ABA chemical pretreatment close to the 14C age limit. *Radiocarbon* 55: 534–544.
Santos, G.M., J.R. Southon, K.C. Druffel-Rodriguez, S. Griffin, and M. Mazon. 2004. Magnesium perchlorate as an alternative water trap in AMS graphite sample preparation: A report on sample preparation at KCCAMS at the University of California, Irvine. *Radiocarbon* 46: 165–173.
Santos, G.M., M. Mazon, J.R. Southon, S. Rifai, and R. Moore. 2007a. Evaluation of iron and cobalt powders as catalysts for C-14-AMS target preparation. *Nuclear Instruments and Methods in Physics Research Section B: Beam Interactions with Materials and Atoms* 259: 308–315.
Santos, G.M., J.R. Southon, S. Griffin, S.R. Beaupré, and E.R.M. Druffel. 2007b. Ultra small-mass AMS ^{14}C sample preparation and analyses at KCCAMS/UCI Facility. *Nuclear Instruments & Methods in Physics Research B* 259: 293–302.
Santos, G.M., A. Alexandre, H.G. Coe, P.E. Reyerson, J.R. Southon, and C.N. De Carvalho. 2010a. The Phytolith C puzzle: A tale of background determinations and accuracy tests. *Radiocarbon* 52: 113–128.
Santos, G.M., J.R. Southon, N.J. Drenzek, L.A. Ziolkowski, E. Druffel, X. Xu, D. Zhang, S. Trumbore, T.I. Eglinton, and K.A. Hughen. 2010b. Blank assessment for ultra-small radiocarbon samples: chemical extraction and separations versus AMS. *Radiocarbon* 52: 1322–1335.
Santos, G.M., A. Alexandre, J.R. Southon, K.K. Treseder, R. Corbineau, and P. Reyerson. 2012. Possible source of ancient carbon in phytolith concentrates from harvested grasses. *Biogeosciences* 9: 1873–1884.
Santschi, P.H., L.D. Guo, M. Baskaran, S. Trumbore, J. Southon, T.S. Bianchi, B. Honeyman, and L. Cifuentes. 1995. Isotopic evidence for the contemporary origin of high molecular weight organic matter in oceanic environments. *Geochimica et Cosmochimica Acta* 59: 625–631.
Schiff, S.L., R. Aravena, S.E. Trumbore, M.J. Hinton, R. Elgood, and P.J. Dillon. 1997. Export of DOC from forested catchments on the Precambrian Shield of Central Ontario: Clues from C-13 and C-14. *Biogeochemistry* 36: 43–65.
Schmidt, M.W.I., M.S. Torn, S. Abiven, T. Dittmar, G. Guggenberger, I.A. Janssens, M. Kleber, I. Kögel-Knabner, J. Lehmann, D.A.C. Manning, P. Nannipieri, D.P. Rasse, S. Weiner, and S.E. Trumbore. 2011. Persistence of soil organic matter as an ecosystem property. *Nature* 478: 49–56.
Schrumpf, M., E. Schulze, K. Kaiser, and J. Schumacher. 2011. How accurately can soil organic carbon stocks and stock changes be quantified by soil inventories? Biogeosciences Discussions 8.
Scott, E. 2003a. Section 2: The results. *Radiocarbon* 45: 151–157.
Scott, E. M. 2003b. The Third International Radiocarbon Intercomparison (TIRI) and the Fourth International Radiocarbon (FIR)—1999–2002—Results, analysis and conclusions. *Radiocarbon* 45:VII–X.
Scott, E., C. Bryant, G. Cook, and P. Naysmith. 2003. Is there a fifth international radiocarbon intercomparison (VIRI)? *Radiocarbon* 45: 493–495.
Scott, E.M., C. Bryant, I. Carmi, G. Cook, S. Gulliksen, D. Harkness, J. Heinemeier, E. McGee, P. Naysmith, G. Possnert, H. van der Plicht, and M. van Strydonck. 2004. Precision and accuracy in applied C-14 dating: some findings from the Fourth International Radiocarbon Inter-comparison. *Journal of Archaeological Science* 31: 1209–1213.

Scott, E.M., G.T. Cook, P. Naysmith, C. Bryant, and D. O'Donnell. 2007. A report on phase 1 of the 5th International Radiocarbon Intercomparison (VIRI). *Radiocarbon* 49: 409–426.

Scott, E.M., G.T. Cook, and P. Naysmith. 2010. The Fifth International Radiocarbon Intercomparison (VIRI): An assessment of laboratory performance in stage 3. *Radiocarbon* 52: 859–865.

Shang, C., and H. Tiessen. 1998. Organic matter stabilization in two semiarid tropical soils: Size, density, and magnetic separations. *Soil Science Society of America Journal* 62: 1247–1257.

Sharp, J.H. 1973. Total organic carbon in seawater-comparison of measurements using persulfate oxidation and high temperature combustion. *Marine Chemistry* 1: 211–229.

Six, J., K. Paustian, E.T. Elliott, and C. Combrink. 2000. Soil structure and organic matter I. Distribution of aggregate-size classes and aggregate-associated carbon. *Soil Science Society of America Journal* 64: 681–689.

Slota, P.J., A.J.T. Jull, T.W. Linick, and L.J. Toolin. 1987. Preparation of small samples for C-14 accelerator targets by catalytic reduction of CO. *Radiocarbon* 29: 303–306.

Sofer, Z. 1980. Preparation of carbon-dioxide for stable carbon isotope analysis of petroleum fractions. *Analytical Chemistry* 52: 1389–1391.

Sollins, P., C. Swanston, M. Kleber, T. Filley, M. Kramer, S. Crow, B.A. Caldwell, K. Lajtha, and R. Bowden. 2006. Organic C and N stabilization in a forest soil: evidence from sequential density fractionation. *Soil Biology & Biochemistry* 38: 3313–3324.

Song, Z., H. Wang, P.J. Strong, and F. Guo. 2014. Phytolith carbon sequestration in China's croplands. *European Journal of Agronomy* 53: 10–15.

Southon, J., and A. Magana. 2010. A comparison of cellulose extraction and ABA pretreatment methods for AMS 14C dating of ancient wood. *Radiocarbon* 52: 1371–1379.

Spycher, G., and J. Young. 1979. Water-dispersible soil organic-mineral particles: II. Inorganic amorphous and crystalline phases in density fractions of clay-size particles. *Soil Science Society of America Journal* 43: 328–332.

Stafford, T.W., A.J.T. Jull, K. Brendel, R.C. Duhamel, and D. Donahue. 1987. Study of bone radiocarbon dating accuracy at the University of Arizona NSF accelerator facility for radioisotope analysis. *Radiocarbon* 29: 24–44.

Stafford, T.W., P.E. Hare, L. Currie, A.T. Jull, and D. Donahue. 1990. Accuracy of North American human skeleton ages. *Quaternary Research* 34: 111–120.

Swanston, C.W., M.S. Torn, P.J. Hanson, J.R. Southon, C.T. Garten, E.M. Hanlon, and L. Ganio. 2005. Initial characterization of processes of soil carbon stabilization using forest stand-level radiocarbon enrichment. *Geoderma* 128: 52–62.

Taylor, R.E., D.G. Smith, and J.R. Southon. 2001. The Kennewick skeleton: Chronological and biomolecular contexts. *Radiocarbon* 43: 965–976.

Thurman, E.M., and R.L. Malcolm. 1981. Preparative isolation of aquatic humic substances. *Environmental Science and Technology* 15: 463–466.

Torn, M.S., and J. Southon. 2001. A new C-13 correction for radiocarbon samples from elevated-CO_2 experiments. *Radiocarbon* 43: 691–694.

Trumbore, S. E., and J. W. Harden. 1997. Input, accumulation and turnover of carbon in soils of the BOREAS northern study area. *JGR Atmospheres* 102:28805–28816.

Trumbore, S.E., and S.H. Zheng. 1996. Comparison of fractionation methods for soil organic matter C-14 analysis. *Radiocarbon* 38: 219–229.

Turteltaub, K.W., and J.S. Vogel. 2000. Bioanalytical applications of accelerator mass spectrometry for pharmaceutical research. *Current Pharmaceutical Design* 6: 991–1007.

Vargas, R., S.E. Trumbore, and M.F. Allen. 2009. Evidence of old carbon used to grow new fine roots in a tropical forest. *New Phytologist* 182: 710–718.

Vogel, J.S. 1992. A rapid method for preparation of biomedical targets for AMS. *Radiocarbon* 34: 344–350.

Vogel, J.S., J.R. Southon, D.E. Nelson, and T.A. Brown. 1984. Performance of catalytically condensed carbon for use in Accelerator Mass-Spectrometry. *Nuclear Instruments & Methods in Physics Research Section B-Beam Interactions with Materials and Atoms* 5: 289–293.

Vogel, J.S., D.E. Nelson, and J.R. Southon. 1987. ^{14}C background levels in an accelerator mass spectrometry system. *Radiocarbon* 29: 323–333.

Wacker, L., S.M. Fahrni, I. Hajdas, M. Molnar, H.-A. Synal, S. Szidat, and Y.L. Zhang. 2013. A versatile gas interface for routine radiocarbon analysis with a gas ion source. *Nuclear Instruments & Methods in Physics Research, Section B: Beam Interactions with Materials and Atoms* 294: 315–319.

Wahlen, M., N. Tanaka, R. Henry, B. Deck, J. Zeglen, J.S. Vogel, J.R. Southon, A. Shemesh, R. Fairbanks, and W.S. Broecker. 1989. Carbon-14 in methane sources and in atmospheric methane: The contribution from fossil carbon. *Science* 245: 286–290.

Walker, B., S. Beaupre, T. Guilderson, E. Druffel, and M. McCarthy. 2011. Large-volume ultrafiltration for the study of radiocarbon signatures and size vs. age relationships in marine dissolved organic matter. *Geochimica et Cosmochimica Acta* 75: 5187–5202.

Wang, W., and D.E. Pataki. 2010. Spatial patterns of plant isotope tracers in the Los Angeles urban region. *Landscape Ecology* 25: 35–52.

Williams, P.M., H. Oeschger, and P. Kinney. 1969. Natural radiocarbon activity of dissolved organic carbon in North-East Pacific Ocean. *Nature* 224: 256–258.

Xu, X.M., S.E. Trumbore, S.H. Zheng, J.R. Southon, K.E. McDuffee, M. Luttgen, and J.C. Liu. 2007. Modifying a sealed tube zinc reduction method for preparation of AMS graphite targets: Reducing background and attaining high precision. *Nuclear Instruments & Methods in Physics Research Section B-Beam Interactions with Materials and Atoms* 259: 320–329.

Zimov, S.A., Y.V. Voropaev, I.P. Semiletov, S.P. Davidov, S.F. Prosiannikov, F.S. Chapin, M.C. Chapin, S. Trumbore, and S. Tyler. 1997. North Siberian lakes: A methane source fueled by Pleistocene carbon. *Science* 277: 800–802.

The manufacturer's authorised representative in the EU is Springer Nature Customer Service Centre GmbH, Europaplatz 3, 69115 Heidelberg, Germany. If you have any concerns regarding our products, please contact ProductSafety@springernature.com

Printed and bound by CPI Group (UK) Ltd, Croydon, CR0 4YY
01/10/2026
02225872-0010